Observing Geohazards from Space

Special Issue Editor
Francesca Cigna

MDPI • Basel • Beijing • Wuhan • Barcelona • Belgrade

MDPI

Special Issue Editor
Francesca Cigna
Italian Space Agency (ASI)
Italy

Editorial Office
MDPI AG
St. Alban-Anlage 66
Basel, Switzerland

This edition is a reprint of the Special Issue published online in the open access journal *Geosciences* (ISSN 2076-3263) from 2017–2018 (available at: http://www.mdpi.com/journal/geosciences/special_issues/observing_geohazards).

For citation purposes, cite each article independently as indicated on the article page online and as indicated below:

Lastname, F.M.; Lastname, F.M. Article title. *Journal Name* **Year**, *Article number*, page range.

First Edition 2018

ISBN 978-3-03842-775-9 (Pbk)
ISBN 978-3-03842-776-6 (PDF)

Cover photo courtesy of European Space Agency (ESA)

Table of Contents

About the Special Issue Editor

Francesca Cigna holds a PhD in Earth Sciences, MSc and BSc in Environment and Territory Engineering, with extensive experience in Earth observation and remote sensing data and techniques. At the Italian Space Agency (ASI), Dr Cigna works as researcher in Earth observation and data analytics, and leads research on time series modelling and big data, with particular focus on satellite Synthetic Aperture Radar (SAR), differential Interferometric SAR (InSAR) for geohazard applications, including natural and man-made hazards and risks, shallow geological processes, ground instability and land changes, landscape archaeology and cultural heritage. She collaborates with the Committee for Earth Observation Satellites (CEOS)—Working Group Disasters. Dr Cigna is Associate Editor for Elsevier—International Journal of Applied Earth Observation and Geoinformation, and Editorial Board Member for Nature—Scientific Reports , and regularly serves as reviewer for international journals in the field of Earth observation, Earth and planetary sciences, geology, natural hazards, geophysics and GIS.

Preface to "Observing Geohazards from Space"

With a wide spectrum of imaging capabilities, Earth Observation (EO) offers several opportunities for the geoscience community to map and monitor natural and human-induced Earth hazards from space.

The objective of this book is to collect scientific contributions on the development, validation, and implementation of satellite EO data, processing methods, and applications for mapping and monitoring of geohazards such as slow moving landslides, ground subsidence and uplift, and active and abandoned mining-related ground movements.

The book includes research papers published in 2017–2018 in the Special Issue 'Observing Geohazards from Space' of MDPI Geosciences, providing a number of novel case studies demonstrating how EO and remote sensing data can be used to detect and delineate land instability and geological hazards in different environmental contexts and using a range of spatial resolutions and image processing methods.

In my Editorial, I provide an overview of the topics covered in the 12 Special Issue papers, and an analysis of the distribution of EO and remote sensing data and methods used in each work, as well as the target geohazards mapped or monitored. The first three papers focus on the use of satellite and aerial optical imagery, and digital elevation models for the analysis of landslide hazards, with case studies in Austria, Italy and Spain. The following seven research papers exploit satellite radar interferometry to showcase the potential of this method and its advanced multi-temporal approaches to characterise subsidence and mining-related ground movements and hazards in Belgium, Italy, Slovakia, Spain, and the UK. The last two papers finally explore the crucial aspect of technical feasibility and, in particular, the assessment of the potential of radar interferometry for a nationwide analysis of geohazards, and to sense pre-failure landslide movements and feed into failure forecasting methods.

Francesca Cigna

Special Issue Editor

geosciences

MDPI

Editorial

Observing Geohazards from Space

Francesca Cigna

Italian Space Agency (ASI), Via del Politecnico s.n.c., 00133 Rome, Italy; francesca.cigna@asi.it

Received: 27 January 2018; Accepted: 5 February 2018; Published: 8 February 2018

Abstract: With a wide spectrum of imaging capabilities—from optical to radar sensors, low to very high resolution, continental to local scale, single-image to multi-temporal approaches, yearly to sub-daily acquisition repeat cycles—Earth Observation (EO) offers several opportunities for the geoscience community to map and monitor natural and human-induced Earth hazards from space. The Special Issue "Observing Geohazards from Space" of *Geosciences* gathers 12 research articles on the development, validation, and implementation of satellite EO data, processing methods, and applications for mapping and monitoring of geohazards such as slow moving landslides, ground subsidence and uplift, and active and abandoned mining-induced ground movements. Papers published in this Special Issue provide novel case studies demonstrating how EO and remote sensing data can be used to detect and delineate land instability and geological hazards in different environmental contexts and using a range of spatial resolutions and image processing methods. Remote sensing datasets used in the Special Issue papers encompass satellite imagery from the ERS-1/2, ENVISAT, RADARSAT-1/2, and Sentinel-1 C-band, TerraSAR-X and COSMO-SkyMed X-band, and ALOS L-band SAR missions; Landsat 7, SPOT-5, WorldView-2/3, and Sentinel-2 multi-spectral data; UAV-derived RGB and near infrared aerial photographs; LiDAR surveying; and GNSS positioning data. Techniques that are showcased include, but are not limited to, differential Interferometric SAR (InSAR) and its advanced approaches such as Persistent Scatterers (PS) and Small Baseline Subset (SBAS) methods to estimate ground deformation, Object-Based Image Analysis (OBIA) to identify landslides in high resolution multi-spectral data, UAV and airborne photogrammetry, Structure-from-Motion (SfM) for digital elevation model generation, aerial photo-interpretation, feature extraction, and time series analysis. Case studies presented in the papers focus on landslides, natural and human-induced subsidence, and groundwater management and mining-related ground deformation in many local to regional-scale study areas in Austria, Belgium, Italy, Slovakia, Spain, and the UK.

Keywords: natural hazards; landslides; subsidence; ground deformation; mining; object-based classification; InSAR; aerial photography; LiDAR; photogrammetry

1. An Overview of the Special Issue

Geohazards, such as landslides, volcanoes, earthquakes, tsunamis, ground subsidence, and uplift, pose significant risks to human life and property. Even when not with catastrophic consequences, or when characterised by relatively slow onset and evolution, these processes may be damaging over long periods. Human interaction with the environment may also combine with natural processes and thus trigger acceleration or exacerbate the impacts of geohazards on the built environment.

With a wide spectrum of imaging capabilities—from optical to radar sensors, low to very high resolution, continental to local scale, single-image to multi-temporal approaches, yearly to sub-daily acquisition repeat cycles—Earth Observation (EO) now offers several opportunities for the geoscience community to map and monitor natural and human-induced Earth hazards from space.

The Special Issue "Observing Geohazards from Space" of *Geosciences* encompasses 12 open access papers presenting research studies based on the exploitation of a broad range of EO data and techniques, as well as focusing on a well-assorted sample of geohazard types.

Table 1 summarizes the distribution of data and techniques used in each paper and the target geohazard types detected or monitored, and also provides the DOI and links to each paper to ease direct access by the readers.

Table 1. Overview of Earth Observation (EO) data, methods, and geohazard types that are discussed in the 12 open access research papers composing the Special Issue "Observing Geohazards from Space" of *Geosciences*. Access links to each paper are also provided via their unique DOI numbers. Papers are sorted in ascending order according to their publication date.

Paper Reference & DOI with Access Link	EO and Remote Sensing Data	Processing and Analysis Methods	Geohazard Types
Novellino et al. [1] 10.3390/geosciences7020019	Sentinel-1 satellite SAR	Intermittent SBAS InSAR	mining-related subsidence and uplift
Solari et al. [2] * 10.3390/geosciences7020021	COSMO-SkyMed and RADARSAT-2 satellite SAR	SBAS InSAR	urbanization-related subsidence
Bonì et al. [3] 10.3390/geosciences7020025	ERS-1/2, RADARSAT-1, ENVISAT, ALOS and COSMO-SkyMed satellite SAR	StaMPS, SPN, SqueeSAR™ and IPTA InSAR	groundwater management-related subsidence and uplift
Cencetti et al. [4] 10.3390/geosciences7020030	Google and Bing satellite optical data, aerial and ortho-photographs	multi-temporal photo-interpretation, feature extraction	landslide dams, river erosion and debris transport
Fernández et al. [5] 10.3390/geosciences7020032	aerial panchromatic and RGB-NIR photographs, LiDAR, GNSS	photogrammetry, DEM generation	landslides
Moretto et al. [6] 10.3390/geosciences7020036	satellite SAR data (simulated)	InSAR (simulated) post-processing	landslides
Hölbling et al. [7] * 10.3390/geosciences7020037	Landsat 7, WorldView-2/3, SPOT-5 and Sentinel-2 satellite optical data	object-based image analysis OBIA, photo-interpretation	landslides
Cigna et al. [8] 10.3390/geosciences7030051	ERS-1/2 satellite SAR, small UAV aerial RGB-NIR photographs, GNSS	SBAS InSAR, SfM photo-interpretation, DEM generation	mine collapse, landslides, natural compaction
Gee et al. [9] * 10.3390/geosciences7030085	ERS-1/2, ENVISAT and Sentinel-1 satellite SAR	Intermittent SBAS InSAR	mining-related subsidence and uplift
Czikhardt et al. [10] 10.3390/geosciences7030087	Sentinel-1 satellite SAR, LiDAR, UAV aerial photographs	StaMPS InSAR, SfM, DEM generation	landslides, mining-related subsidence
Declercq et al. [11] 10.3390/geosciences7040115	ERS-1/2, ENVISAT, TerraSAR-X and Sentinel-1 satellite SAR	StaMPS InSAR, SARProZ PS	groundwater management-related uplift, natural compaction
Matano et al. [12] 10.3390/geosciences8010008	ERS-1/2, RADARSAT-1 and ENVISAT satellite SAR	PS-InSAR™, PSP-DIFSAR	natural compaction and human-induced subsidence, tectonic uplift

* Invited Feature Papers.

1.1. Data, Methods and Geohazard Domains

In their Invited Feature Paper, Hölbling et al. [7] present a spatial accuracy assessment of landslide detection and delineation via semi-automated Object-Based Image Analysis (OBIA). The authors process a set of multi-spectral EO data from the Landsat 7, SPOT-5, WorldView-2/3, and Sentinel-2 missions by using image segmentation and OBIA in five study areas in the Alps in Austria and Italy. Comparison of the OBIA-generated landslide maps with the outcomes from visual image interpretation of the same EO data allows them to identify and discuss in detail the potentials and limitations of the two approaches (visual and semi-automated) and to demonstrate that both methods produce similar results when the same data basis is used.

Keeping the focus on landslide analysis with EO and remote sensing data, by means of an assorted, 60-year long temporal stack of satellite imagery, ortho-photographs, and aerial photographs, Cencetti et al. [4] study the Costantino landslide (Italy) that dammed the Bonamico River Valley in 1973 and caused the formation of a 175,000 m^2 lake. A combined approach of photo interpretation, analysis of climate records, and assessment of wet/drought periods allows the authors to examine the evolution of the lake surface and landslide dam and, in particular, to reconstruct the natural filling of the lake via river debris transport and accumulation, as well as dam overtopping and failure, and to sense the lake extinction in recent years.

Going back to similar historical times, Fernández et al. [5] use a 54-year long multi-temporal set of photogrammetric flights for the generation of Digital Terrain Models (DTM) to analyse the evolution of the Almegíjar rock slide in Spain. The authors exploit an aerial digital camera with four spectral bands (i.e., visible RGB, and near-infrared NIR), a Light Detection and Ranging (LiDAR) sensor, and in-flight orientation based on Global Navigation Satellite System (GNSS) onboard a 2010 flight to generate a high-precision DTM that they then use as a reference for comparison with other elevation models generated based on historical panchromatic flights. Differential DTMs, cross sections, and volumetric changes are computed to identify depletion and accumulation areas and to estimate landslide movement rates and activity.

Moving from optical to Synthetic Aperture Radar (SAR) satellite imagery, the Invited Feature Paper by Gee et al. [9] presents a multi-mission study of land motion and groundwater level change in areas of abandoned mining in the UK using a 21-year long series of ERS-1/2, ENVISAT and Sentinel-1 satellite SAR data in C-band. The authors employ a modified version of the Small BAseline Subset (SBAS) technique allowing extraction of ground motion information across most land cover types, and identify surface depressions in proximity to former collieries and broad areas experiencing regional scale uplift. The latter often occurs in previously mined areas where groundwater levels rise following the cessation of groundwater pumping after mine closure.

The extensive spatial coverage, regular temporal sampling, and free data availability of Sentinel-1 are further discussed in the paper by Czikhardt et al. [10], who provide an in-depth multi-temporal study of recent ground stability and movements for a 440 km^2 study region in Slovakia characterised by high landslide susceptibility and intensive brown coal mining. The authors combine two open-source processing tools, i.e., the Sentinel Application Platform (SNAP) and Stanford Method for Persistent Scatterers (StaMPS), and compare their satellite-based observations with ground truth data from borehole inclinometers and terrestrial levelling to estimate the accuracy of the Sentinel-1 InSAR results. The use of consistent Sentinel-1 stacks in both ascending and descending modes, along with motion decomposition along the vertical and horizontal components, allows the authors to demarcate distinct small-areas of spatially clustered motion due to the presence of slope instability and severe land subsidence due to undermining. This result is also confirmed based on the comparison of Unmanned Aerial Vehicle (UAV) and LiDAR-derived 3D elevation models.

With a similar combination of EO and remote sensing techniques, Cigna et al. [8] discuss how to enhance the understanding of historical ground stability and motion in areas of geotechnical infrastructure and abandoned salt mine workings by means of satellite InSAR and tailored small UAV surveying. With a 3000 km^2 study area in Co. Antrim (UK), the authors utilise ERS-1/2 SBAS InSAR to achieve a baseline ground instability scenario over wide areas encompassing several elements at risk and strategic infrastructure. To detect localised deformation and indicators of ground instability, the InSAR-derived baseline is integrated with analysis of very high resolution ortho-photographs and 3D surface models generated via Structure-from-Motion (SfM). Links to access the 3D model and ortho-photographs for a salt mine collapse site within the study site are also provided as appendices to the paper.

With a focus on land subsidence, in their Invited Feature Paper Solari et al. [2] apply the well-established SBAS algorithm to quantify recent (2011–2014) subsidence rates and temporal evolution in the urban area of Pisa (Italy) using two stacks of RADARSAT-2 C-band and COSMO-SkyMed X-band satellite SAR imagery. From their InSAR analysis at the local and building scales in relation to information on the recent evolution of the city, the authors detect an apparent correlation between subsidence rates and location of modern urban development areas. Their study demonstrates that the construction of new buildings acts as the trigger for accelerated consolidation of highly compressible layers onto which the urban fabric of the city is built.

With an analogous objective, the paper by Declercq et al. [11] provides a 25-year long (1992–2017) study of ground instability for the city of Brussels (Belgium) through combination of C- and X-band satellite SAR data from the ERS-1/2, ENVISAT, TerraSAR-X, and Sentinel-1 missions. A total of ~ 300 scenes are processed with Persistent Scatterers (PS) methods to extract deformation time series and detect the evolution of ground deformation. The authors find evidence of ongoing, slow ground deformations in particular in the historic centre, where land uplift occurs as a result of a reduction of groundwater extraction from deep aquifers and consequent groundwater level rise related to deindustrialization, as confirmed by piezometric data. Natural settlement of soft alluvial deposits along river valleys, possibly increased by the load of buildings, is also detected.

Looking at a whole alluvial plain susceptible to land subsidence and coastal hazards, Matano et al. [12] use almost two decades of InSAR information (1992–2010) derived from the ERS-1/2, ENVISAT and RADARSAT-1 C-band SAR missions to detect subsidence trends of the Volturno River Coastal Plain (Italy). The authors interpret and post-process six datasets of Permanent Scatterers InSAR (PS-InSAR) and Persistent Scatterer Pairs (PSP) results to derive maps of vertical deformation and identify significant subsidence rates in the central axial sectors and the river mouth area, and some uplift related to tectonic activity in the eastern part of the plain. The analysis of subsoil stratigraphy allows them to infer that subsidence is mainly due to natural compaction of the fluvial and palustrine deposits, with anthropogenic influences such as groundwater exploitation and urbanization in some sectors of the study area acting as an additional geohazard factor.

Bonì et al. [3] tackle the use of satellite InSAR methods and SAR data from the methodological point of view, and present three case studies in Italy, Spain, and the UK, where approaches to analysing and interpreting InSAR results for land subsidence detection, characterisation, and modelling are exploited with a focus on the relationship of surface deformation with groundwater level change and exploitation of aquifer resources. Using various stacks of C-, L- and X-band satellite SAR imagery, as well as different InSAR image processing methods (i.e., SqueeSAR^{TM}, StaMPS, Stable Points Network or SPN, and Interferometric Point Target Analysis or IPTA), the authors showcase InSAR time series post-processing methods such as principal component analysis to identify and separate linear, non-linear, and seasonal movements, transient and inelastic aquifer deformation, and accelerating and decelerating subsidence trends.

Looking at the applicability of satellite InSAR methods for geohazard mapping and monitoring, an approach to mapping a priori the feasibility of SAR imaging and InSAR monitoring of land deformation is presented by Novellino et al. [1]. With a focus on the Sentinel-1 SAR constellation, the authors build upon the InSAR feasibility assessment developed in [13] for the ERS-1/2 and ENVISAT missions and run an analysis of Sentinel-1 data availability, topographic distortions, and land cover constraints toward the use of a novel SBAS algorithm to sense ground stability and motion across a whole country. The aspect of achievable standard deviation in the derived InSAR velocities is also discussed to showcase the importance of consistent time stacks of SAR data as inputs for advanced InSAR methods to generate robust geohazard information.

With a specific look at landslide phenomena, on the other hand, Moretto et al. [6] explore the potential of satellite InSAR methods to sense pre-failure landslide movements and to feed into failure forecasting methods. For 56 landslide sites with monitoring data available in the scientific literature, the authors collate pre-failure deformation information from on site and remote monitoring instruments, such as inclinometers, ground-based InSAR, and total station, and resample the frequency of these records to simulate satellite acquisition parameters (e.g., revisit times and phase ambiguity). Their analysis reveals that InSAR could successfully support monitoring of landslide tertiary creep, and in some cases, the time of failure as well. However, for the purposes of risk management and early warning, it is extremely important for future satellite SAR missions to provide high data sampling frequencies (i.e., short revisit times) to capture pre-failure landslide dynamics.

1.2. Statistics, Bibliometrics and Impact

The 12 research papers were published in the Special Issue between the end of March 2017 and early January 2018, with an average time of about two months from first submission to online publication. Each manuscript was assessed via rigorous peer-reviewing from two or more esteemed experts in the respective field [14].

Based on MDPI's article metrics powered by TrendMD, since the publication of the first paper in March 2017 and as of the beginning of 2018 the Special Issue received more than 10,000 views totally, and is now reached by ~35 readers/day on average. A widespread geographic distribution of readers across the globe is observed, with apparent peaks in the USA, Asia and Europe.

Overall, the published papers already received five citations in the indexed literature in the first few months after publication, proving the immediate impact of the published research. In particular, three out of five of the latter citations were received by the first paper of the Special Issue that was published in March 2017 [1], and four out of five citations were made by papers published in other scientific journals in the field of remote sensing, EO, and Earth sciences.

2. Further Reading

Readers interested in the use of geospatial and remote sensing data and methods for the investigation of natural hazards and disasters, in addition to this Special Issue, can also refer to papers published in other recent Special Issues of *Geosciences* such as "Mapping and Assessing Natural Disasters Using Geospatial Technologies" collecting articles published in 2016–2017 [15], "Advances in Remote Sensing and GIS for Geomorphological Mapping" published in 2015–2016 [16], "Remote Sensing and Geosciences for Archaeology" published in 2017 [17], and "Geological Mapping and Modeling of Earth Architectures" published in 2014–2016 [18].

In these Issues, several examples of exploitation of EO and other geospatial data and applications are presented, including use of airborne hyperspectral data, detection of hotspots and thermal anomalies, GIS-based hazard and risk assessment, and geological, structural, and geomorphological mapping. Similarly, other natural hazards and surface processes are covered, such as floods, droughts, fires, tectonics, karst, and ground dissolution, as well as human-induced hazards and threats to urban areas and heritage assets.

Acknowledgments: The Guest Editor would like to acknowledge all the authors for contributing to the Special Issue and the anonymous peer reviewers for assessing the submitted manuscripts and greatly helping the authors to enhance the scientific quality of their papers. Sincere gratitude goes to the Editorial Board and Office of *Geosciences*, especially Assistant Editor Veronica Wang, for the invaluable help and assistance provided at all stages of the design, management, and publication of this Special Issue.

Author Contributions: Following on from the invitation of Jesus Martínez-Frías, Editor-in-Chief of *Geosciences*, Francesca Cigna designed this Special Issue, served as Lead Guest Editor, and wrote this Editorial.

Conflicts of Interest: The author declares no conflict of interest.

References

1. Novellino, A.; Cigna, F.; Brahmi, M.; Sowter, A.; Bateson, L.; Marsh, S. Assessing the feasibility of a national InSAR ground deformation map of great britain with sentinel-1. *Geosciences* **2017**, *7*, 19. [CrossRef]

2. Solari, L.; Ciampalini, A.; Raspini, F.; Bianchini, S.; Zinno, I.; Bonano, M.; Manunta, M.; Moretti, S.; Casagli, N. Combined Use of C- and X-Band SAR Data for Subsidence Monitoring in an Urban Area. *Geosciences* **2017**, *7*, 21. [CrossRef]

3. Bonì, R.; Meisina, C.; Cigna, F.; Herrera, G.; Notti, D.; Bricker, S.; McCormack, H.; Tomás, R.; Béjar-Pizarro, M.; Mulas, J.; et al. Exploitation of satellite A-DInSAR time series for detection, characterization and modelling of land subsidence. *Geosciences* **2017**, *7*, 25. [CrossRef]

4. Cencetti, C.; Di Matteo, L.; Romeo, S. Analysis of Costantino Landslide Dam Evolution (Southern Italy) by Means of Satellite Images, Aerial Photos, and Climate Data. *Geosciences* **2017**, *7*, 30. [CrossRef]

5. Fernández, T.; Pérez, J.L.; Colomo, C.; Cardenal, J.; Delgado, J.; Palenzuela, J.A.; Irigaray, C.; Chacón, J. Assessment of the Evolution of a Landslide Using Digital Photogrammetry and LiDAR Techniques in the Alpujarras Region (Granada, Southeastern Spain). *Geosciences* **2017**, *7*, 32. [CrossRef]

6. Moretto, S.; Bozzano, F.; Esposito, C.; Mazzanti, P.; Rocca, A. Assessment of Landslide Pre-Failure Monitoring and Forecasting Using Satellite SAR Interferometry. *Geosciences* **2017**, *7*, 36. [CrossRef]

7. Hölbling, D.; Eisank, C.; Albrecht, F.; Vecchiotti, F.; Friedl, B.; Weinke, E.; Kociu, A. Comparing Manual and Semi-Automated Landslide Mapping Based on Optical Satellite Images from Different Sensors. *Geosciences* **2017**, *7*, 37. [CrossRef]

8. Cigna, F.; Banks, V.J.; Donald, A.W.; Donohue, S.; Graham, C.; Hughes, D.; McKinley, J.M.; Parker, K. Mapping ground instability in areas of geotechnical infrastructure using satellite InSAR and small UAV surveying: A case study in Northern Ireland. *Geosciences* **2017**, *7*, 51. [CrossRef]

9. Gee, D.; Bateson, L.; Sowter, A.; Grebby, S.; Novellino, A.; Cigna, F.; Marsh, S.; Banton, C.; Wyatt, L. Ground motion in areas of abandoned mining: Application of the intermittent SBAS (ISBAS) to the Northumberland and Durham Coalfield, UK. *Geosciences* **2017**, *7*, 85. [CrossRef]

10. Czikhardt, R.; Papco, J.; Bakon, M.; Liscak, P.; Ondrejka, P.; Zlocha, M. Ground Stability Monitoring of Undermined and Landslide Prone Areas by Means of Sentinel-1 Multi-Temporal InSAR, Case Study from Slovakia. *Geosciences* **2017**, *7*, 87. [CrossRef]

11. Declercq, P.-Y.; Walstra, J.; Gérard, P.; Pirard, E.; Perissin, D.; Meyvis, B.; Devleeschouwer, X. A Study of Ground Movements in Brussels (Belgium) Monitored by Persistent Scatterer Interferometry over a 25-Year Period. *Geosciences* **2017**, *7*, 115. [CrossRef]

12. Matano, F.; Sacchi, M.; Vigliotti, M.; Ruberti, D. Subsidence Trends of Volturno River Coastal Plain (Northern Campania, Southern Italy) Inferred by SAR Interferometry Data. *Geosciences* **2018**, *8*, 8. [CrossRef]

13. Cigna, F.; Bateson, L.B.; Jordan, C.J.; Dashwood, C. Simulating SAR geometric distortions and predicting Persistent Scatterer densities for ERS-1/2 and ENVISAT C-band SAR and InSAR applications: Nationwide feasibility assessment to monitor the landmass of Great Britain with SAR imagery. *Remote Sens. Environ.* **2014**, *152*, 441–466. [CrossRef]

14. Geosciences Editorial Office. Acknowledgement to Reviewers of Geosciences in 2017. *Geosciences* **2018**, *8*, 33. [CrossRef]

15. Pu, R. A Special Issue of Geosciences: Mapping and Assessing Natural Disasters Using Geospatial Technologies. *Geosciences* **2017**, *7*, 4. [CrossRef]

16. Propastin, P.; Sheng, Y. Geosciences Special Issue "Advances in Remote Sensing and GIS for Geomorphological Mapping". Available online: http://www.mdpi.com/journal/geosciences/special_issues/geomorphological-mapping (accessed on 11 January 2018).

17. Tapete, D. Remote Sensing and Geosciences for Archaeology. *Geosciences* **2018**, *8*, 41. [CrossRef]

18. McCaffrey, K. Geosciences Special Issue "Geological Mapping and Modeling of Earth Architectures" Available online: http://www.mdpi.com/journal/geosciences/special_issues/geological-mapping (accessed on 11 January 2018).

MDPI

Article

Comparing Manual and Semi-Automated Landslide Mapping Based on Optical Satellite Images from Different Sensors

Daniel Hölbling [1],*, Clemens Eisank [2], Florian Albrecht [1], Filippo Vecchiotti [3], Barbara Friedl [1], Elisabeth Weinke [1] and Arben Kociu [3]

[1]   Department of Geoinformatics–Z_GIS, University of Salzburg, Schillerstrasse 30, 5020 Salzburg, Austria;
        florian.albrecht@sbg.ac.at (F.A.); barbara.friedl@sbg.ac.at (B.F.); elisabeth.weinke@sbg.ac.at (E.W.)
[2]   GRID-IT—Gesellschaft für angewandte Geoinformatik mbH, Technikerstraße 21a, 6020 Innsbruck, Austria;
        eisank@grid-it.at
[3]   Geological Survey of Austria (GBA), Neulinggasse 38, 1030 Vienna, Austria;
        filippo.vecchiotti@geologie.ac.at (F.V.); arben.kociu@geologie.ac.at (A.K.)
*   Correspondence: daniel.hoelbling@sbg.ac.at; Tel.: +43-662-8044-7581

Academic Editors: Francesca Cigna and Jesus Martinez-Frias
Received: 31 March 2017; Accepted: 11 May 2017; Published: 19 May 2017

Abstract: Object-based image analysis (OBIA) has been increasingly used to map geohazards such as landslides on optical satellite images. OBIA shows various advantages over traditional image analysis methods due to its potential for considering various properties of segmentation-derived image objects (spectral, spatial, contextual, and textural) for classification. For accurately identifying and mapping landslides, however, visual image interpretation is still the most widely used method. The major question therefore is if semi-automated methods such as OBIA can achieve results of comparable quality in contrast to visual image interpretation. In this paper we apply OBIA for detecting and delineating landslides in five selected study areas in Austria and Italy using optical Earth Observation (EO) data from different sensors (Landsat 7, SPOT-5, WorldView-2/3, and Sentinel-2) and compare the OBIA mapping results to outcomes from visual image interpretation. A detailed evaluation of the mapping results per study area and sensor is performed by a number of spatial accuracy metrics, and the advantages and disadvantages of the two approaches for landslide mapping on optical EO data are discussed. The analyses show that both methods produce similar results, whereby the achieved accuracy values vary between the study areas.

Keywords: landslides; remote sensing; semi-automated mapping; object-based image analysis (OBIA); manual mapping; visual interpretation; spatial accuracy metrics; Alps

1. Introduction

Optical Earth Observation (EO) data has proven beneficial for mapping the location and spatial extent of landslides at various spatial scales, ranging from local to regional [1]. The interpretation of optical EO data for landslide recognition may be conducted using manual, semi-automatic, and automatic approaches [2].

Traditionally, manual visual image interpretation for landslide recognition has been based on aerial photographs and recently extended to high resolution (HR) and very high resolution (VHR) optical satellite imagery (e.g., WorldView-2/3, QuickBird) [2] and is still the most widely used method for landslide detection and inventory preparation. Imagery with lower resolutions (e.g., SPOT, Landsat) has also been successfully employed for delineating single, larger landslides [3], yet VHR optical data has become the best choice for landslide recognition on the basis of landform analysis over larger areas [1]. The comparison between landslide maps derived from aerial images and satellite images reveals that both data sources yield results of equal value [4,5]. Manual visual image interpretation of optical imagery turned out to be practicable for landslide mapping, especially where landslides led to significant changes in the land cover, as is usually the case for shallow landslides. Shape, size, pattern, and texture with characteristic contrast in brightness (for black and white photography) or color variation (in case of a multispectral image) are used for manual landslide interpretation [6]. One major advantage related to visual image interpretation is the use of stereo-vision for landslide recognition, as some landslide features are detectable only employing the third dimension [2]. However, manual interpretation reveals several drawbacks; it is resource and time consuming and the quality of the resulting landslide maps strongly depends on the experience and skills of the investigator [7–9]. The variability in human visual interpretation tasks across time and individuals is a common but understated problem and has also been investigated for remote sensing image interpretation of features in urban areas [10], in rural areas [11], and in slums [12]. This subjectivity issue is difficult to measure, but influences the reliability of landslide mapping [13,14], which further depends on the persistence of the landslide morphology, the quality and scale of the imagery, the complexity of the geological and geomorphological setting, the presence of vegetation coverage, and the final map scale [9]. It has been recognized that, when more than one compilation of landslide inventories is closely evaluated, different operators produce unequal results in terms of the number and density of landslides, which also vary in geometry and extent [7,8]. Ardizzone et al. [14] compared three independently produced landslide inventory maps and conclude that the detail and precision of landslide maps primarily depend on the experience and skills of the investigator and only secondly on the quality of the remote sensing data and base maps used.

During the past decade, advances in computer science and machine intelligence and the ever increasing number of satellite missions acquiring VHR data led to the development of novel technologies, fostering the semi-automated interpretation of EO data. The current approaches can be split into two categories [1,15], pixel-based and object-based, both containing techniques applicable to single and multi-temporal images and frequently making use of additional data, e.g., digital elevation models (DEMs) [16]. Pixel-based methods take into account the spectral information associated with each pixel without considering its actual neighborhood [2]. Considering the resolution of HR/VHR data and the size and spatial distribution of landslides, pixel-based methods tend to be additionally sensitive to errors [17], frequently resulting in salt-and-pepper classifications, with single pixels being demarcated as landslides.

Due to these circumstances, there has been a trend towards object-based landslide recognition in recent years, as demonstrated by a range of studies [8,16–29]. Object-based image analysis (OBIA) provides a set of innovative tools for semi-automatically delineating and classifying landslides based on EO data. Object-based algorithms rely on the concepts of image segmentation and classification. Regarding landslide recognition, landslides or landslide features are composed of aggregations of homogeneous pixels rather than spatially uncorrelated cells and are then classified according to specific characteristics. By supporting the use of a plethora of properties during the classification process, e.g., spectral, spatial, textural, and contextual, OBIA is somewhat mimicking human perception and thereby proves its potential for producing more accurate and realistic landslide maps than pixel-based procedures [15,17,18]. OBIA therefore can be considered a specific type of model of the human perception process [30]. This model type performs well in producing an understandable encoding of explicit expert knowledge in rulesets for the production of repeatable image classification results, e.g., as demonstrated by Eisank et al. [31]. At the same time, OBIA is also a simplification of the human perception process and has certain disadvantages. In practice, OBIA workflows are a tradeoff between keeping a low complexity and covering all details and tend to ignore special cases in favor of generally applicable classification rules. Therefore, like manual visual image interpretation, OBIA-derived classification results are not free of error either. Consequently, the accuracy of semi-automated landslide classifications should be documented, but, at the same time, the reliability of the reference should be critically scrutinized. Attainable accuracy values vary considerably between studies, often due to the varying complexity and appearance of landslides. Accuracy values reach up to 95% for the detection of landslide scars and range between 73% and 93.7% for the detection of landslide-affected areas and between 44.8% and 90% for the identification of different landslide types such as debris slides, rock slides, or debris flows [15–22,25–27]. Object-based landslide mapping routines are often tailored to specific study areas and data and thus reveal a lack of transferability and robustness. Also the selection of suitable segmentation parameters for the creation of "meaningful" image objects, which in turn affects the accuracy of image classification as well as the application of user driven thresholds, remain critical issues [27].

In this study, we aim to compare the results of two independently performed landslide mapping approaches; semi-automated detection of landslides by OBIA and manual mapping of landslides. The analyses are based on optical satellite imagery obtained from different sensors, i.e., Landsat 7, SPOT-5, WorldView-2, WorldView-3, and Sentinel-2. Based on five selected case studies in the Alps, spatial accuracy values are calculated to evaluate if the results are comparable when the same data basis is used for analyses.

2. Materials and Methods

2.1. Study Areas

Five landslide prone areas in the Alps with different geomorphic characteristics were selected as test sites for the development and comparison of the mapping approaches. Two study sites are located in the federal state of Salzburg, Austria, two in the federal state of Vorarlberg, Austria, and one in Südtirol-Alto Adige Autonomous Province, Italy (Figure 1). The Italian test site is divided into two sub-areas located in the Gader Valley.

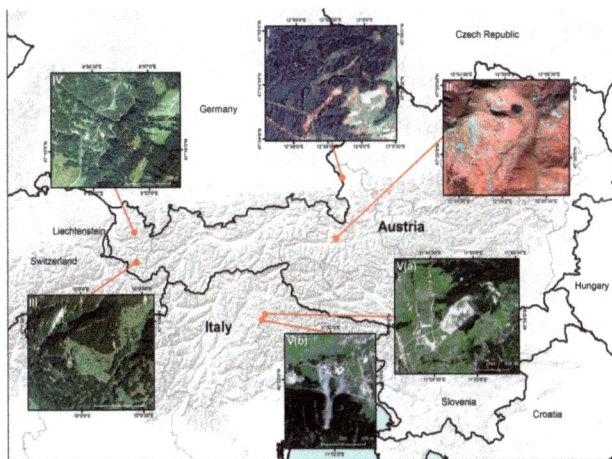

Figure 1. Study areas in Austria and Italy. Study areas I (Fürwag/Haunsberg) and II (Pinzgau) are located in the federal state of Salzburg, Austria; study areas III (Montafon) and IV (Bregenzerwald) are located in the federal state of Vorarlberg, Austria; and study area V (Gader Valley) is in Südtirol-Alto Adige Autonomous Province, Italy. The Italian study area in the Gader Valley is divided in two sub-areas (V(a) and V(b)).

2.1.1. Study Areas I and II in the Federal State of Salzburg

Study areas I and II (Salzburg; Figure 1) are located in a geologic domain that encompasses Quaternary until Permian units. They are respectively [32]:

- The quaternary alluvial and moraine deposits
- The tertiary Eocene-Miocene Molasse
- The Jura-mid Eocene continental margin Helvetic Zone
- The Lower Cretaceous-Eocene Penninic Rhenodanubian Flysch
- The Permo-Mesozoic Northern Calcareous Alps (NCA)

The Permo-Mesozoic Northern Calcareous Alps were deposited in a reef-submerged environment corresponding to the opening of the Penninic Sea. At the very end of the Cretaceous Period, on the active margin of the Austroalpine microplate, as a result of the subduction, an accretionary wedge was formed, and a new deep water flysch-like deposition facies, the Rhenodanubian Flysch, took place [32].

The erosional debris of the Alpine uplift occurring between the Alpine orogenic front and the European foreland formed the tertiary Molasse. During the Würmian, the extent of the Salzach glacier reached the Inn basin to the north. The Salzach valley, an over-deepened Alpine valley, was filled mainly with lacustrine and delta sediments. The melting outlet-glacier deposited big terminal moraines, and the melt water runoff caused strong lateral erosion along the foot of the slopes, which locally provoked deep-seated gravitational slope deformations [33].

The "Fürwag Landslide" (study area I) is located at the Haunsberg, approximately 10 km north of the city of Salzburg, and covers an area of approximately 170 ha. The landslide-prone area consists of various mudslides, which have been formed at different times. The sandstones, shales, and the Flysch, in particular the Zementmergelserie (marl cement series), of this pre-alpine zone are very prone to mass movements [34]. The latest sliding processes occurred between 1999 and 2003, and have almost reached the valley floor, representing a constant threat to major infrastructure facilities [35].

In the Pinzgau region (study area II) the greywacke zone is especially susceptible to sliding processes. The Central Alps are dominated by different metamorphic rock types, leading to a great variety of mass movement types [34]. In June 2013 parts of this region have been declared a disaster zone due to torrential rains, which caused severe flooding and numerous landslides. A mountainous area (the Hundstein Mountain) characterized by a high relief and sparse vegetation, where several geomorphological phenomena occur, has been selected as study area in the Pinzgau region.

2.1.2. Study Areas III and IV in the Federal State of Vorarlberg

The major tectonic units of study areas III and IV (Vorarlberg; Figure 1), can be subdivided from North to South into three groups [36]:

- Molasse
- Helvetic Nappes and Vorarlberg Flysch (Eastern Alps origin)
- Northern Calcareous Alps and Silvretta Crystalline (Western Alps origin)

The gneisses, schists, and amphibolites of the Silvretta Crystalline of the Variscan age are the oldest rocks in the area. The Helvetic limestones and marls (as well as the NCA) were formed in a shallow sea in Mesozoic times, whereas the deep turbiditic origin of the Vorarlberg Flysch dates back to the Late Cretaceous-Early Tertiary Period. The Molasse was deposited instead in the Alpine foreland basin as a residual product of the mountain building process and consists of conglomerate sandstones, claystone and marls of mixed marine and non-marine nature [36].

In the Montafon valley (study area III) tectonic deformation and weakening of rock strength during Alpine orogenesis favors the occurrence of landslides [37]. Additionally, glacial erosion led to the steepening of slopes, triggering tensional fissures and accelerating a number of geomorphological processes, e.g., rock slides and rock falls [38].

In Bregenzerwald (study area IV), the valley slopes were oversteepened by glacial erosion, leading to increased slope instability after deglaciation [39]. In May 2005, intense and prolonged rainfall induced inundations and numerous landslides, leading to severe infrastructure and property damages. Prominent examples of major landslides in the Bregenzerwald area are the Doren landslide and the landslide Rindberg/Sibratsgfäll [40,41].

2.1.3. Study Area V in Südtirol—Alto Adige Autonomous Province

The major lithological formations of study area V (Südtirol-Alto Adige Autonomous Province; Figure 1) are [42]:

- Schliar dolomites (detritic massive dolomite of Anisian age)
- Wengen formation (marls and calciturbidites mixed with volcaniclastic conglomerates of Ladinian age)
- S. Cassian formation (grey mudstones and calciturbidites of Carnian age)

The geology of the Dolomite area is complex being the result of the dismantling of the Hercinian mountains and Permo-Trias rifting tectonics in Permian times, which gave way to the Athesian magmatism [42]. In Anisian times, the massive detritic Schliar dolomite was formed as a consequence of a rifting event. After another upper Ladinian paroxistic volcanic phase, the Wengen formation was formed. The successive recovery of carbonate sedimentation allowed the right conditions for the sedimentation of the S. Cassiano Formation [43].

The Gader Valley (study areas V(a) and V(b)) is prone to gravitational processes such as landslides due to its geological and geomorphological conditions, particularly the upper parts of the valley [44]. In December 2012, after long and intensive rainfall, a large landslide destroyed several buildings and endangered critical infrastructure in Badia. Further large landslides are located near the village of Corvara. The Corvara landslide, which affects an area of more than 2.5 km^2 and has an overall volume of more than 300 million m^3 [45], repeatedly causes damages to infrastructure. The most recent landslide happened in April 2014 at the border between Corvara and Colfosco. Some areas on its western part were reactivated in August 2016.

2.2. Data

For each study area, selected subsets of optical satellite images with different properties from different sensors were used (Table 1). Digital elevation models (DEMs), and particularly the derived slope information, were used as supplementary data. The database for Montafon and Bregenzerwald was also complemented by a terrain roughness index (TRI) layer, computed as the sum of change in elevation between a grid cell and its eight neighboring grid cells [46]. For the study areas in Salzburg, a DEM derived from airborne laser scanning (ALS), which is freely available at a resampled resolution of 10 m from the geodata portal of Austria (www.geoland.at), was used. For the two study areas in Vorarlberg, an ALS-based DEM at 5 m spatial resolution provided by the federal government of Vorarlberg was available. The ALS campaigns over the Austrian Alps were conducted between 2006 and 2013 by the regional governments. In Südtirol-Alto Adige Autonomous Province, a DEM at 5 m resolution (acquired between 2004 and 2006) was provided by the Autonomous Province of Bolzano via their geodata portal (http://geoportal.buergernetz.bz.it/default.asp).

Table 1. Selected database for each study area.

Study Area	Size (km^2)	Optical Sensor	Acquisition Date	Spectral Resolution [1]	Spatial Resolution (m)	DEM Resolution (m)
I	4.7	Landsat 7	28 July 2002	1× pan 7× multispectral (blue, green, red, nir, swir-1, thermal, swir-2)	15 30	10
II	2.8	SPOT-5	10 September 2011	3× multispectral (green, red, nir)	2.5	10
III	1.4	WorldView-2	29 August 2015	1× pan 4× multispectral (blue, green, red, nir)	0.5 2	5
IV	1.8	WorldView-3	13 August 2015	1× pan 4× multispectral (blue, green, red, nir)	0.5 2	5
V(a)	3	Sentinel-2	27 August 2016	13× multispectral (coastal aerosol, blue, green, red, red edge 1-3, nir, red edge 4, water vapour, swir cirrus, swir 1-2)	10 (blue, green, red, nir) 20 (red edge 1-4, swir 1-2) 60 (other)	5
V(b)	1.2					

[1] refers to the available panchromatic (pan) and multispectral bands for each dataset.

The image selection process relied primarily on the availability of cloud-free data acquired in the summer season, possibly soon after a major landslide-triggering event or after the reactivation of landslide processes. Regarding DEM information, the supply of multi-temporal and post-event data was limited; the available data sets were therefore considered even if the acquisition date is not always consistent with the respective optical images. Pre-processing of the satellite images was implemented with ERDAS IMAGINE software (Hexagon Geospatial, Stockholm, Sweden) and included:

- Pansharpening if a panchromatic band was provided along with multispectral bands.
- Orthorectification supported by rational polynomial coefficients (RPCs) and/or ground control points (GCPs) to achieve sub-pixel geolocation accuracies.

2.3. Object-Based Landslide Mapping

For semi-automated landslide mapping, object-based approaches were developed and applied for each dataset and study area. Efforts were made to follow similar workflows; however, due to the different characteristics of the satellite images and study areas and because of the varying complexity and appearance of landslides, the transferability was limited. Thus, the classification rules were adapted for each test case. Moreover, two different but interchangeable software products, eCognition 9.2 (Trimble Geospatial, Sunnyvale, CA, USA) and InterIMAGE 1.43 (Pontifical Catholic University of Rio de Janeiro (PUC-Rio) and Brazilian National Institute for Space Research (INPE), Rio de Janeiro, Brazil), were used to develop the knowledge-based image analysis workflows. The two products are similar in their logic and functionality, meaning that a workflow implemented in eCognition can also be implemented in InterIMAGE and vice versa. Since the major objective of this study is the evaluation of OBIA mapping results in comparison to manual mapping results, rather than the development of sophisticated mapping routines, the implemented OBIA approaches were kept as simple as possible to produce acceptable results in a fast and efficient manner.

Table 2 shows the software that was used for mapping landslides in each study area, the segmentation method and the segmentation parameters that were used for creating image objects, and the main classification parameters for identifying landslide objects. Slightly different segmentation and classification parameters were used for mapping the two large landslides in study area V. Thus, the sub-areas (V(a), a landslide near the villages of Badia and La Villa, and V(b), a landslide near the village of Corvara) were treated separately during OBIA mapping. In general, the Normalized Difference Vegetation Index (NDVI) and the slope information were most useful for the detection of landslide bodies. For refining the classification, further parameters (e.g., brightness, length/width ratio, relation to neighboring objects, compactness, perimeter/area ratio) were used additionally or in combination with each other. Thresholds were selected based on (1) an exploratory analysis of image object properties (e.g., mean slope, mean NDVI) and (2) recommendations in the literature [17,31].

Table 2. Software, the segmentation method and parameters (SP—Scale Parameter, S—Shape criterion, C—Compactness criterion), and the main classification parameters used for object-based landslide mapping for each case study.

Study Area	Software	Segmentation Method and Parameters	Main Classification Parameters
I	eCognition 9.2 (Trimble Geospatial)	Multiresolution segmentation: SP: 25; S: 0.1; C: 0.5	Mean diff. to neighbors (NDVI) < 0 Mean slope > 10 Length/Width > 4
II	eCognition 9.2 (Trimble Geospatial)	Multiresolution segmentation: SP: 15; S: 0.3; C: 0.5	Mean NDVI < 0.04 Mean slope > 10 Mean brightness > 85
III	InterIMAGE 1.43	Threshold segmentations based on NDVI and brightness layer (Threshold = Mean of layer)	Mean slope > 35° Mean TRI > 1.5 Size > 100 m^2 Perimeter/Area Ratio (P/A) < 0.55 Mean NDVI < 0.55

Table 2. *Cont.*

Study Area	Software	Segmentation Method and Parameters	Main Classification Parameters
IV	InterIMAGE 1.43	1. Threshold segmentations based on NDVI and brightness layer (Threshold = Mean of layer) 2. Multiresolution segmentation based on slope; SP: 400; S: 0.5; C: 0.5	Same as III, except: Mean slope > 15° No use of Mean NDVI
V (a)	eCognition 9.2 (Trimble Geospatial)	Multiresolution segmentation: SP: 150; S: 0.5; C: 0.5	Mean NDVI < 0.4 or < 0.5 (depending on fine adjustment in combination with other features) Mean slope > 10
V (b)	eCognition 9.2 (Trimble Geospatial)	Multiresolution segmentation: SP: 150; S: 0.3; C: 0.8	Mean NDVI < 0.4 Mean slope > 8 Length/Width > 1.6 Shape index < 3.6

2.4. Manual Landslide Mapping

Three different types of features were manually mapped. First, points at the central area of landslide scars were located. Second, the same DEMs as for semi-automated mapping were used to recognize and digitize the crown scars (if possible) and at last, for most of the mass movements detected, polygons were drawn. Some basic information was recorded and stored in a GIS database assigned with several attribute fields covering the following aspects:

- Technical aspects connected to the intrinsic characteristics of the sensors adopted (Table 3)
- Geomorphologic and physical aspects of the phenomena mapped (Table 4)
- Statistical information showing, for example, the number of mapped landslides and their size (Table 5)

Table 3. Database fields that describe how the remote sensing image quality influences the manual landslide mapping.

Band Combination	Digitalization Level of Detail	Geometric Quality	Completeness Quality	Landslide Classification Quality
R-G-B	Mapping Scale (varies between 1:1.000–1:20,000) depends on spatial resolution	accurate (1)	complete (1)	certain (1)
		low degree of inaccuracy (2)	high degree of completeness (2)	low degree of uncertainty (2)
		high degree of inaccuracy (3)	low degree of completeness (3)	high degree of uncertainty (3)
		inaccurate (4)	incomplete (4)	uncertain (4)

The technical aspects concerning the scale used for the manual mapping can be distinguished between objective and subjective ones. The first group contains aspects such as the band combination (most of the time false color composite infrared was used) and the spatial resolution, which influenced the digitalization scale. For the second group, a broader distinction was proposed; for subjective factors, which are synthesized in three quality fields (Table 3), scores from 1 (best) to 4 (worst) were assigned to each polygon created.

The first quality field is related to the geometry that relies directly on the image resolution; for example, landslides mapped on Landsat 7 result in a lower geometric accuracy than landslides mapped on SPOT-5 and WorldView-2/3 images. The completeness quality field represents a purely subjective index, being related to the ability of the operator to map a phenomenon as completely as possible. In this case, the landscape context and local characteristics such as forest presence, which obliterated parts of the phenomena, or the re-growing of vegetation, which concealed previous landslide scars and deposits, played a very important role. Concerning the landslide classification quality, the best score was reached when a VHR image was available for the manual mapping.

Visual interpretation partly allowed distinguishing between different subtypes and parts of landslides (Table 4), depending on the specific geographical and geological settings of the five study areas. If a landslide scar could be identified, then it was represented as polyline; when the scar could not be recognized due to data constraints, the accumulation and the transport areas were mapped instead.

The presence of a DEM in order to delineate the exact boundaries of the landslides was useful; even if it was mainly used for the confirmation of the scar extension highlighted in the VHR images (by means of the hillshade and the contour lines). Bulges identification on the deposition area were also supported by the digital elevation data, whereas, since the trail left at the transport section of a landslide is enhanced by a difference in local re-growing of vegetation, only the optical images could be used for identifying these landslide sub-parts.

The landslide activity field was derived from additional alternative data sources (e.g., a public image Web Map Service (WMS) showing older aerial photography for the study areas I, II, and V), published landslide data such as the multi-temporal landslide inventory of Vorarlberg [47] for study areas III and IV, and from unpublished material such as the Austrian mass movement cadastre GEORIOS [48] for study area II, since it was not possible to obtain this information from satellite imagery from one point in time. Objects stored in the GEORIOS database were digitized from geological maps published at different scales (ranging between 1:50,000 and 1:5,000). Although this information represents historical evidence of mass movements, these kind of products only represent active (at the time of the map release) phenomena that are larger than the surrounding geological feature [49] and care should be taken when using this information in order to produce detailed (scale 1:10,000) landslide inventories.

Table 4. Database fields that describe the geomorphological aspects of the manually digitized landslides.

Landslide Type	Landslide Subtype	Landslide Accumulation	Landslide Transport	Landslide Activity [1]
Landslide process type classification [50]	Landslide process subtype classification [50,51]	accumulation area complete (1) incomplete (2) absent (3)	transport area complete (1) incomplete (2) absent (3)	active (1) probably still active (2) dormant (3) not active (4)

[1] based on additional archived data and literature.

2.5. Comparison Methods

For comparing the semi-automated mapping results from OBIA to the mapping results of the visual landslide interpretation, the traditional thematic accuracy assessment approach for land use/land cover image classifications has been employed and adapted [52–54]. It has been applied for landslide classifications by Hölbling et al. [23]. Accordingly, we calculated the landslide area that was mapped by both methods (i.e., the overlapping area). Based on this overlapping area, the respective producer's (overlap expressed as share of the manual mapping area) and user's (as share of the OBIA mapping area) accuracies were computed. Furthermore, the difference in the landslide area mapped by both methods was computed and expressed as a ratio relative to the manual mapping area.

Beyond the area-based comparison, we performed a comparison of the number of polygons mapped as landslides. The difference in the count of landslide polygons between the OBIA result and the manual mapping result was expressed as a ratio relative to the manual mapping count.

Additionally, a set of object-by-object comparison metrics was calculated, which has proven to be useful in the geomorphological context of object-based landform delimitation [55]. Spatial accuracy metrics were used to quantify the spatial match between the manually digitized landslide polygons (considered reference areas) and the landslide polygons extracted with OBIA workflows (considered test areas). In total, five established spatial accuracy metrics were selected [55–57]: (1) Quality Rate (QR), (2) Area-Fit-Index (AFI), (3) Over-Segmentation Rate (OR), (4) Under-Segmentation Rate (UR), and (5) a compound metric D, which combines OR and UR. All these metrics rely on area proportions, i.e., the area of manual polygons, the area of OBIA polygons, and the spatial overlap between manual and OBIA polygons. Except for AFI, the possible range of values is from 0 to 1. A value of 0 indicates a perfect spatial match between the manual and the OBIA area, i.e., the areas have exactly the same spatial extent. In contrast, if the manual and OBIA polygons do not produce any spatial overlap, a value of 1 is obtained. A value around 0.5 can be interpreted such that the majority of manual polygons are overlapped by OBIA polygons but that the spatial extent of the overlapping polygons is different and that some non-overlapping polygons exist. Only the values of AFI can be negative. For the presented landslide mapping tests, it is therefore desirable to obtain metric values as close as possible to 0. The computation of the metrics was automated in the open source software QGIS (Open Source Geospatial Foundation, Beaverton, OR, USA), based on a tool that has been developed previously for use in object-based software [55]. Metrics were computed (1) globally, i.e., for all landslides per study site, and (2) locally, i.e., separately for each landslide. The formulas for and mathematical notations of the spatial accuracy metrics can be found in Clinton et al. [56]. It is also important to mention that there exists a mathematical relation between UA and UR, as well as PA and OR; the total (not percentage) values of each pair (UA-UR, PA-OR) sum up to 1.

Next to the calculation of spatial accuracy metrics, a non-spatial accuracy metric, the so-called Miss Rate (MR), was used. MR quantifies the completeness of the semi-automated landslide extraction process. It is computed as the ratio between the number of undetected manual polygons and the total number of manual polygons per study site [55]. The range of values is between 0 and 1. The closer the values are to 0, the more complete is the OBIA landslide extraction.

3. Results

3.1. Landslide Mapping Results

3.1.1. OBIA Landslide Mapping Results

Figures 2–6 show the OBIA landslide mapping results in comparison to the manual mapping results for each study area. There is good coincidence between most of the results; however, several disparities can be observed that are often related to the ability of a human interpreter to delineate a landslide body as whole, even if parts of the landslide are covered by vegetation. This is the case, for example, for the large landslide near the villages of Badia and La Villa in Südtirol-Alto Adige Autonomous Province (study area V(a)). Similarly, landslides that are still visible on the images but are already overgrown by grass are difficult to detect semi-automatically due to the missing spectral difference to the surrounding areas. This can be recognized in the southwestern part of study area IV, where not all landslides, and particularly landslide tails, could be detected. The detection of landslides by OBIA was most challenging in study area II, since it was hardly possible to differentiate landslides from other processes using the SPOT-5 image and the DEM only. Statistics and detailed information about the accuracy of the semi-automated mapping in comparison to the manual mapping are given in Section 3.2.

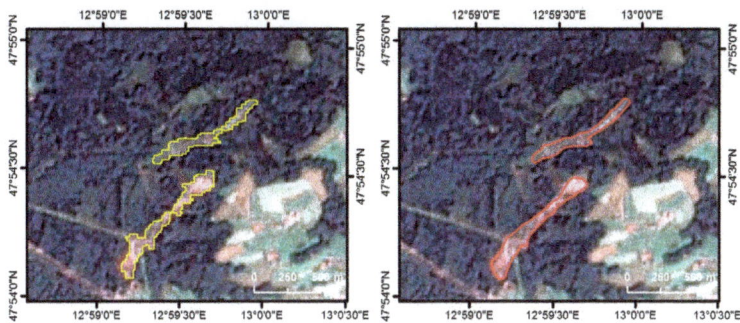

Figure 2. Semi-automatically mapped landslides using the object-based image analysis (OBIA) approach (in yellow; **left**) and manually mapped landslides (in red; **right**) for study area I based on the Landsat 7 image.

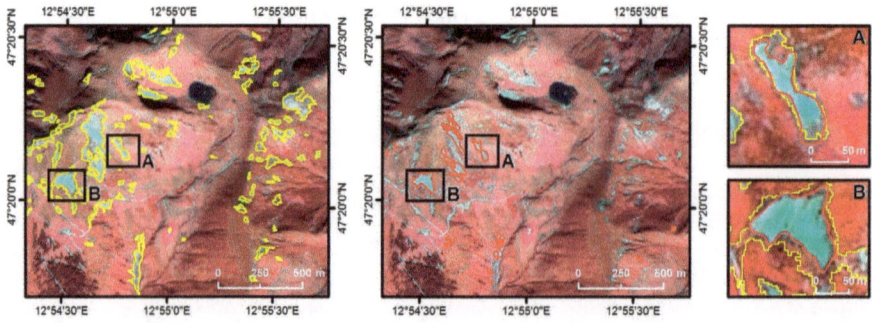

Figure 3. Semi-automatically mapped landslides using the object-based image analysis (OBIA) approach (in yellow; **left**) and manually mapped landslides (in red; **middle**) for study area II based on the SPOT-5 image. Two subsets demonstrating the variation of the mapping results for selected landslides are shown on the **right**. The black rectangles (A, B) indicate the location of the subsets.

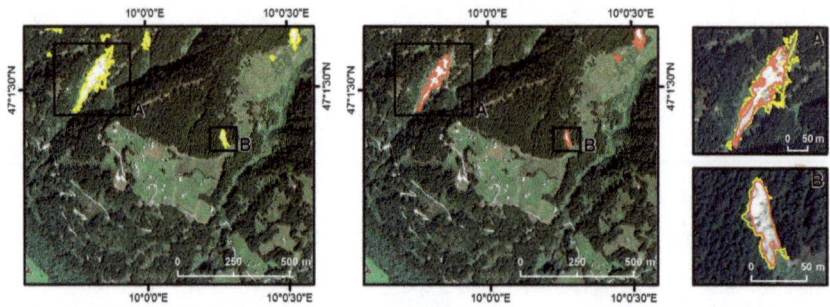

Figure 4. Semi-automatically mapped landslides using the object-based image analysis (OBIA) approach (in yellow; **left**) and manually mapped landslides (in red; **middle**) for study area III based on the WorldView-2 image. Two subsets demonstrating the variation of the mapping results for selected landslides are shown on the **right**. The black rectangles (A, B) indicate the location of the subsets.

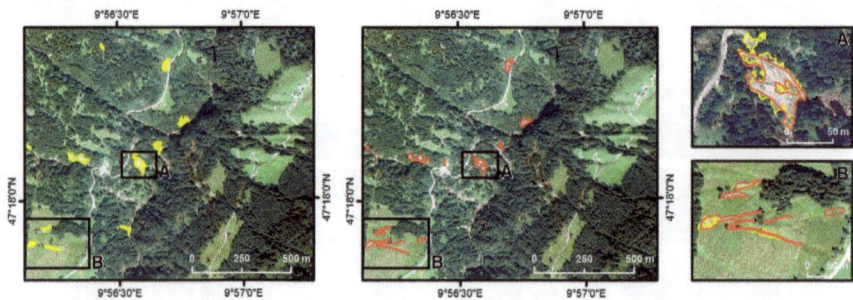

Figure 5. Semi-automatically mapped landslides using the object-based image analysis (OBIA) approach (in yellow; **left**) and manually mapped landslides (in red; **middle**) for study area IV based on the WorldView-3 image. Two subsets demonstrating the variation of the mapping results for selected landslides are shown on the **right**. The black rectangles (A, B) indicate the location of the subsets.

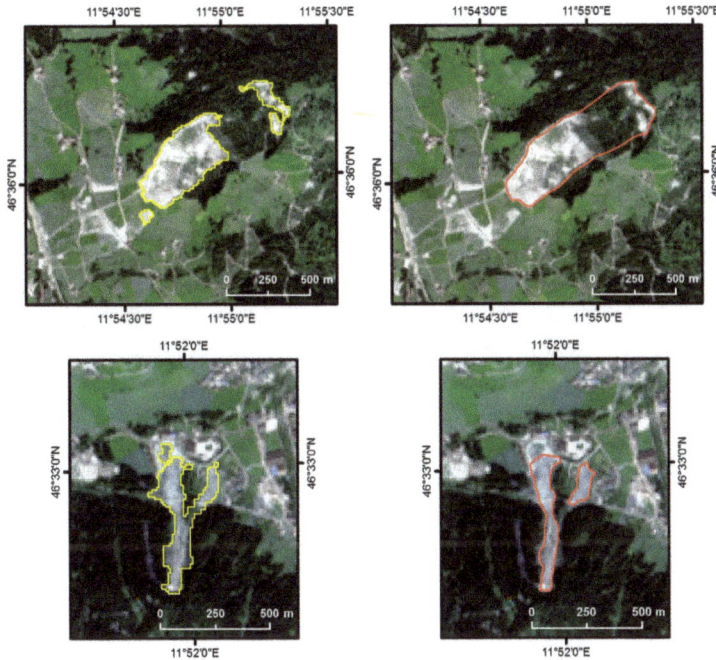

Figure 6. Semi-automatically mapped landslides using the object-based image analysis (OBIA) approach (in yellow; **left**) and manually mapped landslides (in red; **right**) for study areas V(a) **(top)** and V(b) **(below)** based on the Sentinel-2 image.

Table 5 presents some basic statistics for the OBIA mapping results for each study area.

Table 5. Statistics of the OBIA mapping results for each of the five study areas.

Statistical Property	Study Area I	Study Area II	Study Area III	Study Area IV	Study Area V
Number of landslides	2	85	9	11	5
Area affected by landslides (km^2)	0.148	0.203	0.019	0.011	0.250
Smallest mapped landslide (m^2)	59700	75	135	244	4200
Largest mapped landslide (m^2)	87900	38741	12577	2400	148800
Average size of mapped landslides (m^2)	73800	2393	2110	992	49960

3.1.2. Manual Landslide Mapping Results

Table 6 shows the statistics of the manual landslide mapping results for each study area. Probably the most notable case is the Pinzgau study area (II). This is a special area of interest, being a mountainous area characterized by a high relief and sparse vegetation, where several geomorphological phenomena such as snow scars induced by avalanches, gully erosions connected to torrent activity, and rock glacier deposits could be perceived and mapped as mass movements. Surprisingly, a high landslide density can be recognized. Since the differentiation of landslides from other mass movement types only based on the SPOT-5 image and the 10 m DEM was hardly possible in this case, even by visual interpretation, a 5 m DEM and aerial photography from 2011 (accessible through a free WMS service) were additionally used for manual landslide mapping. Furthermore, some GEORIOS polygons were used as guidance for the identification of landslide subtypes.

Table 6. Statistics of the manual mapping results for each of the five study areas.

Statistical Property	Study Area I	Study Area II	Study Area III	Study Area IV	Study Area V
Number of landslides	2	18	5	14	3
Area affected by landslides (km^2)	0.140	0.031	0.011	0.010	0.320
Smallest mapped landslide (m^2)	54855	247	66	30	10781
Largest mapped landslide (m^2)	85339	8374	7200	1882	270966
Average size of mapped landslides (m^2)	70097	1830	2157	709	106533

The three quality fields were introduced in order to trace back the logic used for the task of manual landslide mapping and to communicate the perceived difficulties encountered during the mapping process. The three quality indices are directly correlated to the spatial resolution of the used satellite images. For each of the analyzed images, the achievable score varies. The best score reached among all the quality categories is:

- Score of "1" for study area III and IV (1 m resolution)
- Score of "2" for area II (2.5 m resolution) and area V (10 m resolution)
- Score of "3" for study area I (15 m resolution)

In general, when considering all quality measures, the best results were achieved on the VHR images (including area II with 2.5 m resolution), always bearing in mind the subjectivity introduced by the interpreter.

3.2. Accuracy Assessment

A variety of different metrics was used to quantitatively capture the agreement between the OBIA mapping results and the manual delineation. Comparisons were performed in all study areas and for all datasets.

3.2.1. Comparison of Overlapping Areas

The metrics according to the general accuracy assessment approach for image classifications, as presented in Hölbling et al. [23], are summarized in Table 7. Concerning the number of landslides, the best agreement occurred in study area I (Fürwag), where both approaches detected two large landslides. The highest discrepancy occurred in study area II (Pinzgau), where a large number of objects were considered landslides only by the OBIA result but not by the manual delineation. In the other study areas, the OBIA and manual results deviated from each other by two to four landslides. Since the numbers of landslides were relatively low, with a count between three and 14, the difference by percentage still showed high values between −20% and 80%. The comparison of the landslide affected area reveals similar tendencies: again, study area I showed the best agreement with a difference of only 5.28%, and study area II showed the highest discrepancy, with OBIA detecting an affected area that is almost five times as large as the manual mapping. In study area III (Montafon), the difference in the landslide-affected area was 76% and reflected the agreement already shown by the difference in landslide numbers. For study area IV (Bregenzerwald) and study area V (Gader Valley), the OBIA deviation from the manually mapped area is relatively low, with 9.91% and −21.8%., but here the difference shows a contrasting direction related to the number of landslides. The producer's and user's accuracy values in study area I and IV were very similar. For study area II, the values show that most of the landslide affected areas identified with manual mapping have also been captured by OBIA (producer's accuracy 90.6%) but that many more areas were classified that were left out by manual mapping (user's accuracy 15.6%). Study areas III and V(b) show a similar situation to study area II but not as pronounced. Study area V(a) showed the opposite behavior because a significant portion of a landslide has been included in the manual mapping that was not covered by the OBIA result.

Table 7. Comparison of OBIA mapping results and manual mapping results per study area.

Comparison Metric	Study Area I	Study Area II	Study Area III	Study Area IV	Study Area V
Number of landslides (OBIA)	2	85	9	11	5
Number of landslides (manual)	2	18	5	14	3
Difference in numbers OBIA—manual (count; %)	0 0%	+67 372%	+4 80%	−3 −21%	+2 +67%
Landslide affected area (OBIA, km^2)	0.148	0.203	0.0190	0.0109	0.250
Landslide affected area (manual, km^2)	0.140	0.0351	0.0108	0.00993	0.320
Area difference OBIA—manual (%)	5.28%	480%	76 %	9.91%	−21.8%
Overlap area (km^2)	0.118	0.0318	0.0104	0.00691	0.208
Producer's Accuracy (%)	84.3%	90.6%	95.8%	69.6%	V(a): 60.6% V(b): 90.0%
User's Accuracy (%)	80.1%	15.6%	54.4%	63.3%	V(a): 91.2% V(b): 62.7%

3.2.2. Completeness of the Semi-Automated Extraction

Relatively low MR values in the range from 0 to 0.35 were obtained (Figure 7), demonstrating that at least two thirds of the manual polygons were also extracted by the OBIA method. For the Gader Valley and Fürwag, even all reference polygons were detected with the semi-automated workflows. This is indicated by a MR value of 0. However, for these areas, the total number of reference polygons was small ($n = 1$–2) and the spatial extent relatively large. Usually, the combination of low reference number and large reference areas increases the success of the automated landslide detection, which is also supported by the presented results.

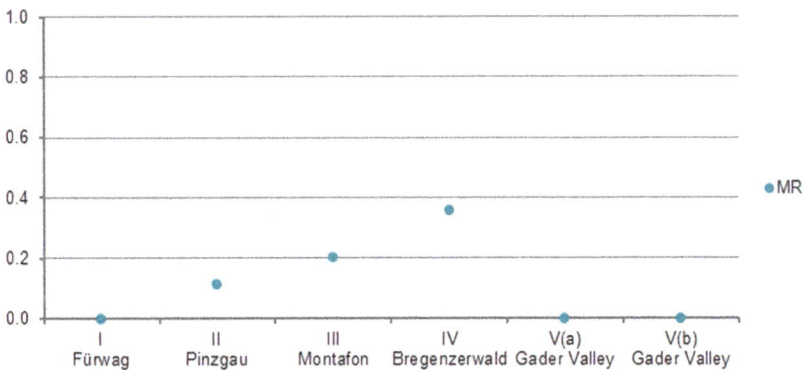

Figure 7. Completeness of the automated landslide extraction, quantified by the values of Miss Rate (MR).

3.2.3. Spatial Accuracy of the Semi-Automated Extraction

Generally, global spatial accuracy values are below 0.5, indicating good to high spatial agreements between the reference and automated landslide polygons for the selected study sites (Figure 8). Only for the Pinzgau area higher values, and hence, lower spatial accuracies were achieved. For Pinzgau, the high values of QR, D, and UR show that there is a high difference between the size of reference and test polygons. As a result, small spatial overlaps are produced. Since OR is close to 0 and AFI is negative (Figures 8 and 9), it can be concluded that the extracted landslide polygons are far larger than the reference polygons and/or that landslides were extracted at the wrong locations, i.e., the automated extraction over-estimates the spatial extent of the reference. To a lesser extent, the afore-mentioned conclusion also holds true for the Montafon study site (III). From Figures 8 and 9, it is also obvious that study area V(a) is the only area where the value of UR is lower than OR and where AFI is positive. This means that the extracted landslide polygon is smaller than the reference polygon ($n = 1$), i.e., the extraction under-estimates the reference. The best spatial agreement between reference and test polygons was obtained for the Fürwag area (*n* = 2). The spatial accuracy values are relatively low, and the deviation between them is minimal. These characteristics are achieved if the over- and under-estimated areas are relatively small and the area proportions of the overlap, test, and reference polygons are similar.

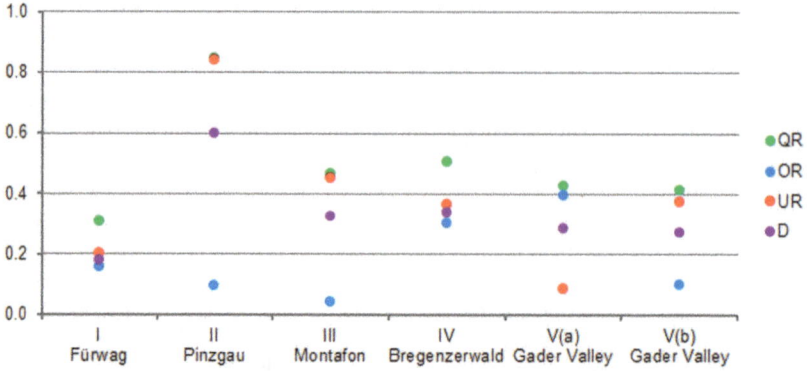

Figure 8. Global spatial accuracies for the selected study sites. Results presented for four metrics; Quality Rate (QR), Under-Segmentation Rate (UR), Over-Segmentation Rate (OR), and the UR-OR compound metric D.

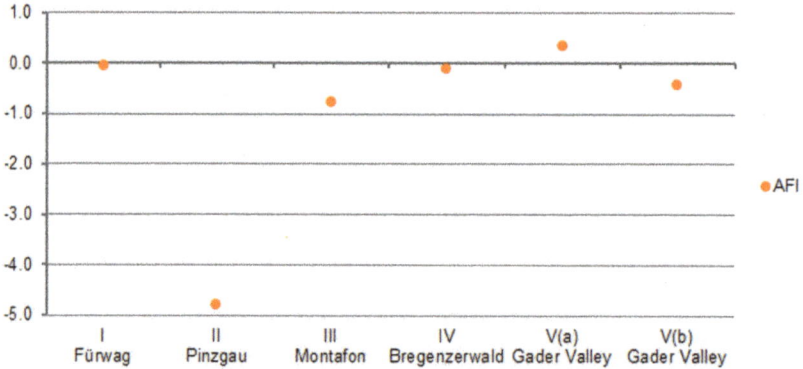

Figure 9. Global values of Area-Fit-Index (AFI) for the selected study sites.

Figure 10 depicts the spatial accuracy signatures of individual landslides per study site. The signatures are obtained by plotting the four metrics, QR, UR, OR, and D; each on one side of the X- and Y-axis. In case of high spatial agreement, the shape of the graph is diamond-like and the diamond is positioned as close as possible around the zero point. The afore-mentioned characteristics are best met for the two Fürwag landslides. When looking at the deviation of spatial accuracy signatures for the landslides per study area, it can be concluded that the lowest values are obtained for Fürwag and Montafon, whereas for Pinzgau and Bregenzerwald higher deviations are depicted in Figure 10. Higher deviations indicate a higher variation of results for a study area, i.e., some reference landslides are under-estimated by the test polygons, while others are over-estimated.

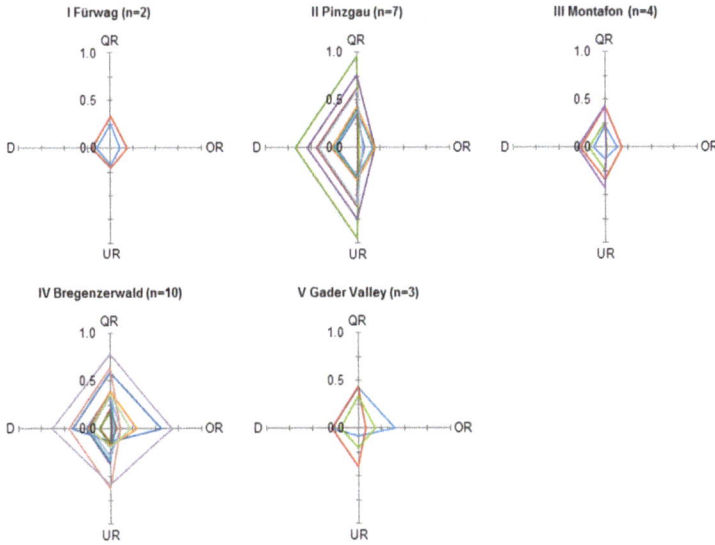

Figure 10. Landslide-specific spatial accuracy values per study site I–V. Each graph in I–V depicts the spatial accuracy signature of a detected landslide. Results presented for four metrics; Quality Rate (QR), Under-Segmentation Rate (UR), Over-Segmentation Rate (OR), and the UR-OR compound metric D.

4. Discussion

Landslide maps produced with semi-automated methods are commonly compared to manually mapped landslides in order to quantitatively assess the accuracy of results. Visual interpretation, however, while still the most common method for mapping landslides on optical remote sensing data, has some disadvantages: it is a time-consuming process, depends on various factors, and cannot constitute a completely "true" reference. However, often it is the only available reference data [8,23]. A "true" reference, i.e., ground truth reference data, would require the reference dataset to be of a significantly higher accuracy and to be collected independently from the tested dataset [58]. The manual delineation and the OBIA mapping are two independently performed processes, but they employ the same satellite images, and thus their results cannot be considered fully independent. For the same reason, the manual interpretation did not result in landslide delineation with a significantly higher accuracy. Consequently, manual mapping implies some degree of uncertainty that cannot be avoided, even if created by an expert. Therefore, if some of the disagreement between the two compared datasets can be attributed to the uncertainty in the manual delineation, the accuracy of the OBIA results may be larger than the calculated accuracy values suggest. At least they would fall in the same range of acceptable uncertainty that commonly used manual delineations imply. Of course, this could only be confirmed if a valid ground truth reference dataset was available that also allows the accuracy of the manual delineation to be assessed. In the current practice of assessing the accuracy of OBIA results with manual delineations, only strong disagreements can clearly be attributed to mapping errors, e.g., in the case of study area II where a user's accuracy value of only 15.6% occurred. The conceptual problem remains that the manual mapping does not completely fulfill the standards as a reference for a classical accuracy assessment. Due to the uncertainty inherent in the manual mapping, we suggest to use the term "agreement" instead of "accuracy" when comparing the two classifications. The concept of agreement for evaluating the quality of OBIA landslide mapping results also applies to other fields of applications where a fully reliable reference is missing. The approach of assessing the agreement may not yield as sharp a quality indicator as the classical approach of an accuracy assessment, but it can still be highly valuable. The selection of appropriate quality indicators is a subject for a necessary discussion within the landslide community. It enables the design of a validation approach for EO-based landslide mapping that conforms to the Quality Assurance Framework for Earth Observation (QA4EO; http://QA4EO.org/).

The accuracy assessment achieved a number of different metrics that were used for assessing the accuracy of the OBIA landslide maps. In addition to classical PA and UA, which were computed separately for each site, this study proposes to integrate spatial accuracy metrics for quantifying the spatial agreement between manual and automated landslide polygons on an object-by-object basis. The use of landslide-specific accuracies allows a more detailed interpretation of the results. For instance, the deviation of landslide accuracy values within a study area may give clues about the stability/robustness of the automated extraction workflow. The best results where all measures show similar values were achieved for study area I, with PA and UA of approximately 80% and QR, UR, OR, and D of approximately 0.2. However, for area I only two reference polygons were extracted. The results of this study confirm that lower accuracies are achieved for areas where a larger number of landslides have to be mapped by OBIA methods. We also compared delineations of single landslides that performed equally well or even better. However, what makes it difficult to interpret the calculated accuracy values is the lack of information about the accuracy thresholds accepted within the landslide community. Such information would be highly valuable for estimating the quality of landslide maps derived from EO data with semi-automated methods.

The results in this study obviously show that the very high spatial resolution of the optical satellite image does not necessarily guarantee a high quality of the automated landslide mapping. The relative quality of semi-automated landslide mapping can be better for Landsat 7 data than for WorldView-2/3 images, as exemplified by the Fürwag test case. However, when drawing such conclusions, one has to keep in mind that there is a direct relationship between the spatial resolution of the optical satellite image and the reliability of the reference polygons digitized therefrom. It is not surprising that VHR images allow the mapping of smaller landslides and to some degree also the differentiation of subtypes, particularly during manual mapping. On the other hand, Landsat 7 and Sentinel-2 proved their applicability for identifying and delineating larger landslides, whereby the agreement between both results was high (apart from the special case of study area V(a), as described below). Especially Sentinel-2 can be considered a relevant data source for semi-automated landslide mapping and landslide change detection, also because of its high revisiting frequency.

Both methods produced acceptable and comparable results. However, the differentiation of landslide types and sub-parts with OBIA was not possible with the available data. Very high resolution DEM data from different dates would be helpful for that, but data availability was limited. Also, optical data acquired directly after a landslide triggering event and another image taken several weeks or months later could be helpful for the semi-automated differentiation of source, transport, and accumulation area, since landslide tails tend to revegetate faster than landslide scars [23]. These difficulties become obvious when looking at the results from the Pinzgau study area, where different mass movement processes are present. While the OBIA approach classified too many areas as landslides due to similar spectral properties, only a few landslides were identified by manual mapping. In such a case, ground truthing would be necessary to create a reliable landslide inventory. An example that requires expert interpretation can be found in the southwestern part of the Bregenzerwald study site where some small shallow landslides were manually mapped, which could not be identified by OBIA despite the use of WorldView-3 since they were already grassed over. Study area V(a) is an interesting case because a significant portion of a landslide has been included in the manual mapping that was not covered by the OBIA mapping. This part of the landslide is covered by vegetation. In the manual interpretation, the inclusion of this part was logically inferred by expert knowledge, an element that is difficult to implement in automated mapping workflows. Eisank et al. [31] proposed a knowledge-based landslide mapping system, but further research is needed in this direction to compete with a human interpreter.

Two OBIA software products with a different range of functionalities were used for semi-automated landslide mapping, both delivering acceptable results for the presented case studies. OBIA workflows were developed in such a way that they can potentially be set up in any of the two software products, but each image analysis tool consists of different implemented approaches for image segmentation and classification. As a consequence, the workflows may produce different results, particularly related to the object delineations. Further studies could focus on a systematic comparison of the applicability of different software products to identify the most adequate tool for each sensor and landscape characteristic. However, this also depends on the data used and on the preferences and skills of the software user.

By combining both approaches, i.e., starting with the semi-automated detection of landslide candidates and then refining this classification manually, the whole mapping process could be improved while achieving adequate mapping results. Particularly large-scale mapping tasks would probably benefit from a degree of automation [59]. By implementing such a combined approach in an interactive and easy-to-use web service intended for practitioners with at least a certain knowledge in landslide recognition, landslide mapping could become faster and more efficient [60,61]. Such a web service can be relevant, especially for users and practitioners who might not be familiar with sophisticated image analysis software.

5. Conclusions

In this study we compared two landslide mapping approaches, OBIA and manual mapping, in five selected test areas with various geological and geomorphological settings in the Alps. Satellite images with different resolutions from different sensors were used. The analyses show that both methods produce similar results, whereby the achieved accuracy values vary between the study areas. It is important to note that using manual mapping for reference is critical, and the associated uncertainty should be considered when interpreting the accuracy values. While visual expert interpretation has advantages for differentiating landslides subtypes and for delineating single landslides as single objects and thus to identify the number of mapped landslides, it is a subjective and time-consuming task. This subjectivity issue is difficult to measure but influences the reliability of the landslide mapping. We proposed a scheme for including different quality measures in manual mapping. This might be useful for evaluating the reliability of manual mapping as reference data, even if some of these measures are subjective and rely on the interpreter. On the other hand, the OBIA approach is more transparent due to the use of replicable classification rules, but the expert knowledge required for the mapping of complex landslides or landslides that are partly covered by vegetation is difficult to implement in automated mapping workflows. Semi-automated landslide mapping approaches show high potential and would find even more acceptance if a set of generally agreed quality indicators and accuracy thresholds for the validation of EO-based landslide maps were commonly used.

Acknowledgments: This research has been supported by the Austrian Research Promotion Agency FFG in the Austrian Space Applications Program (ASAP 11) through the project "Land@Slide" (contract no: 847970).

Author Contributions: Daniel Hölbling had the idea for the paper and performed the object-based landslide mapping for study areas I, II, and V. Clemens Eisank did the OBIA mapping for study areas IV and V and calculated the spatial accuracy values. Florian Albrecht contributed to the accuracy assessment and the interpretation of results. Filippo Vecchiotti performed the visual landslide interpretation and provided the geological background information. Barbara Friedl wrote major parts of the introduction and contributed to the discussion. Elisabeth Weinke revised the manuscript and contributed to the discussion and conclusion. Arben Kociu provided valuable background information and revised the manuscript. All authors contributed to writing the paper.

References

1. Van Westen, C.J.; Castellanos, E.; Kuriakose, S.L. Spatial data for landslide susceptibility, hazard, and vulnerability assessment: An overview. *Eng. Geol.* **2008**, *102*, 112–131. [CrossRef]
2. Scaioni, M.; Longoni, L.; Melillo, V.; Papini, M. Remote Sensing for Landslide Investigations: An Overview of Recent Achievements and Perspectives. *Remote Sens.* **2014**, *6*, 9600–9652. [CrossRef]
3. Singhroy, V. Remote sensing of landslides. In *Landslide Hazard and Risk*; Glade, T., Anderson, M.G., Crozier, M.J., Eds.; Wiley & Sons: West Sussex, UK, 2005; pp. 469–492. ISBN 9780471486633.
4. Fiorucci, F.; Cardinali, M.; Carlà, R.; Rossi, M.; Mondini, A.C.; Santurri, L.; Ardizzone, F.; Guzzetti, F. Seasonal landslide mapping and estimation of landslide mobilization rates using aerial and satellite images. *Geomorphology* **2011**, *129*, 59–70. [CrossRef]
5. Gao, J.; Maro, J. Topographic controls on evolution of shallow landslides in pastoral Wairarapa, New Zealand, 1979–2003. *Geomorphology* **2010**, *114*, 373–381. [CrossRef]
6. Morgan, J.L.; Gergel, S.H.; Coops, N.C. Aerial Photography: A Rapidly Evolving Tool for Ecological Management. *BioScience* **2010**, *60*, 47–59. [CrossRef]
7. Galli, M.; Ardizzone, F.; Cardinali, M.; Guzzetti, F.; Reichenbach, P. Comparing landslide inventory maps. *Geomorphology* **2008**, *94*, 268–289. [CrossRef]
8. Hölbling, D.; Friedl, B.; Eisank, C. An object-based approach for semi-automated landslide change detection and attribution of changes to landslide classes in northern Taiwan. *Earth Sci. Inform.* **2015**, *8*, 327–335. [CrossRef]
9. Guzzetti, F.; Mondini, A.C.; Cardinali, M.; Fiorucci, F.; Santangelo, M.; Chang, K.-T. Landslide inventory maps: New tools for an old problem. *Earth-Sci. Rev.* **2012**, *112*, 42–66. [CrossRef]
10. Van Coillie, F.M.B.; Gardin, S.; Anseel, F.; Duyck, W.; Verbeke, L.P.C.; De Wulf, R.R. Variability of operator performance in remote-sensing image interpretation: The importance of human and external factors. *Int. J. Remote Sens.* **2014**, *35*, 754–778. [CrossRef]
11. Albrecht, F.; Lang, S.; Hölbling, D. Spatial accuracy assessment of object boundaries for object-based image analysis. In Proceedings of the GEOBIA 2010—Geographic Object-Based Image Analysis, Ghent, Belgium, 29 June–2 July 2010; Addink, E., Van Coillie, F.M.B., Eds.; ISPRS Vol. No. XXXVIII-4/C7, Archives ISSN No 1682-1777.
12. Kohli, D.; Stein, A.; Sliuzas, R. Uncertainty analysis for image interpretations of urban slums. *Comput. Environ. Urban Syst.* **2016**, *60*, 37–49. [CrossRef]
13. Guzzetti, F.; Cardinali, M.; Reichenbach, P.; Carrara, A. Comparing Landslide Maps: A Case Study in the Upper Tiber River Basin, Central Italy. *Environ. Manag.* **2000**, *25*, 247–263. [CrossRef]
14. Ardizzone, F.; Cardinali, M.; Carrara, A.; Guzzetti, F.; Reichenbach, P. Impact of mapping errors on the reliability of landslide hazard maps. *Nat. Hazards Earth Syst. Sci.* **2002**, *2*, 3–14. [CrossRef]
15. Moosavi, V.; Talebi, A.; Shirmohammadi, B. Producing a landslide inventory map using pixel-based and object-oriented approaches optimized by Taguchi method. *Geomorphology* **2014**, *204*, 646–656. [CrossRef]
16. Stumpf, A.; Kerle, N. Object-oriented mapping of landslides using Random Forests. *Remote Sens. Environ.* **2011**, *115*, 2564–2577. [CrossRef]
17. Martha, T.R.; Kerle, N.; Jetten, V.; van Westen, C.J.; Kumar, K.V. Characterising spectral, spatial and morphometric properties of landslides for semi-automatic detection using object-oriented methods. *Geomorphology* **2010**, *116*, 24–36. [CrossRef]
18. Barlow, J.; Franklin, S.; Martin, Y. High spatial resolution satellite imagery, DEM derivatives, and image segmentation for the detection of mass wasting processes. *Photogramm. Eng. Rem. S.* **2006**, *72*, 687–692. [CrossRef]
19. Behling, R.; Roessner, S.; Kaufmann, H.; Kleinschmit, B. Automated Spatiotemporal Landslide Mapping over Large Areas Using RapidEye Time Series Data. *Remote Sens.* **2014**, *6*, 8026–8055. [CrossRef]
20. Blaschke, T.; Feizizadeh, B.; Hölbling, D. Object-Based Image Analysis and Digital Terrain Analysis for Locating Landslides in the Urmia Lake Basin, Iran. *IEEE J. Sel. Top. Appl.* **2014**, *7*, 4806–4817. [CrossRef]
21. Heleno, S.; Matias, M.; Pina, P.; Sousa, A.J. Semiautomated object-based classification of rain-induced landslides with VHR multispectral images on Madeira Island. *Nat. Hazards Earth Syst. Sci.* **2016**, *16*, 1035–1048. [CrossRef]

22. Hölbling, D.; Füreder, P.; Antolini, F.; Cigna, F.; Casagli, N.; Lang, S. A Semi-Automated Object-Based Approach for Landslide Detection Validated by Persistent Scatterer Interferometry Measures and Landslide Inventories. *Remote Sens.* **2012**, *4*, 1310–1336. [CrossRef]
23. Hölbling, D.; Betts, H.; Spiekermann, R.; Phillips, C. Identifying Spatio-Temporal Landslide Hotspots on North Island, New Zealand, by Analyzing Historical and Recent Aerial Photography. *Geosciences* **2016**, *6*, 48. [CrossRef]
24. Kurtz, C.; Stumpf, A.; Malet, J.-P.; Gançarski, P.; Puissant, A.; Passat, N. Hierarchical extraction of landslides from multiresolution remotely sensed optical images. *ISPRS J. Photogramm.* **2014**, *87*, 122–136. [CrossRef]
25. Lahousse, T.; Chang, K.-T.; Lin, Y. Landslide mapping with multiscale object-based image analysis—A case study in the Baichi watershed, Taiwan. *Nat. Hazards Earth Syst. Sci.* **2011**, *11*, 2715–2726. [CrossRef]
26. Lu, P.; Stumpf, A.; Kerle, N.; Casagli, N. Object-oriented change detection for landslide rapid mapping. *IEEE Geosci. Remote Sens.* **2011**, *8*, 701–705. [CrossRef]
27. Martha, T.R.; Kerle, N.; van Westen, C.J.; Jetten, V.; Kumar, K.V. Segment Optimization and Data-Driven Thresholding for Knowledge-Based Landslide Detection by Object-Based Image Analysis. *IEEE Geosci. Remote Sens.* **2011**, *49*, 4928–4943. [CrossRef]
28. Martha, T.R.; Kerle, N.; van Westen, C.J.; Jetten, V.; Kumar, K.V. Object-oriented analysis of multi-temporal panchromatic images for creation of historical landslide inventories. *ISPRS J. Photogramm.* **2012**, *67*, 105–119. [CrossRef]
29. Rau, J.-Y.; Jhan, J.-P.; Rau, R.-J. Semiautomatic object-oriented landslide recognition scheme from multisensor optical imagery and DEM. *IEEE Trans. Geosci. Remote Sens.* **2014**, *52*, 1336–1349. [CrossRef]
30. Lang, S. Object-based image analysis for remote sensing applications: modeling reality—Dealing with complexity. In *Object-Based Image Analysis*; Blaschke, T., Lang, S., Hay, G.J., Eds.; Springer: Heidelberg/Berlin, Germany; New York, NY, USA, 2008; pp. 1–25. ISBN 978-3-540-77058-9.
31. Eisank, C.; Hölbling, D.; Friedl, B.; Chen, Y.C.; Chang, K.T. Expert knowledge for object-based landslide mapping in Taiwan. *South-East. Eur. J. Earth Obs. Geomat.* **2014**, *3*, 347–350.
32. Wagreich, M.; Lukeneder, A.; Egger, H. Cretaceous History of Austria. In Proceedings of the International Meeting on Correlation of Cretaceous Micro—And Macrofossils—Berichte der Geologischen Bundesanstalt, Vienna, Austria, 16–18 April 2008; Volume 74, pp. 12–30.
33. Vecchiotti, F.; Kociu, A. Geohazard Description for Salzburg, Public Pangeo FP7 Final Report, 2013. Available online: http://www.pangeoproject.eu/eng/coverage_map (accessed on 30 March 2017).
34. Embleton-Hamann, C. Geomorphological hazards in Austria. In *Geomorphology for the Future*; Kellerer-Pirklbauer, A., Keiler, M., Embleton-Hamann, C., Stötter, J., Eds.; Innsbruck University Press: Innsbruck, Austria, 2007; pp. 33–56; ISBN 978-3-902571-18-2.
35. Fiebiger, G. Die Rutschung Fürwag im Norden von Salzburg/Österreich. Prozess und Maßnahmen. In Proceedings of the International Congress Interpraevent, Matsumoto, Japan, 14–18 October 2002; pp. 629–639.
36. Friebe, J.G. *Geologie der Österreichischen Bundesländer. Vorarlberg. Erläuterungen der Geologischen Karte Vorarlberg 1:100,000*; Geologische Bundesanstalt: Vienna, Austria, 2007.
37. Seijmonsbergen, A.C.; Anders, N.; Bouten, W. Geomorphological change detection using object-based feature extraction from multi-temporal LiDAR data. In Proceedings of the 4th GEOBIA, Rio de Janeiro, Brazil, 7–9 May 2012; Feitosa, R., Costa, G., Almeida, C., Eds.; pp. 484–489.
38. Dorren, L.K.; Maier, B.; Seijmonsbergen, A.C. Improved Landsat-based forest mapping in steep mountainous terrain using object-based classification. *Forest Ecol. Manag.* **2003**, *183*, 31–46. [CrossRef]
39. De Graaff, L.W.S.; Seijmonsbergen, A.C. Postglacial landslides and their impact on Pleistocene lake floor deposits in the Balderschwang Valley as witnessed by geomorphological, sedimentological and geophysical evidence (Vorarlberg, Austria). *Vorarlb. Naturschau* **2001**, *9*, 237–251.
40. Ghuffar, S.; Szekely, B.; Roncat, A.; Pfeifer, N. Landslide Displacement Monitoring Using 3D Range Flow on Airborne and Terrestrial LiDAR Data. *Remote Sens.* **2013**, *5*, 2720–2745. [CrossRef]
41. Jaritz, W.; Supper, R.; Wöhrer-Alge, M. Beurteilung geogener Gefahren in Hinblick auf eine Risikominderung in der Gemeinde Sibratsgfäll (Österreich). In Proceedings of the 11th Interpraevent Congress, Dornbirn, Austria, 26–30 May 2008; pp. 171–182.

42. Brandner, R.; Gruber, A.; Morelli, C.; Mair, V. Field trip 1—Pulses of Neotethys-Rifting in the Permomesozoic of the Dolomites. In *Geo.Alp*; Institute of Geology—University of Innsbruck: Innsbruck, Austria, 2016; Volume 13, pp. 7–70.

43. Gianolla, P.; Andreetta, R.; Furin, S.; Furlanis, S.; Riva, A. Geology of the Dolomites. In *Nomination of the Dolomites for Inscription on the World Natural Heritage List Unesco*; Artimedia: Trento, Italy, 2008; pp. 3–77.

44. Larcher, V.; Notarnicola, C.; Piacentini, D.; Pinter, T.; Schneiderbauer, S.; Soldati, M.; Strada, C. Analyse flachgründiger Massenbewegungen mittels Verwendung zweier statistischer Methoden im Gadertal (Südtirol). In Proceedings of the COGeo, Salzburg, Austria, 11 June 2010; Marschallinger, R., Wanker, W., Zobl, F., Eds.; pp. 1–11.

45. Corsini, A.; Pasuto, A.; Soldati, M.; Zannoni, A. Field monitoring of the Corvara landslide (Dolomites, Italy) and its relevance for hazard assessment. *Geomorphology* **2005**, *66*, 149–165. [CrossRef]

46. Riley, S.J.; De Gloria, S.D.; Elliot, R. A terrain ruggedness that quantifies topographic heterogeneity. *Interm. J. Sci.* **1999**, *5*, 23–27.

47. Tilch, N. Identifizierung gravitativer Massenbewegungen mittels multitemporaler Luftbildauswertung in Vorarlberg und angrenzender Gebiete. *Jahrb. Geol. Bundesanstalt* **2014**, *154*, 21–39.

48. Kociu, A.; Kautz, H.; Tilch, N.; Grösel, K.; Horst, H.; Reischer, J. Massenbewegungen in Österreich. *Jahrb. Geol. Bundesanstalt* **2007**, *147*, 215–220.

49. Wills, C.J.; McCrink, T.P. Comparing Landslide Inventories: The Map Depends on the Method. *Environ. Eng. Geosci.* **2002**, *8*, 279–293. [CrossRef]

50. Hungr, O.; Leroueil, S.; Picarelli, L. The Varnes classification of landslide types, an update. *Landslides* **2014**, *11*, 167–194. [CrossRef]

51. Cruden, D.M.; Varnes, D.J. Landslide Types and Processes. In *Landslides: Investigation and Mitigation*; Turner, A.K., Schuster, R.L., Eds.; Sp. Rep. 247; Transportation Research Board, National Research Council, National Academy Press: Washington, DC, USA, 1996; pp. 36–75. ISBN 978-0309062084.

52. Congalton, R.G.; Green, K. *Assessing the Accuracy of Remotely Sensed Data—Principles and Practices*, 2nd ed.; CRC Press: Boca Raton, FL, USA, 2008; p. 183. ISBN 978-1420055122.

53. Foody, G.M. Status of land cover classification accuracy assessment. *Remote Sens. Environ.* **2002**, *80*, 185–201. [CrossRef]

54. Pontius, R.G.; Millones, M. Death to Kappa: Birth of quantity disagreement and allocation disagreement for accuracy assessment. *Int. J. Remote Sens.* **2011**, *32*, 4407–4429. [CrossRef]

55. Eisank, C.; Smith, M.; Hillier, J. Assessment of multiresolution segmentation for delimiting drumlins in digital elevation models. *Geomorphology* **2014**, *214*, 452–464. [CrossRef] [PubMed]

56. Clinton, N.; Holt, A.; Scarborough, J.; Yan, L.; Gong, P. Accuracy Assessment Measures for Object-based Image Segmentation Goodness. *Photogramm. Eng. Rem. S.* **2010**, *76*, 289–299. [CrossRef]

57. Belgiu, M.; Draguţ, L. Comparing supervised and unsupervised multiresolution segmentation approaches for extracting buildings from very high resolution imagery. *ISPRS J. Photogramm.* **2014**, *96*, 67–75. [CrossRef] [PubMed]

58. Lang, S.; Albrecht, F.; Kienberger, S.; Tiede, D. Object validity for operational tasks in a policy context. *J. Spat. Sci.* **2010**, *55*, 9–22. [CrossRef]

59. Hölbling, D.; Betts, H.; Spiekermann, R.; Phillips, C. Semi-automated landslide mapping from historical and recent aerial photography. In Proceedings of the 19th AGILE Conference on Geographic Information Science, Helsinki, Finland, 14–17 June 2016; p. 5.

60. Hölbling, D.; Eisank, C.; Friedl, B.; Weinke, E.; Kleindienst, H.; Kociu, A.; Vecchiotti, F.; Albrecht, F. EO-based landslide mapping: From methodological developments to automated web-based information delivery. In Proceedings of the 13th Congress Interpraevent—Extended Abstracts, Lucerne, Switzerland, 30 May–2 June 2016; pp. 102–103.

61. Weinke, E.; Albrecht, F.; Hölbling, D.; Eisank, C.; Vecchiotti, F. Verfahren zur Implementierung eines Kartierungsdienstes für Rutschungen auf Basis von Fernerkundungsdaten und Nutzereinbindung. *AGIT J. Angew. Geoinform.* **2016**, 46–55. [CrossRef]

geosciences

MDPI

Article

Analysis of Costantino Landslide Dam Evolution (Southern Italy) by Means of Satellite Images, Aerial Photos, and Climate Data

Corrado Cencetti *, Lucio Di Matteo and Saverio Romeo

Department of Physics and Geology, University of Perugia, Via Pascoli snc, 06123 Perugia, Italy;
lucio.dimatteo@unipg.it (L.D.M.); saverio.romeo@studenti.unipg.it (S.R.)
* Correspondence: corrado.cencetti@unipg.it; Tel.: +39-075-5840-303

Academic Editors: Francesca Cigna and Jesús Martínez Frías
Received: 16 March 2017; Accepted: 13 April 2017; Published: 19 April 2017

Abstract: Large landslides, triggered by earthquakes or heavy rainfall, often obstruct the river's flow to form landslide dams, causing upstream inundations, and downstream flooding. In Italy, landslide dams are rather widespread along in Alps and Apennines: although the identification of past events is a complex task, some hundreds of landslide dams are identified in the literature. In order to assess the formation and evolution of landslide dams, several studies suggested the employment of geomorphological indexes. In this framework, the knowledge of site-specific time-space evolution can be useful in the understanding of the landslide dams phenomena. The present work focuses on a landslide dam that occurred in January 1973, which totally dammed the Bonamico River Valley (Southern Italy): the lake reached an area of about 175,000 m^2, a volume of about 3.6×10^6 m^3 and a maximum depth of 40 m. During 1973–2008, the lake surface gradually decreased and nowadays it is completely extinct by filling. By using satellite and aerial images, the paper discusses the evolution of the lake surface and the causes of the lake extinction. The use of a climate index (i.e., standardized precipitation index at different time scale) indicates that in recent decades the alternance of drought and heavy rainfall periods affected the inflow/outflow dynamics, the filling of lake due to the solid transport of the Bonamico River, and the failure of the landslide dam.

Keywords: Costantino Lake; landslide dam; remote sensing

1. Introduction

Landslide dams recur in areas affected by active tectonics (typically areas involved in recent orogenesis), where the rising trend is the main cause of hydrographic network entrenchment and consequent formation of narrow valleys, delimited by steep slopes. This is the main geological-structural factor predisposing to obstruction of riverbeds by landslides [1–3]. In addition, earthquakes, even of high magnitude, frequently affect these areas so that, among the causes of occlusion of the riverbeds, seismically-induced landslides represent a high percentage [1,4,5].

The landslide damming phenomenon produces a significant increase in risk compared to that due to landslide movement alone. In fact, the formation of a lake (dam lake) can cause both flooding on the upstream portion of the barrier and abnormal flooding downstream of the occlusion in the case of sudden collapse of the dam (e.g., rapid emptying of the lake, earthquake, piping). Within this framework, even small landslides mobilizing a reduced volume of material can occlude the riverbed, changing the morphological conditions. Overall, the approach of the use of fixed return periods to manage the hydraulic risk of downstream flooding produced by the overtopping or the failure of the landslide dam is not appropriated. In fact in these conditions, downstream flooding is not related to the recurrence time of river peak discharge due to the occasional character of the phenomenon [6–8].

In Italy, geological and structural conditions produced the formation of two chains (Alps and Apennines), frequently affected by landslide dam phenomena. Censuses conducted in Alps [9], in the northern Apennines [10,11] and, more recently, in the whole national territory [12], showed the presence of about 300 main occlusions of riverbeds due to landslides. However, in the zoning of hydrogeological risk, promoted in Italy by specific laws (e.g., Hydrogeological Assessment Plans—PAI), the two types of risk (landslide and hydraulic risks) are always addressed separately, without considering what can be generated as a result of the interference between gravitational slope processes and dynamics of riverbeds. The scientific literature mainly focused on the following fundamental aspects of landslide dams:

- the description of case histories (surely the most treated issue, with a countless number of cases described around the world);
- the evaluation of the landslide dam hazards [2,13–17];
- methods and criteria for the prediction of the dam formation and break [2,13,18–22], even with the introduction of indexes related to the dam break hazard, based on morphological parameters [10,20,23–27];
- the prediction of the consequences due to dam break [10,28–32], even on the basis of modeling [33–38];
- the mitigation measures of hydraulic risk resulting from the collapse of a natural dam [1,39,40].

As reported by [18,41], many landslide dams fail shortly after formation and overtopping is by far the most common cause of failure: among the 73 landslide-dam failures documented by these papers, 27% of the landslide dams failed less than one day after formation, and about 50% failed within 10 days.

Moreover, in the Mediterranean region, the increase of length and frequency of drought periods, and extreme rainfalls affect the inflow and outflow dynamics of lakes as well as the erosion rates in river basin and solid transport. These factors—coupled with geotechnical properties of dam materials—influence the survival time of the lake and dam stability. The present work takes as reference the landslide dam that occurred in January 1973, which totally dammed the Bonamico River Valley (Southern Italy), producing a small lake that survived for about 36 years. In detail, the paper aims to study and discuss the evolution of the lake surface, analyzing causes that conditioned its evolution up to the drastic surface reduction produced by filling and overtopping/dam failure phenomena.

2. Study Area

The Costantino landslide dam is located in the Bonamico basin along the *Arco Calabro-Peloritano* in the Southern Ionian side of Calabrian territory, Southern Apennines (Italy) (Figure 1). The Bonamico is a typical Mediterranean basin: it has a maximum altitude of 1956 m a.s.l. (Montalto Mount, Figure 1) and a mean altitude of about 780 m a.s.l. The high relief energy is due to the tectonic uplift during the Pleistocene. The length of the main river channel is 29.68 km with a mean gradient of 6% [42]. Although the basin has a total area of about 137 km², its upper part (subtended by the Costantino landslide dam) covers an area of about 100 km². The Bonamico basin, as most basins in Calabria territory, show intense hydraulic and geomorphological dynamics. From a geomorphological point of view, the Bonamico River is a *fiumara*, a braided river with high gradient, delimited by very steep slopes and characterized by coarse-grained alluvium [43,44]. The hydraulic regime of *fiumara* is typically torrential, with the predominance of debris production by mass-movement and erosion on slopes. For these reasons, the transport capacity of *fiumara* torrents is very high. According to the landslide map proposed by [45], the upper part of the basin is characterized by the presence of more than 20 landslides.

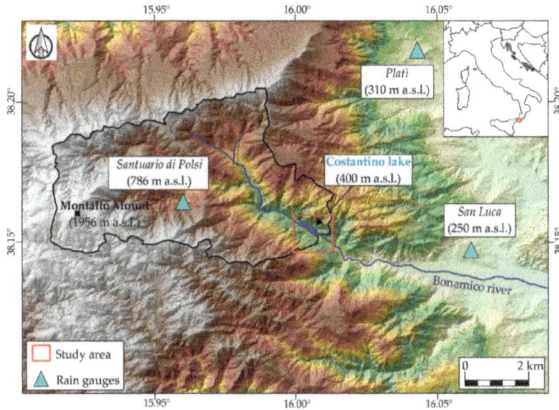

Figure 1. Location map of the study area.

The Costantino landslide, classifiable as debris/earth translational slide, occurred during the night between 3 January and 4 January in 1973 (volume of about 23×10^6 m^3 [46–48]: the displaced landslide materials—coming from the slope on right bank of the Bonamico river—reached the opposite slope and occluded the riverbed producing a lake. According to [46,47]), the lake after three minor flood waves had a maximum depth of 40 m with a volume of water of about 3.6×10^6 m^3. As reported by [48], "the lake lasted only a few hours before breaching occurred, and no significant damage was recorded at the nearby villages and towns". After about one month (4 February 1973), filtration and erosion phenomena along the lake outlet lowered the lake to a depth of about 20 m. This was the first huge mobilization of landslide materials downstream [47]: this event is clearly shown in Figure 2 that documents the formation of a deep canyon downstream the lake [49]. By this time, the threshold has suffered a temporary stop in its evolution, probably for armoring phenomena, as often occurs in such cases, when the material constituting the natural dam is very heterogeneous and characterized by low sorting [50].

According to the official geological map [51], the Costantino landslide is hosted in metamorphic rocks characterized by schists with garnet and muscovite (unit of *Madonna di Polsi*). The weathering processes and the high degree of jointing in the rock mass decomposed and transformed the rocks into soils, strongly reducing the overall resistance of the rock mass [46,48].

Figure 2. Photos taken immediately after the landslide events of January 1973. On the left the Bonamico river valley, on the right the diffuse erosion along the Costantino Lake outflow and on the landslide dam body. Photos taken from [49] (used with permission of the copyright holder).

3. Materials and Methods

3.1. Climatic Characterization and Assessment of Wet/Drought Periods

Climatic characterization of the Bonamico catchment at the Costantino landslide dam (Figure 1) was carried out on different time-scale taking into account the most reliable meteorological stations located close to the study area. In detail, three available meteorological stations were considered: Platì, San Luca, and Santuario di Polsi, located 8, 5, and 4.5 km away from the Costantino Lake, respectively. Daily analysis was carried out to investigate the rainfall amount/intensity responsible for the landslide occurrence in January 1973 and to examine the accumulated rainfall during the days before the sudden reduction of the lake surface (January 2009). Monthly analysis helped us to study the evolution of the Costantino Lake surface during the last decades also considering wet/drought periods. As commonly occurs in other locations along the Apennines, only a few stations registered meteorological data up to the latest years, especially those located at higher altitudes [52,53]. Table 1 shows the characteristics of the available meteorological stations in the study area, two of which are located within the Bonamico basin (San Luca and Santuario di Polsi, Figure 1). Data sets were collected from the rain gauge network of the *Centro Funzionale of Calabria Region*.

Table 1. Characteristics of the available meteorological stations (for locations, see Figure 1).

Station	Code	Type	Elevation (m a.s.l.)	Period	Missing Data
Platì	2230	Rain gauge	310	1920–2016	4%
San Luca	2260	Rain gauge	250	1924–2016	5% (no data during 2001–2004)
Santuario di Polsi	2250	Rain gauge	786	1928–2005	27% (no data during 1973–1992)

San Luca rain gauge (code 2260) was operating during 1924–2016 with less than 5% of missing data, with a total lack of data during 2001–2004. The Platì rain gauge (code 2230) represents the most reliable rain gauge in the study area (missing data less than 4% during 1920–2016 period). Gaps in time-series were filled by applying multiple regressions based on the best-correlated data series ($R^2 > 0.75$) of the nearest rain gauges. Double mass analysis [54] was used to investigate the consistency of precipitation records in the study area. In detail, a straight double mass curve between Platì and San Luca rain gauges was obtained, indicating a consistent and reliable precipitation record. Data of the Santuario di Polsi rain gauge (code 2250) are not continuous during 1921–2016, since no data have been registered between 1974–1991 period and after 2004. Due to the high gaps in the data time series, rainfall data of this station have been used only to check the amount of daily rain accumulation before the landslide occurrence.

The Mann-Kendall test (MK) [55,56], commonly used to detect monotonic trends in series of data (e.g., climate data, hydrological data, etc.), was used to check the presence of significant rainfall trends on differing time-scales (annual, seasonal, etc.) by analyzing the long-time series of Platì rain gauge. To check the occurrence and intensity of wet/droughts periods, we used the Standardized Precipitation Index, SPI [57,58], which was declared to be an official meteorological drought index by the World Meteorological Organization [59]. This index was also used by [60] to assess drought occurrence in Calabria. As reported by [61], robust relationships were found between the SPI time-scales on river discharges and reservoir storages in complex hydrological systems in mountainous regions and in the Mediterranean region. In the SPI computation, at least 30 years of data are needed [57,62]. In this study, the SPI analysis was carried out on the 1950–2015 period (65 years): this time-span was chosen because we are mainly focused on the analysis of wet/drought periods preceding and following the occurrence of the Costantino landslide dam. Rainfall data passed the test for gamma probability distribution according to Kolmogorov-Smirnov and Chi-square tests. The SPI computation was carried out on varying time-scales with the DrinC software [63]. According to [57,58], seven severity classes—from severe drought (SPI $\leq$ 2) to severe wet (SPI $\geq$ 2)—have been proposed (Table 2). The lack of any long time-series of temperature data did not allow the use of index based on monthly precipitation (P) and potential evapotranspiration (PET), such as the Reconnaissance Drought Index (RDI) [64].

Table 2. Standardized Precipitation Index (SPI) classification (adapted from [57]).

SPI Values	Classification	Abbreviation
≥2.0	Extremely wet	EW
1.5–1.99	Very wet	VW
1.0–1.49	Moderately wet	MW
0.99–0	Normal	N
0––0.99	Near Normal	NN
−1.0––1 49	Moderately drought	MD
−1.5––1.99	Severe drought	SD
≤−2.0	Extremely drought	ED

3.2. Aerial and Satellite Imagery Database

In order to check the evolution of lake dam, a multi-temporal image analysis was carried out. A similar analysis, based only on visual analysis of three recent satellite images (SI) images (2005, 2008, and 2012), was performed by [12]. In the present work, several SI, aerial photos (AP), and orthophotos (OP) from different sources were collected: SI are available for viewing on Google (www.google.com) and Bing (www.bing.com), AP were bought from Istituto Geografico Militare, while OP were collected through the Web Map Service (WMS) made available by Ministero dell'Ambiente e della Tutela del Territorio e del Mare (http://www.pcn.minambiente.it/GN/accesso-ai-servizi/servizi-di-visualizzazione-wms). Overall, the analysis covered the 1983–2015 time-span also considering the conditions before the landslide occurrence (*a* in Table 3). Although among the available images, one has a very low resolution (*e* in Table 3): we believe that the study of this image is useful to increase the understanding of temporal evolution of the lake surface.

Table 3. Characteristics of the imagery database used in the study area: AP: aerial photos; OP: orthophotos; SI: satellite images. (Source: Ministero dell'Ambiente e della Tutela del Territorio e del Mare, Istituto Geografico Militare, Google, Bing).

Labels of images	Type	Season	Date
a	AP black/white	Summer	July 1955
b	AP black/white	Summer	July 1983
c	OP black/white	Winter	March 1989
d	OP black/white	Summer	August 1996
e	SI colour	Spring	May 2003
f	SI colour	Winter	March 2005
g	SI colour	Summer	August 2005
h	SI colour	Summer	July 2008
i	SI colour	Spring	March 2009
l	SI colour	Winter	March 2010
m	SI colour	Spring	April 2012
n	SI colour	Spring	April 2015

4. Results

4.1. Climatic Analysis and Assessment of Wet/Drought Periods

To check the presence of significant rainfall trends in the study area (both positive and negative), a statistical analysis was carried out on monthly data of the Platì station which represents one of the most reliable stations of the Calabria region as it has a long-term dataset (about 96 years of rainfall observations). The data show a statistically significant negative precipitation trend both at annual scale

(−2.60 mm/year, MK test > 95%) and autumn/winter seasons (−3.65 mm/6-months, from October to March—MK test > 98%). Conversely, positive rainfall trends have been detected during summer months (+0.8 mm/three months, from July to September—MK test > 98%). Overall, the results are consistent with those of previous studies [65–69]. As reported by [69], the negative trend of annual rainfall data is a typical feature of all the long aggregation rainfalls observed in the Calabria Region and especially on the Southern Ionian side of Calabrian territory. Moreover, also the general significant increase of rainfall during summer months and the significant decrease of rainfall in autumn/winter months is well spatially distributed on the Southern Ionian side of Calabrian territory. In this context, results from the Platì rain gauge can be representative of climatic trend in the upper part of the Bonamico basin.

According to [70], "the paradoxical increase of extreme rainfall in spite of a decrease in the totals is not present in this part of southern Italy". This is also confirmed—at monthly scale—by the SPI index. As shown in Figure 3 in the last three decades (1986–2016) the number and frequency of prolonged drought periods are increased while the prolonged wet and very wet periods are—according to the index used—not so numerous as those recognized in the 1950–1985 period. In detail, for the last three decades, five severe droughts (−2.0 ≤ SPI < −1.5) occurred. Conversely, there were only two main prolonged drought periods between 1950 and 1986 (Figure 3).

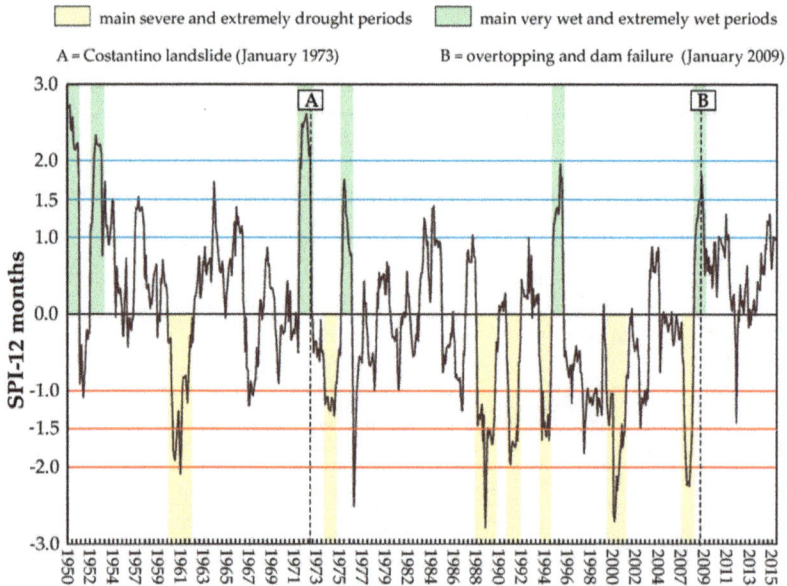

Figure 3. Twelve-month SPI calculated from meteorological data of Platì station. The SPI computation was carried out with DrinC software [63].

4.2. Evolution of Lake Surface and of Landslide Dam

The temporal evolution of the lake surface can be inferred by the imagery database in Table 3. The analysis was carried out in GIS environment by using the ArcGIS Earth (ESRI, Redlands, CA, USA), Google Earth (Google, Mountain View, CA, USA), and Quantum GIS software (OSGeo). As shown in Figures 4 and 5, the lake surface gradually decreased from about 100,000 (July 1983) to about 60,000 m² (July 2008). After 2008, a sudden decrease of the lake surface was observed by using satellite images: the lake surface passed in few months from about 60,000 (July 2008) to 30,000 m² (March 2009), reaching about 10,000 m² (after April 2012). Figure 6a shows the large amount of sediment in the area previously occupied by the lake. Figure 6b shows a picture taken during spring 2012: the reduction of lake surface has nowadays allowed the vegetative colonization of aquatic and riparian plants along the river and the small surviving lake.

Figure 4. Evolution of lake surface by using the imagery database in Table 3.

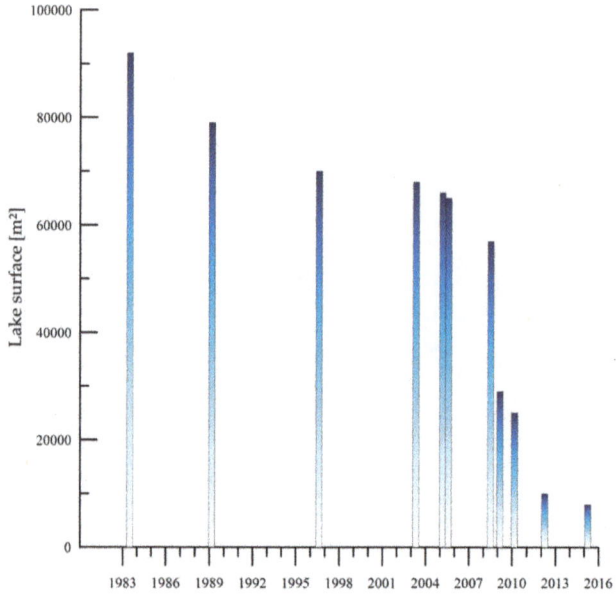

Figure 5. Evolution of lake surface during the 1983–2015 period as resulted from the analysis of the imagery database in Table 3.

Figure 6. Photos of Costantino Lake in 2012. (**a**) Detail of sediment accumulation in the area originally occupied by the lake; (**b**) vegetative colonization along the river and the small surviving lake.

5. Discussion on Origin and Evolution of the Costantino Lake

The Costantino Lake was one of the few examples of formed-stable landslide dam documented in Southern Italy [12], which gradually disappeared due to erosion/filling and overtopping/dam failure processes. The landslide originating the dam was triggered by heavy rainfalls that occurred between 21 December 1972 and 4 January 1973. First rainfall events were spatially distributed with peaks of 300–350 and 200 mm/day, in the upper part (data of Santuario di Polsi) and in the medium part (data of San Luca and Platì) of the Bonamico basin, respectively. Although the Platì station is not located within the Bonamico basin, the rainfall data from this station were taken into consideration to better understand rainfall event. In the few days before the landslide triggering, rainfalls mainly affected the medium part of the Bonamico basin with peaks of 200–300 mm/day (Figure 7a), by focusing in the area where the landslide has taken place.

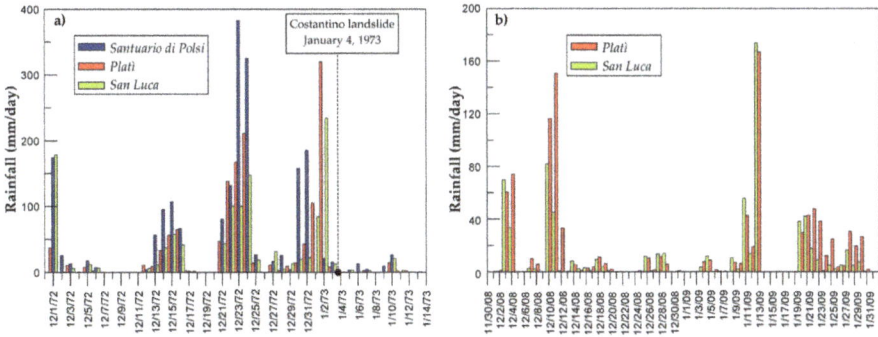

Figure 7. Daily rainfalls registered by the available rain gauges during: (**a**) the occurrence of the Costantino landslide in January 1973; (**b**) the overtopping/dam failure in December 2008–January 2009.

The lake originating from the landslide dam evolved over time (about 36 years) experiencing a gradual reduction of the lake surface up to 2008 [42], based on extrapolation of the amount of solid material deposited in the Costantino Lake, suggesting an average sediment transport of about 10,000 m^3/year: this estimation is referred to the two decades before the publication of the research (about 20 years after the landslide occurrence). Since such data are not available after 1992, we studied the evolution of the lake surface by investigating, in a qualitative way, the effect of rainfall on average annual erosion, taking into account the significant decreasing trend of autumn/winter and annual rainfall in the study area. In other words, the use of empirical equations—based on rainfall data—can be useful when no lengthy series of sediment transport measurements are available.

$$FI = \frac{P_{imax}^{\;2}}{P} \tag{1}$$

$$MFI = \sum_{i=1}^{12} \left(\frac{p_i^{\;2}}{P} \right) \tag{2}$$

where:

p_{imax} = rainfall in the rainiest month of the year i (mm);
P_i = yearly rainfall (mm);
p_i = mean monthly rainfall (mm);
P = mean annual rainfall (mm).

We calculated the Fournier Index (FI) [71] and the Modified Fournier Index (MFI) [72], defined by Equations (1) and (2), respectively. Both indexes allow understanding, qualitatively at least, of the effect of rainfall on erosion. Based on FI and MFI values, different rainfall erosivity classes are defined (Table 4). Although the risk of erosion is not only linked to the rainfall but also to other factors related to the vegetation dynamic, land use, etc. [73], the decrease of the two indexes should reduce the erosion rates.

Table 4. The erosivity classes by Fournier index (FI) and modified Fournier index (MFI). * from [72]; ** from [74].

Erosivity Classes	FI *	Erosivity Classes	MFI **
Very low	0–20	Very low	0–60
Low	20–40	Low	60–90
Moderate	40–60	Moderate	90–120
Severe	60–80	High	120–160
Very severe	80–100	Very high	>160
Extremely severe	>100		

Taking into account the monthly rainfall data of Platì and San Luca stations, FI and MFI values were computed five year by five year (Figure 8), starting from 1951. It is interesting to highlight as the FI index after the occurrence of the Costantino landslide dam (1973), decreased on both stations of about 70 mm, passing from about 165 to 95 mm. Overall, the peak of FI recorded in 1969–1974 is placed in a period characterized by a general decrease of the index since 1951–1956. By comparing the 1951–1986 and 1987–2016 periods, the MFI index has decreased of about 17% on both stations (306 to 257 mm in Platì station and 246 to 203 mm in San Luca station). If other conditions in the catchment remained the same (topographic factors, land use, etc.), as it is likely, a decrease in such indexes implied a decrease in erosion in mountain areas, with a subsequent decrease in sediment yield to the alluvial plains [75].

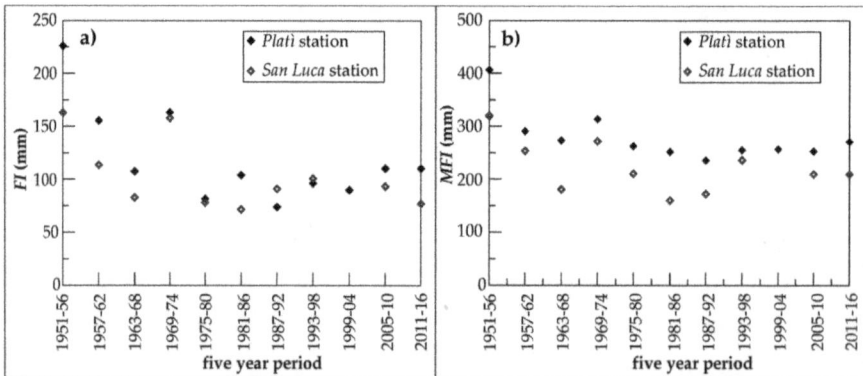

Figure 8. (a) Fournier index (FI); and (b) modified Fournier index (MFI) for Platì and San Luca stations, five year by five year, starting from 1951.

Despite the increase of droughts periods highlighted in the last three decades by SPI (Figure 3), the erosivity classes for the Bonamico basin remained very severe/very high. The analysis here presented is based only on two meteorological stations, which are located at an altitude much lower than that of the mean altitude of the Bonamico basin; moreover, the station of Platì is located out of the Bonamico basin (about 8 km in line of sight from the Costantino Lake). We believe that without reliable rainfall data at high elevation, the analysis may be considered as a qualitative representation of rainfall effects on erosion and therefore on sediment transport to the Costantino Lake. Overall, considering that the FI index remains almost constant (between 80 and 100 mm) after 1975 (Figure 8), the estimation of sediment transport made by [42], could be representative—at least as order of magnitude—even for the period after 1992, on the basis of only rainfall data. The high sediment transport played a key role in the lake filling, especially up to 2008: as illustrated in Figure 5a gradual decrease of the lake surface during 1983–2008 is mainly due to sediment accumulation in the lake produced by solid transport. After these phases, a sudden reduction of lake surface after January 2009 was observed (Figure 5). It is linked to overtopping and dam failure phenomena. A rainfall event occurred in January 2009, also documented by [76], affected the mountains in the southern-eastern part of Calabria territory with three days of rainfall with a 100-year return period (>200 mm/day). This event was also registered—although in lesser amounts—at the Platì and San Luca rain gauges (Figure 7b) located at low altitude (about 240 mm/3-days on 11–13 January 2009). The effect of the flood related to this rainfall event is well recognized by comparing the satellite images of July 2008 (Figure 4h) and March 2009 (Figure 4i): the vegetation along the lake outflow and on the landslide dam body has been removed and dam materials have been highly eroded. Overtopping and dam failure produced the lowering of the threshold elevation of lake outlet with a consequent reduction of the lake surface.

6. Conclusions

The study, taking as reference the Costantino landslide dam, emphasizes the use of Earth Observation (EO) through multi-temporal high-resolution images as useful tool for understanding, both quantitatively and qualitatively, changes occurring on dam lakes and landslide bodies. On the basis of the results of the study carried out, we can draw the following conclusions:

- The use of imagery dataset of different nature (aerial photos, orthophotos, and satellite images) allowed to study the evolution of the Costantino dam lake. The lake surface between 1983 and 2008 gradually decreased of 40% (from about 100,000 to 60,000 m^2). After January 2009, a sudden decrease of the lake surface was observed in few months, leaving a very small lake of less than 8000 m^2.

- Climatic data from the most reliable rain gauges in the Costantino landslide dam area indicate a significant decreasing trend in annual and autumn-winter rainfalls, with some significant increase in summer months. The SPI index shows a notable increase in the length and frequency of droughts. The most severe droughts have occurred during the last two decades: this problem is common to many other mountain areas in the Mediterranean basin.

- The use of basic empirical equations shows that the general reduction of rainfalls inevitably affects the erosion rates on the Bonamico catchment: the MFI index decreased of about 17% in the last three decades. This result agrees with those obtained in other areas of the Mediterranean basin (e.g. [75–78]), but is in contrast to findings in the Mediterranean Iberian peninsula, where some increase (+8.8%) of MFI was observed (e.g. [73]).

- Despite the general decrease of both precipitation and rain erosivity index, the erosivity classes within the Bonamico basin remained 'very severe/very high' in the last decades. This indicates that the erosion and sediment transport processes played a key role in the filling up of the Costantino dam lake, at least up to 2008.

- The use of multi-temporal image observations coupled with analysis of daily rainfalls enabled individuation causes of sudden changes in lake surface after January 2009: overtopping/dam failure processes related to heavy rainfall and floods that occurred in January 2009 deeply affected the threshold of the lake outlet. Satellite images visually show the result of these processes, which can be summarized in the removal of vegetation and the mobilization of dam materials along the riverbed downstream.

In conclusion, the results of the present work indicate that Earth Observation data are fundamental in studying and monitoring land processes such as the landslide dams. It should be emphasized that, for a proper understanding of the evolution of this type of landslides and related lakes, an accurate and reliable hydro-meteorological network should be assured. This is a key point in the next few decades, since those data often are missing, especially in mountainous areas.

Acknowledgments: Authors wish to thank Giancarlo Parisi who kindly provided the most recent photos of Costantino Lake.

Author Contributions: All authors contributed equally to the research.

Conflicts of Interest: The authors declare no conflict of interest.

References

1. Schuster, R.L. *Landslide Dams: Processes, Risk, and Mitigation*; Geotechnical Special Publication; American Society of Civil Engineering: New York, NY, USA, 1986; Volume 3, p. 164.
2. Costa, J.E.; Schuster, R.L. The formation and failure of natural dams. *Geol. Soc. Am. Bull.* **1988**, *100*, 1054–1068. [CrossRef]
3. Evans, S.G. The formation and failure of landslide dams: An approach to risk assessment. *Ital. J. Eng. Geol. Environ.* **2006**, *1*, 15–19.
4. Schuster, R.L. Landslide Dams in the Western United States. In Proceedings of the 4th International Conference and Field Workshop on Landslides, Tokyo, Japan, 23–31 August 1985; pp. 411–418.
5. Costa, J.E.; Schuster, R.L. *Documented Historical Landslide Dams from Around the World*; Open-File Report; United States Department of the Interior Geological Survey: Vancouver, WA, USA, 1991; pp. 91–239.
6. Cencetti, C.; Conversini, P.; Ribaldi, C.; Tacconi, P. Studio dei sistemi alveo—Pianura fluviale in relazione alle interazioni con fenomeni franosi di versante. *Mem. Soc. Geol. Ital.* **2001**, *56*, 249–263.
7. Cencetti, C.; Conversini, P.; Marchesini, I.; Ribaldi, C.; Tacconi, P. Pericolosità dei fenomeni franosi che interferiscono con i sistemi alveo-pianura fluviale: Un approccio probabilistico. *Ital. J. Eng. Geol. Environ.* **2002**, *1*, 49–60.
8. Cencetti, C.; De Rosa, P.; Minelli, A. A sensitivity analysis on main factors involved in the landslide dam phenomena. *Ital. J. Eng. Geol. Environ.* **2011**, *1*, 61–72.
9. Pirocchi, A. Laghi di sbarramento per frana nelle Alpi: Tipologia ed evoluzione. *Ric. Sci. Educ. Perm.* **1991**, *93*, 127–136.
10. Canuti, P.; Casagli, N.; Ermini, L. Inventory of landslide dams in the Northern Apennine as a model for induced flood hazard forecasting. In *Managing Hydro-Geological Disasters in a Vulnerate Environment*; Andah, K., Ed.; CNR-GNDCI Publication 1900; CNR-GNDCI-UNESCO (IHP): Perugia, Italy, 1998; pp. 189–202.
11. Casagli, N.; Ermini, L. Geomorphic analysis of landslide dam of Northern Appenine. *Trans. Jpn. Geomorphol. Union* **1999**, *20*, 219–249.
12. Tacconi Stefanelli, C.; Catani, F.; Casagli, N. Geomorphological investigations on landslide dams. *Geoenviron. Dis.* **2015**, *2*, 1–15. [CrossRef]
13. Dunning, S.A.; Petley, D.N.; Rosser, N.J. Landslide dams: Causes, catastrophic failure prediction and downstream consequences. *Geophys. Res. Abstr.* **2005**, *7*, 06893.
14. Hungr, O. Prospects for prediction of landslide dam geometry using empirical and dynamic models. *Ital. J. Eng. Geol. Environ.* **2006**, *1*, 151–155.
15. Kuo, Y.S.; Tsang, Y.C.; Chen, K.T.; Shieh, C.L. Analysis of landslide dam geometries. *J. Mt. Sci.* **2011**, *8*, 544–550. [CrossRef]
16. Dal Sasso, S.F.; Sole, A.; Pascale, S.; Sdao, F.; Bateman Pinzòn, A.; Medina, V. Assessment methodology for the prediction of landslide dam hazard. *Nat. Hazards Earth Syst. Sci.* **2014**, *14*, 557–567. [CrossRef]
17. Cencetti, C.; De Rosa, P.; Fredduzzi, A. Evaluation of landslide dams hazard and risk: An application in Upper Tiber Valley (central Italy). *Rend. Online Soc. Geol. Ital.* **2015**, *35*, 54. [CrossRef]
18. Costa, J.E. *Floods from Dam Failures*; Open-File Report n. 85-560; United States Department of the Interior Geological Survey: Vancouver, WA, USA, 1987; pp. 1–59.
19. Ermini, L.; Casagli, N. Criteria for a preliminary assessment of landslide dam evolution. In *Landslides—Proceedings of 1st European Conference on Landslides*; Rybar, J., Stemberk, J., Wagner, P., Eds.; Balkema: Prague, Czech Republic, 2002; pp. 157–162.

20. Ermini, L. Gli sbarramenti d'alveo da frane: Criteri speditivi per la stesura di scenari evolutivi derivanti dalla loro formazione. In *Atti del 1° Congresso Nazionale AIGA (Chieti, 19–20 Febbraio 2003)*; Rendina Editore: Rome, Italy, 2003; pp. 355–367.

21. Dong, J.J.; Tung, Y.H.; Chen, C.C.; Liao, J.J.; Pan, Y.W. Logistic regression model for predicting the failure probability of a landslide dam. *Eng. Geol.* **2011**, *117*, 52–61. [CrossRef]

22. Peng, M.; Zhang, L.M. Breaching parameters of landslide dams. *Landslides* **2012**, *9*, 13–31. [CrossRef]

23. Swanson, F.J.; Oyagi, N.; Tominaga, M. Landslide dam in Japan. In *Landslide Dam: Processes Risk and Mitigation*; Schuster, R.L., Ed.; Geotechnical Special Publication; American Society of Civil Engineers: New York, NY, USA, 1986; Volume 3, pp. 131–145.

24. Moore, I.D.; Grayson, R.B.; Ladson, A.R. Digital terrain modelling: A review of hydrological, geomorphological, and biological applications. *Hydrol. Process.* **1991**, *5*, 3–30. [CrossRef]

25. Ermini, L.; Casagli, N. Prediction of the behaviour of landslide dams using a geomorphological dimensionless index. *Earth Surf. Processes Landf.* **2003**, *28*, 31–47. [CrossRef]

26. Fan, X.; van Westen, C.J.; Korup, O.; Gorum, T.; Xu, Q.; Dai, F.; Huang, R.; Wang, G. Transient water and sediment storage of the decaying landslide dams induced by the 2008 Wenchuan earthquake, China. *Geomorphology* **2012**, *171*, 58–68. [CrossRef]

27. Tacconi Stefanelli, C.; Segoni, S.; Casagli, N.; Catani, F. Geomorphic indexing of landslide dams evolution. *Eng. Geol.* **2016**, *208*, 1–10. [CrossRef]

28. Pingyi, Z.; Tianchi, L. Flash Flooding Caused by Landslide Dam Failure. In *ICIMOD, Mountain Flash Floods*; ICIMOD: Kathmandu, Nepal, 2000; Newsletter No. 38.

29. Schuster, R.L. Outburst debris flows from failure of natural dams. In Proceedings of the 2nd International Conference on Debris-Flow Hazard Mitigation, Taipeh, Taiwan, 16–18 August 2000.

30. Li, M.H.; Hsu, M.H.; Hsieh, L.S.; Teng, W.H. Inundation Potentials Analysis for Tsao-Ling Landslide Lake Formed by Chi-Chi Earthquake in Taiwan. *Nat. Hazards* **2002**, *25*, 289–303. [CrossRef]

31. Korup, O.; Strom, A.L.; Weidinger, J.T. Fluvial response to large rock-slope failures: Examples from the Himalayas, the Tien Shan, and the southern alps in New Zealand. *Geomorphology* **2006**, *78*, 3–21. [CrossRef]

32. Yoshino, K.; Uchida, T.; Shimizu, T.; Tamura, K. Geomorphic changes of a landslide dam by overtopping erosion. *Ital. J. Eng. Geol. Environ.* **2011**, 797–804. [CrossRef]

33. Fread, D.L. *DAMBRK: The NWS Dam-Break Flood Forecasting Model*; Hydrologic Research Laboratory, National Weather Service: Silver Spring, MD, USA, 1984.

34. Fread, D.L. *BREACH: An Erosion Model for Earthen Dam Failures*; Hydrologic Research Laboratory, National Weather Service: Silver Spring, MD, USA, 1987.

35. Fread, D.L. *The NWS Dambrk Model: Theoretical Background/User Documentation*; National Weather Service, NOAA: Silver Spring, MD, USA, 1991.

36. Walder, J.S.; O'Connor, J.E. Methods for predicting peak discharge of floods caused by failure of natural and earthen dams. *Water Resour. Res.* **1997**, *33*, 2337–2348. [CrossRef]

37. Tabata, S.; Ikeshima, T.; Inoue, K.; Mizuyama, T. Study on prediction of peak discharge in floods caused by landslide dam failure. *J. Jpn. Soc. Eros. Control Eng.* **2001**, *54*, 73–76.

38. Cencetti, C.; Fredduzzi, A.; Marchesini, I.; Naccini, M.; Tacconi, P. Some considerations about the simulation of the breach channel erosion on landslide dams. *Comput. Geosci.* **2006**, *10*, 201–219. [CrossRef]

39. Ishikawa, Y.; Irasawa, M.; Kuang, S.F. Study on prediction and countermeasures of flood disasters caused by landslide dam failure. *J. Jpn. Soc. Eros. Control Eng.* **1992**, *45*, 14–23.

40. Schuster, R.L. Risk-reduction measures for landslide dams. *Ital. J. Eng. Geol. Environ.* **2006**, *1*, 9–13.

41. Schuster, R.L.; Costa, J.E. Effects of landslide damming on hydroelectric projects. In Proceedings of the 5th International Association of Engineering Geology, Buenos Aires, Argentina, 20–25 October 1986; CRC Press: BOca Ratón, FL, USA, 1990; pp. 1295–1307.

42. Ergenzinger, P. A conceptual geomorphological model for the development of a Mediterranean river basin under neotectonic stress (Buonamico basin, Calabria, Italy). In *Erosion, Debris Flows and Environment in Mountain Regions*; Walling, D.E., Davies, T.R., Hasholt, B., Eds.; International Association of Hydrological Sciences (IAHS): Wallingford, UK, 1992; Volume 209, pp. 51–60.

43. Petrucci, O.; Polemio, M. Catastrophic geomorphological events and the role of rainfalls in South-Eastern Calabria (Southern Italy). In Proceedings of the 2nd EGS Plinius Conference on Mediterranean Storms, Siena, Italy, 16–18 October 2000; Editoriale Bios: Cosenza, Italy, 2000; pp. 449–459.

44. Sorriso-Valvo, M.; Terranova, O. The Calabrian fiumara. *Z. Geomorphol.* **2006**, *143*, 105–121.
45. Sorriso-Valvo, M. *Mass Movements and Accelerated Erosion in the Buonamico Basin*; Berliner Geowissenschaftliche Abhandlungen: Berlin, Germany, 1984.
46. Guerricchio, A.; Melidoro, G. Segni premonitori e collassi delle grandi frane della valle della fiumara Buonamico (Aspromonte, Calabria). *Geol. Appl. Idrogeol.* **1973**, *8*, 315–346.
47. Ibbeken, H.; Schleyer, R. *Source and Sediment: A Case Study of Provenance and Mass Balance at an Activeplatemargin (Calabria, Southern Italy)*; Pangaea: Bremerhaven, Germany, 1991.
48. Calcaterra, D.; Parise, M. Weathering in the crystalline rocks of Calabria, Italy, and relationships to landslides. *Eng. Geol. Spec. Geol. Soc. Lond.* **2010**, *23*, 105–130. [CrossRef]
49. Delfino, A. *L'Aspromonte*; Falzea Editore: Reggio Calabria, Italy, 2006; p. 128.
50. Cencetti, C.; De Rosa, P.; Fredduzzi, A. The Landslide Dam of Ventia Creek (Umbria, Central Italy). In *Engineering Geology for Society and Territory—Volume 2. Landslide Processes*; Lollino, G., Giordan, D., Crosta, G.B., Corominas, J., Azzam, R., Wasowski, J., Sciarra, N., Eds.; Springer: Cham, Switzerland, 2015; pp. 1125–1128.
51. ISPRA, CARG Project (Geological Cartography). *Geological Map "Foglio 603 Bovalino"*; ISPRA—Servizio Geologico d'Italia: Rome, Italy, 2005.
52. Cambi, C.; Valigi, D.; Di Matteo, L. Hydrogeological study of data-scarce limestone massifs: The case of Gualdo Tadino and Monte Cucco structures (Central Apennines, Italy). *Boll. Geofis. Teor. Appl.* **2010**, *51*, 345–360.
53. Di Matteo, L.; Valigi, D.; Cambi, C. Climatic characterization and response of water resources to climate change in limestone areas: Some considerations on the importance of geological setting. *J. Hydrol. Eng.* **2013**, *18*, 773–779. [CrossRef]
54. Peterson, T.C.; Easterling, D.R.; Karl, T.R.; Groisman, P.; Nicholls, N.; Plummer, N.; Torok, S.; Auer, I.; Boehm, R.; Gullett, D.; et al. Homogeneity adjustments of in situ atmospheric climate data: A review. *Int. J. Climatol.* **1998**, *18*, 1493–1517. [CrossRef]
55. Mann, H.B. Nonparametric tests against trend. *Econometrica* **1945**, *13*, 245–259. [CrossRef]
56. Kendall, M.G. *Rank Correlation Methods*; Griffin: London, UK, 1975.
57. McKee, T.B.; Doesken, N.J.; Kleist, J. The relationship of drought frequency and duration to time scales. In Proceedings of the 8th Conference of Applied Climatology, Anaheim, CA, USA, 17–22 January 1993; American Meteorological Society: Boston, MA, USA, 1993; pp. 179–184.
58. Edwards, D.C.; McKee, T.B. Characteristics of 20th Century Drought in the United States at Multiple Time Scales. Available online: http://ccc.atmos.colostate.edu/edwards.pdf (accessed on 25 January 2017).
59. WMO. Experts Agree on a Universal Drought Index to Cope with Climate Risks. WMO Press Release No. 872. 2009. Available online: http://www.wmo.int/pages/prog/wcp/agm/meetings/wies09/documents/872_en.pdf (accessed on 5 March 2017).
60. Buttafuoco, G.; Caloiero, T. Drought events at different timescales in southern Italy (Calabria). *J. Maps* **2014**, *10*, 529–537. [CrossRef]
61. Vicente-Serrano, S.M.; López-Moreno, J.I. Hydrological response to different time scales of climatological drought: An evaluation of the standardized precipitation index. *Hydrol. Earth Syst. Sci.* **2005**, *9*, 523–533. [CrossRef]
62. Naresh Kumar, M.; Murthy, C.S.; SeshaSai, M.V.R.; Roy, P.S. On the use of Standardized Precipitation Index (SPI) for drought intensity assessment. *Meteorol. Appl.* **2009**, *16*, 381–389. [CrossRef]
63. Tigkas, D.; Vangelis, H.; Tsakiris, G. DrinC: A software for drought analysis based on drought indices. *Earth Sci. Inform.* **2015**, *8*, 697–709. [CrossRef]
64. Tsakiris, G.; Vangelis, H. Establishing a Drought Index Incorporating Evapotranspiration. *Eur. Water* **2005**, *9*, 3–11.
65. Coscarelli, R.; Gaudio, R.; Caloiero, T. Climatic trends: An investigation for a Calabrian basin (southern Italy). *IAHS Publ.* **2004**, *286*, 255–266.
66. Cotecchia, V.; Casarano, D.; Polemio, M. Characterisation of rainfall trend and drought periods in southern Italy from 1821 to 2001. In *Proceedings of 1st Italian-Russian Workshop New Trends in Hydrology*; Gaudio, R., Ed.; CNR-GNDCI Publ. 2823; CNR: Rome, Italy, 2004.

67. Ferrari, E.; Terranova, O. Non-parametric detection of trends and change point years in monthly and annual rainfalls. In *Proceedings of 1st Italian-Russian Workshop New Trends in Hydrology*; Gaudio, R., Ed.; CNR-GNDCI Publ. 2823; CNR: Rome, Italy, 2004.

68. Polemio, M.; Casarano, D. Climate change, drought and groundwater availability in southern Italy. In *Climate Change and Groundwater*; Dragoni, W., Ed.; Special Publications, No. 288; Geological Society: London, UK, 2008; pp. 39–51.

69. Caloiero, T.; Coscarelli, R.; Ferrari, E.; Mancini, M. Trend detection of annual and seasonal rainfall in Calabria (Southern Italy). *Int. J. Climatol.* **2011**, *31*, 44–56. [CrossRef]

70. Caloiero, T.; Sirangelo, B.; Coscarelli, R.; Ferrari, E. An Analysis of the Occurrence Probabilities of Wet and Dry Periods through a Stochastic Monthly Rainfall Model. *Water* **2016**, *8*, 39. [CrossRef]

71. Fournier, F. *Climat et Erosion*; Presses Universitaires de France: Paris, France, 1960.

72. Arnoldus, H.M.J. An approximation of the rainfall factor in the universal Soil Loss Equation. In *Assessment of Erosion*; FAO Land and Water Development Division, Wiley & Sons: Chichester, UK, 1980; pp. 127–132.

73. De Luis, M.; Gonzalez-Hidalgo, J.C.; Longares, L.A.; Stepanek, P. Seasonal precipitation trends in Mediterranean Iberian Peninsula in second half of XX century. *Int. J. Climatol.* **2009**. [CrossRef]

74. European Environment Agency. *CORINE—Soil Erosion Risk and Important Land Resources in the Southern Regions of the European Community*; Final Report EUR 13233-EN; Publications Office of the European Union: Luxembourg, Luxembourg, 1992.

75. Dragoni, W.; Valigi, D. Some considerations regarding climatic change and specific erosion in Central Italy. In *Geomorphology and Global Environmental Change*; Slaymaker, O., Ed.; John Wiley & Sons: Hoboken, NJ, USA, 2000; pp. 197–208.

76. Gullà, G.; Antronico, L.; Borrelli, L.; Caloiero, T.; Coscarelli, R.; Iovine, G.; Nicoletti, P.G.; Pasqua, A.A.; Petrucci, O.; Terranova, O. Indicazioni conoscitive e metodologiche connesse all'evento di dissesto idrogeologico dell'autunno-inverno 2008–2009 in Calabria. *Geol. Calabria* **2009**, *10*, 4–21.

77. Angulo-Martinez, M.; Beguería, S. Do atmospheric teleconnection patterns influence rainfall erosivity? A comparison between NAO, MO and WEMO in NE Spain, 1955–2006. *J. Hydrol.* **2012**, *450*, 168–179. [CrossRef]

78. Meddi, M. Sediment transport and rainfall erosivity evolution in twelve basins in Central and Western Algeria. *J. Urban Environ. Eng.* **2013**, *7*, 253–263. [CrossRef]

geoscience

MDPI

Article

Assessment of the Evolution of a Landslide Using Digital Photogrammetry and LiDAR Techniques in the Alpujarras Region (Granada, Southeastern Spain)

Tomás Fernández [1,2,*], **José Luis Pérez** [1,2], **Carlos Colomo** [1], **Javier Cardenal** [1,2], **Jorge Delgado** [1], **José Antonio Palenzuela** [3], **Clemente Irigaray** [3] and **José Chacón** [3]

[1] Department of Cartographic, Geodetic and Photogrammetric Engineering, University of Jaén, Campus de las Lagunillas s/n, 23071 Jaén, Spain; jlperez@ujaen.es (J.L.P); cmcj0002@red.ujaen.es (C.C.); jcardena@ujaen.es (J.C.); jdelgado@ujaen.es (J.D.)
[2] Centre for Advanced Studies in Earth Sciences (CEACTierra), University of Jaén, Campus de las Lagunillas s/n, 23071 Jaén, Spain
[3] Department of Civil Engineering, University of Granada, Campus de Fuentenueva s/n, 18071 Granada, Spain; jpalbae@ugr.es (J.A.P.), clemente@ugr.es (C.I.); jchacon@ugr.es (J.Ch.)
* Correspondence: tfernan@ujaen.es; Tel.: +34-53-212843

Academic Editors: Francesca Cigna and Jesus Martinez-Frias
Received: 4 March 2017; Accepted: 20 April 2017; Published: 27 April 2017

Abstract: In this work a detailed analysis of the temporal evolution of the Almegíjar landslide is presented. It is a rock slide located in the Alpujarras region (Granada, Spain) that has developed over the last 30 years. Six datasets and photogrammetric flights corresponding to the years 1956, 1984, 1992, 2001, 2008, and 2010 were surveyed. The more recent flight of 2010 combined an aerial digital camera and a LiDAR sensor and was oriented by means of in-flight data and tie points. This 2010 flight allowed for the generation of a reliable and high-precision Digital Terrain Model (DTM). The other flights were oriented using second-order ground control points transferred from the 2010 flight, and the corresponding DTMs were prepared by automatic matching and subsequent editing from the stereoscopic models. After comparing the DTMs of different dates, it has been observed that the landslide was triggered after 1984 and since then has evolved in an irregular pattern with periods of variable activity. On average, the ground surface dropped more than 8 m in depleted zones and rose nearly 4 m in the accumulation zones, with a velocity catalogued as very slow (about 15–30 cm/year) over a time span corresponding to a degree VIII of diachroneity. The total volume of the mobilized mass of this large contemporary slide was about 300×10^3 m^3.

Keywords: landslide evolution; DTM; digital photogrammetry; LiDAR; Alpujarras region

1. Introduction

This article is an extended and improved version of the conference paper titled "Digital photogrammetry and LiDAR techniques to study the evolution of a landslide" [1], presented at the Eighth International Conference on Geo-information for Disaster Management (Gi4DM), celebrated in Enschede (The Netherlands) in December 2012.

The application of remote sensing techniques to natural hazards and landslide research has been steadily expanding in the last two decades [2,3], with different approaches from the optical spectrum to Synthetic Aperture Radar (SAR) techniques [4,5]. Among these techniques, those based on photogrammetry and Light Detection and Ranging (LiDAR) are adequate for middle- to high-resolution studies.

Regarding photogrammetric techniques, it should be taken into account that the only current approach that enables reconstruction of the landslide kinematics at sufficiently significant temporal and spatial resolution is the use of aerial photographs from historical flights [6,7] and, alternatively, but with less accuracy, the use of historical cartography. Therefore the photogrammetric techniques have been widely employed [1,7–22], sometimes combined with LiDAR [1,23,24], Global Navigation Satellite Systems, GNSS [25,26], digitization of old maps [24], and other techniques such as the electric resistivity tomography [27]. Recently, the use of Unmanned Aerial Vehicles (UAV) has extended the use of photogrammetric techniques to very high resolution studies [28–39].

Most studies based on photogrammetric techniques follow a similar methodology, starting with the orientation of the digital aerial images using photo-triangulation techniques by means of bundle block adjustment [1,7,10,11,13–16,19–23,26,27]. Then it is possible to use a reduced number of ground control points (GCPs), which are generally established by means of GNSS techniques, discarding in each survey any control points that are not adequately identified in the corresponding images.

Once the stereoscopic models are oriented, Digital Elevation Models (DEMs)—Digital Terrain Models (DTMs) or Digital Surface Models (DSMs)—and orthophotographs are generated by image matching techniques. After this step, almost all these studies perform different quantitative analyses, including those to prepare differential DEMs between surveys, profiles, and volumetric calculations. Also, in some studies, displacement vectors between single points are determined [10,11,13–18,21,26,29,30,32,36–38] or correlation algorithms are used with a much larger number of points [20,31,33]. From these analyses, useful qualitative features concerning their evolution are also established [1,7,13–16,20,23,24,27,32].

The quality or positional accuracy of these analyses is easily assessed in most of the aforementioned studies by means of the root mean square error (RMS) at the residuals of the control and/or check points [19,22,29,32,33,37,39,40], including the error propagation to DEMs and their differentials [7,23,38].

On the other hand, LiDAR techniques supply reliable and high-density point clouds from which true Digital Terrain Models (DTMs) can be generated [41–52], as the processes of classification and filtering of ground points work better in LiDAR than in photogrammetric approaches. In this way, DTMs from LiDAR data corresponding to different dates, together with those from digital photogrammetry, may be compared in order to determine vertical ground displacements or calculate volumes [23,24,47,48,51,52]. Also, these DTMs, both photogrammetry and LiDAR-derived, provide the starting steps for geomorphological studies based on different parameters (slope, aspect, roughness, etc.), enabling landforms and landslides to be identified [42,45,46,49–52]. In some cases, data from different surveys are compared in order to assess the ground surface evolution [46]. In addition, accuracy analysis of LiDAR data and models are accomplished [47,48,52].

The methodology of this study starts from a combined camera and LiDAR flight oriented with in-flight information (onboard GNSS and inertial data), which provides a common reference system for both the photogrammetric and the high-resolution LiDAR digital models. Taking this flight as a reference, the other photogrammetric flights were oriented by means of second-order ground control points that were transferred from the reference flight. Once these historical flights were oriented, the DTMs corresponding to each survey were obtained by outdating the reference DSM (from LiDAR data) by means of stereoscopic editing in a photogrammetric workstation. Then differential DTMs and subsequent profiles and volumetric changes were computed. It is a rather efficient methodology that considerably reduces the number of field-surveyed ground control points and allows for the geo-referencing of all data in a common reference system.

2. Geographical and Geological Setting

The study zone is located in the Alpujarras region, a mountainous area in the Sierra Nevada range (Granada province, Southern Spain), near the Mediterranean Sea (Figure 1). The region is widely affected by slope instability processes, giving rise to abundant rock falls, slides, and debris flows resulting from a combination of abrupt relief and metamorphic geological units of the Internal Zone of the Betic Cordillera, rather susceptible to landslides [53–59].

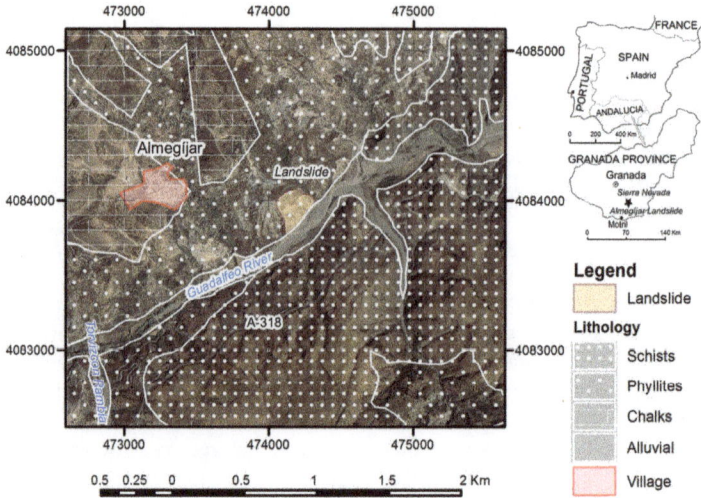

Figure 1. Geographical and geological setting. Coordinates are in ETRS89/UTM30.

The main triggering factor in the region is the combination of torrential rainfalls, the morphological slope, and a river channel evolution controlled by semiarid environmental conditions which gives place to a "rambla" segment in the Guadalfeo River. This channel has a planar bottom in which ephemeral streams alternate with variable flooding controlled by extraordinary intense storms or the quick melting of Sierra Nevada snow during spring rising of temperatures.

The landslide studied is a good example of the many landslides resulting from the slope evolution on the intensely eroded northern margin of the Guadalfeo River [59]. It is settled near the village of Almegíjar (Granada, Spain), but clearly outside the urban area and affecting only terrain free of any buildings or infrastructure. Given the fresh morphological features and the sliding material, it has been classified as a translational rock slide, with a clear upper scarp and a planar crown of 20 m height. The landslide mass is 250 m long, 300 m wide, and 140 m high (Figure 2). Also, some secondary minor scarps are visible, besides rock fall and debris flow in the frontal part of the mass [1,60].

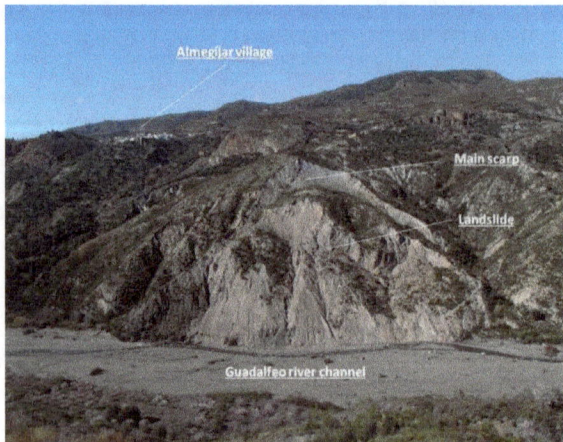

Figure 2. The Almegíjar landslide.

The landslide is considered to be active, although alternating periods of very slow speed with more active periods during ephemeral floods of the "rambla" [53,58,59]. Also, according to [60], the slide reactivation in the period 2009–2010 showed a bulging in the lower sector of the displaced mass, following a mechanism described by [61].

The mass of the Almegíjar slide is composed of an Alpujarride unit of quartzites and quartzphyllites of Permian to Triassic age [62]. Nevertheless, the slide is part of an unstable but dormant slope showing an upward extension of the instability expressed by the appearance of large rock blocks and open cracks affecting the overlying marble unit, with water flows in its bottom infiltrating the mass of phyllite rock.

3. Materials and Methods

Digital photogrammetry and LiDAR techniques represent a significant advance in the study of ground surface evolution derived from landslides and other processes. The results using photogrammetric workstations have offered better and more accurate interpretations of changes on the terrain surface over the past decade [8,13–16].

Thus, landslide inventories and their geomorphic features (scarps, crown, lateral flanks, cracks, etc.) may be prepared by means of 3D digital stereoplotting. In this sense, not only single landslides but also the identification of unstable slopes in a region may be analyzed with these techniques, starting from high-resolution DEMs and 3D inventories [22,44,47]. In addition, if data from different surveys are available, it becomes possible to study landslide evolution. The overlaying of landslide inventories in Geographical Information Systems (GIS) allows the study of landslide evolution in a given region and the identification of reactivated landslide areas as well as the growth or generation of new landslides. Even landslide multitemporal landslides can be obtained from accurate DEMs of different epochs [44,47]. Thus it is possible to determine landslide activity, differentiating between active, dormant, and relict landslides. The concept of landslide diachroneity [63] distinguishes landslides considering the time span of the active periods and the initial triggering data.

For the analysis of single landslides, two approaches are available: the first is based on a comparison between DEMs from different flights, calculating distances and volumes [1,7–11,13–26,29–32,36–38,48,52]; the second approach consists of a comparison between points and other elements identified without any ambiguity between the different surveys, enabling the determination of 3D displacement vectors for accurate identification of the landslide kinematics [10,11,13–18,21,26,29,30,32,36–38]. In this paper the quantitative analysis of ground changes will be approached by comparing different DTMs.

The starting point is a dataset made up by photogrammetric flights and LiDAR data. The methodology, established by the research team based on previous results [1,19], includes a set of tasks summarized in the following steps (Figure 3):

1. Dataset compilation and digitization of images from historical flights.
2. Definition of reference system and orientation of reference data.
3. Orientation of historical photogrammetric flights.
4. Building and editing DTMs.
5. Comparison between models and calculations.

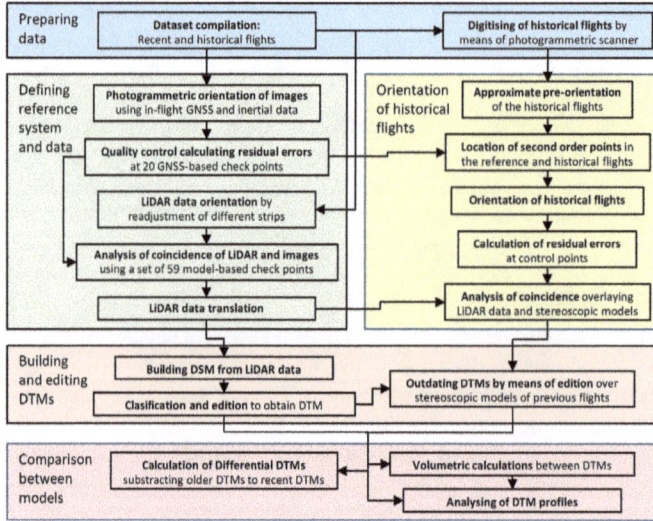

Figure 3. Flow chart of the methodology.

3.1. Dataset Compilation and Digitization of Images from Historical Flights

First, a photogrammetric flight corresponding to the year 2010 was available, combined with a digital camera (Z/I DMC) and a LiDAR sensor (Leica ALS50-II), as well as attached GNSS/IMU systems for direct orientation. The Z/I digital camera offers a resolution of 0.20 m and has four spectral bands, three in the visible spectrum (RGB) and one in the near-infrared (NIR). The LiDAR instrument produces a ground resolution of about 1–1.5 points per m^2. A second photogrammetric flight of 2008 with the same properties of the first one was also available but without LiDAR dataset.

Besides the highly accurate and recent flights, historical images from aerial film cameras were also available, corresponding to four photogrammetric flights of the years 1956, 1984, 1992, and 2001. The first one was the so-called "American flight", as it was undertaken by the U.S. government in 1956, and it was panchromatic at a middle scale (1:33,000). The second one (1984) was a panchromatic flight made by the National Geographic Institute of Spain (IGN) at scale of 1:30,000. The other two (1992 and 2001) were also panchromatic flights at a scale of 1:20,000, made by the Andalusian Cartographic and Statistical Institute (IECA). The main flight features are indicated in Table 1 and their distributions are shown in Figure 4.

Table 1. Properties of the image datasets.

Date	Bands	Format	Pixel [1]	GSD [2]	Camera	Focal Distance [3]	LiDAR
1956	Panchromatic	Film	15	0.60	-	151.42	No
1984	Panchromatic	Film	20	0.80	UAG-II	153.03	No
1992	Panchromatic	Film	15	0.30	UAG-II	153.03	No
2001	Panchromatic	Film	15	0.30	Wild UAO-S	152.75	No
2008	RGB-NIR	Digital	12	0.20	Z/I DMC	120.00	No
2010	RGB-NIR	Digital	12	0.20	Z/I DMC	120.00	Yes

[1] Pixel size is in microns; [2] GSD (ground sample distance) is in meters; [3] Focal distance is in mm.

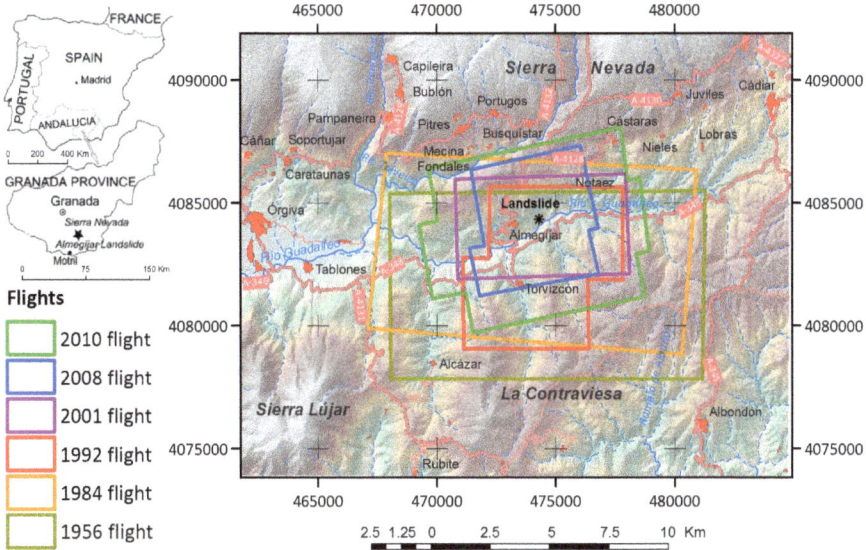

Figure 4. Distribution of available flights. Coordinates are in ETRS89/UTM30.

These historical flights were digitized with a precision photogrammetric scanner Vexcel Ultrascan 5000 at a resolution of 15 microns, which, in turn, implies a ground sample distance (GSD) of 0.60 m in the 1956 flight and 0.30 m in the 1992 and 2001 flights. The 1984 flight was already supplied in digital format after a scanning at 20 microns that provided a GSD of 0.80 m.

The period considered (c. 50 years) is long enough to study short-term changes in the relief and, more precisely, to analyze the landslide evolution in the study zone.

3.2. Definition of Reference System and Orientation of Reference Data

As a starting point for the entire process, the 2010 flight (camera and LiDAR) was used. This flight was oriented by means of in-flight orientation systems (direct orientation). As this dataset was considered to be of the highest quality, both in image resolution and point cloud accuracy, it has been proposed as the reference data and the previous flights have been registered with respect to it.

Thus, it is important to ensure the adjustment and minimal data discrepancies between the photogrammetric flight and the LiDAR point cloud of this 2010 reference flight. The discrepancies between them are originated as a consequence of the independent orientation systems of these sensors—despite the simultaneous recording of data—and thus its different positioning, which may prompt a lack of geometrical coincidence in the data compiled. This raises the need for an optimal adjustment between the two types of data [64].

In accordance with the above, the reference data are compiled and refined by three steps:

1. Photogrammetric orientation using in-flight GNSS and inertial data to ensure the geometrical quality of the whole photogrammetric block. This procedure was performed by means of a photogrammetric workstation and Socet Set 5.2 software (BAE Systems Plc., London, UK) [65] by measuring about 200 tie points and readjusting the orientation parameters. The RMS errors of the resulting model, based on the residual of 20 check points surveyed with GNSS (GNSS-based check points), are 0.258 m in XY and 0.100 m in Z (Table 2).

2. LiDAR data orientation, by readjustment of different strips, which is employed to obtain a single homogeneous point cloud, even in overlapping zones. This task was carried out with the TerraMatch module of TerraSolid software (Terrasolid Ltd., Helsinki, Finland) [66].

3.   Analysis of height coincidence of both datasets. For this analysis an internal control process of the LiDAR point cloud was made using check points (model-based check points) taken from the refined photogrammetric model of the step 1. These points were selected taking into account the following:

- Points on stable and flat surfaces, avoiding high slopes or roughness zones, were selected.
- The points had to be well distributed throughout the study area and the number high enough to get an appropriate redundancy and significant results.

Table 2. Orientation errors.

Date	Photographs Number	GCP/CHK Number	Tie Points Number	RMS Pixel	GCP/CHK RMS [1]		Propagation [1]	
					RMS_{XY}	RMS_Z	PE_{XY}	PE_Z
2010	36	20	195	0.529	0.258	0.100	-	-
2008	9	8 xyz	82	0.246	0.145	0.042	0.296	0.108
2001	3	8 xyz	14	0.363	0.187	0.213	0.319	0.235
1992	4	7 xyz	21	0.423	0.255	0.067	0.363	0.120
1984	3	8 xyz	15	0.621	0.517	0.234	0.578	0.254
1956	3	8 xyz.4 z	14	1.038	0.566	0.269	0.622	0.287

[1] Root Mean Square (RMS) and propagation errors are in meters. For 2010, flight errors are calculated in check points (CHK). For the remaining flights, errors correspond to the residual calculated in ground control points (GCPs). GCP types: full (xyz) and height (z).

Taking into account the good adjustment of the image orientation process, with mean values of about 0.10 m in GNSS-based check points, what really was tested in Step 3 were the errors of the LiDAR data. The vertical accuracy of aerial LiDAR data is well documented in many works [47–49,51,52] although the values present a great dispersion—from 0.16 to 3.26 m [47]—depending on different factors (flight altitude, field of view, beam angle respect to the vertical, coverage reflectance, vegetation, etc.). Furthermore, according to [67], the vertical error in LIDAR increases by 0.1 m per 1000 m of flight altitude, so it reaches values of about 0.10–0.20 m in conventional flights. In this sense, some works, in which the point density is usually higher than 1 point/m², establish a vertical accuracy of 0.10–0.20 m [48,49,52,68]; but in other works, with point densities lower than 1 point/m², the accuracy is about 0.50 m [47,50]. Thus, the higher the flight is, the lower the point accuracy and density, and therefore the vertical errors and uncertainties usually increase with the flight altitude. Meanwhile, in [47] the vertical discrepancies—expressed as RMS—between two LIDAR surveys in a close time interval were of 0.28 m, measured in stable zones (out of a mask drawn with the landslide inventory); however, the calculated uncertainties from comparison of LiDAR data with GNSS-measured points ranges from 0.50 m in leaf-off conditions and 0.75 m in leaf-on conditions. Bearing in mind these previous data, a reference value of accuracy of 0.20 m has been established for the present work, according the flight altitude and LiDAR configuration.

Nevertheless, a set of 59 model-based check points was extracted in a wide area surrounding the studied landslide to test the adjustment of LIDAR data and the photogrammetric model. The distribution of these points, all of them located outside of the landslide limits, is shown in Figure 5. The discrepancies or errors between the datasets range from 0 to near 1 m, with a mean positive value of 0.529 m, as can be observed in Table 3 and Figure 5a,c. Considering the good adjustment of the photogrammetric orientation process, this analysis allows us to infer a systematic error upwards in LiDAR data, corresponding to a translation of about 0.5 m, due to weak absolute georeferencing of the whole LiDAR block. This discrepancy could be eliminated by a simple translation downwards in height of the LiDAR data equal to the mean error. Meanwhile, the standard deviation (0.260 m) was of the same order of expected accuracy of LiDAR [48,49,52,67,68] and remained constant after correction. The discrepancies in the points after correction are also shown in Table 5 and Figure 5b,c. Thus, in the new situation, a refined adjustment has been reached with a zero mean error and a standard deviation around the considered accuracy of LiDAR data.

Table 3. Statistics of errors between LiDAR and photogrammetric data of the 2010 flight.

Statistics of Errors	Before Correction [1]	After Correction [1]
Mean	0.529	0.000
Standard deviation	0.259	0.259
Maximum	0.939	0.410
Minimum	0.045	−0.484
Median	0.564	0.035

[1] Errors are in meters.

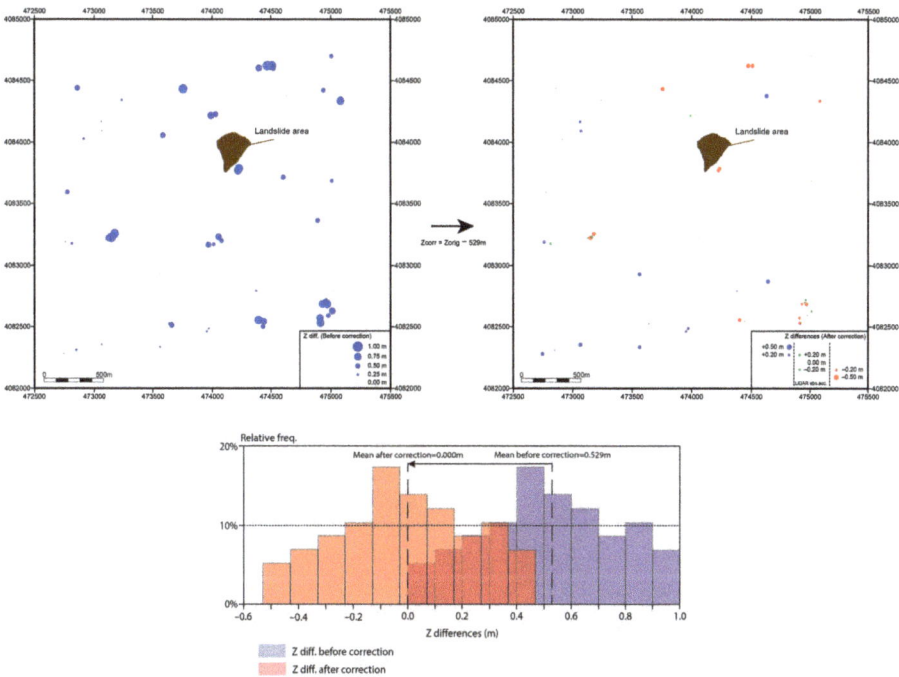

Figure 5. Distribution of errors or discrepancies in height between 2010 LiDAR and photogrammetric data: (**a**) before the adjustment; (**b**) after the adjustment; (**c**) histograms.

3.3. Orientation of Historical Photogrammetric Flights

Once the 2010 reference flight was re-oriented and the agreement between the photogrammetric and LiDAR data was ensured, the historical flights were oriented with respect to the 2010 reference system. The processing of several historical flights corresponding to a broad time span is rather difficult as the field ground control points (GCPs) must be stable, accessible, and unequivocally recognizable, not only in images from the time period considered but also in the field, where the points have to be measured with differential GNSS observations. Thus, this paper proposes the use of GCPs transferred from the reference photogrammetric flight (second-order GCPs) recognizable in flights corresponding to each survey. As the GCPs are directly located on the photogrammetric flights (the one to be oriented and the reference flight), it will be possible to certify in real time whether the point is observable and unequivocally recognizable on both flights. In addition, with a zenithal perspective the observer would have access to the entire study area, without limitation in accessibility or observation. Furthermore, with this procedure the number of required GCPs measured in the field is optimized, as well as the field work and processing time.

The main constraint of this methodology is a theoretical reduction in the accuracy of the coordinates in the second-order GCPs with regard to the accuracy of the first-order GCPs due to error propagation. Nevertheless, as the 2010 photogrammetric flight with higher resolution and a RMS lower than the pixel size was always taken as the reference, the accuracy in the GCPs transferred in this way is proved to be better than the resolution of the historical flights.

The process of transferring points from the reference flight (more recent) to the previous flights to be oriented is made iteratively following the next steps (Figure 3):

1. Approximate pre-orientation of the flight to be oriented in the reference system.
2. Once the two photogrammetric flights are pre-oriented in the same reference system, a set of several common points in both flights are located and uploaded simultaneously in the digital photogrammetric workstation.
3. All these points located and measured in the reference flight are considered second-order GCPs for the flight to be oriented (Figure 6). This process includes full GCPs (known X,Y,Z), horizontal GCPs (known X,Y), and height GCPs (known Z). Table 2 shows the number and type of the second-order GCPs used in the orientation of the historical flights.
4. The flight to be oriented is adjusted and the process is repeated from step 2 until a final proper orientation is attained, that is when the difference between coordinates of GCPs in both datasets was lower than the image resolution of the flight to be oriented.

Figure 6. Example of transferred second-order control point (GCP).

In Table 2, the results of the process of orientation of historical flights are presented together with data from the reference flight. The orientation errors of the 2008 flight are satisfactory but the errors increase in the older flights. Thus, XY errors range from 0.15 m (2008), about 0.20 m (2001 and 1992), 0.50 m (1984), to near 0.60 m (1956). Regarding to heights, Z errors are between 0.04 m for 2008 flight, 0.07 m for 1992 flight and 0.20–0.30 m for 2001, 1984 and 1956 flights. But, as the historical flights were oriented using GCPs obtained from the reference 2010 flight, the precision of any measurement in those flights is influenced by the error propagation from the 2010 flight. Therefore the final precision of these flights can be calculated according to the following expression:

$$\text{Prop}_{YEAR} = (\text{RMS}_{2010}^2 + \text{RMS}_{YEAR}^2)^{0.5} \tag{1}$$

Thus, the propagated error increases in the following way: XY errors range between 0.30–0.36 m for the 2008, 2001, and 1992 flights to near 0.60 m for the 1984 and 1956 flights; and Z errors are between 0.11–0.12 m for 2008 and 1992 flights to 0.24–0.30 m for 2001, 1984, and 1956 flights.

After applying this procedure, the spatial correspondence between the two compared flights is checked out under the premise of finding a total correspondence in stable zones where no modifications were detected between the two surveys. The comparison was made by overlaying the reference 2010 LiDAR data and the stereomodels of the historical flights with the stereoscopic viewing tools of the Socet Set and the photogrammetric workstation. If the orientation was correct, then DTM contours and points derived from LiDAR data had to be adjusted to the ground in the stable areas of the study zone. If significant discrepancies were found, then an orientation re-adjustment would be necessary. In the case studies, the stereoviewing display comparison showed a good coincidence between elevation and stereoscopic models in all the flights, and therefore no readjustments of the flight orientation were necessary.

For this comparison, different types of color compositions were applied in recent flights: true color (Blue-Green-Red) and false color (Green-Red-NIR), which was possible in recent flights (2008 and 2010), since they incorporated the near infrared (NIR) band. With these compositions the different ground features were more visible and identifiable in addition to the element over them, such as vegetation, which is very sensible to detection with near infrared.

3.4. Building and Editing DTMs

A Digital Surface Model (DSM) was obtained from the 2010 LiDAR data. Then a classification was made between ground-points and non-ground-points (vegetation, buildings, artifacts, etc.). This classification process was performed with the software TerraSolid (module TerraScan). Once this classification was finished, the ground surface corresponding to the reference survey was defined by those filtered ground-points, thus obtaining the Digital Terrain Model (DTM).

From the initial 4,240,499 points in the 2010 LiDAR point cloud, 1,943,545 were classified as ground-points while 2,296,954 were non-ground points. Then, the results of this classification were displayed in the stereo photogrammetric workstation together with the reference photogrammetric flight, and edited using the stereoscopic tools supplied by Socet Set. This editing process consisted basically on correcting wrong classified points and stereoplotting breaklines and some typical ground features, such as scarps, water streams, slope changes, etc., which allowed for a more realistic and precise DTM generation.

After preparing a reference DTM from LiDAR data classification, the models corresponding to the historical flights are obtained by editing and outdating the 2010 reference model. Other authors [7,20,21,23,24,26] have generated DEMs corresponding to each survey by a process of image correlation independent for each flight. However, in this work another approach has been adopted for the following two reasons. First, the accuracy of models is consistent with the resolution and the radiometric quality of the aerial images, which in the older flights has led to a lower accuracy in the models. A second reason is the need to build the model again by matching, which in addition requires the editing of all the surveys in the study zone, no matter if there have been any changes in the terrain or not, thereby giving rise to inevitable false ground modifications in the comparison of models corresponding to different surveys and noise in the detection of changes.

For these reasons, an outdating of the reference model—of a higher quality and geometric resolution than those models that would be obtained by a matching technique in each case—is proposed. This outdating process consists of uploading the reference model overlaid to the historical flights in the photogrammetric workstation. Then, by means of the stereoscopic viewing tools is possible to analyze the lack of coincidence between the two datasets. When that lack of coincidence occurs, the 2010 DTM is edited in that zone to adapt the terrain to the ground surface in the stereomodel of the historical flight. In the case of changes affecting large surfaces, it is necessary to launch a matching process exclusively for that area. In ground areas where no change is detected, the model is not modified, thus reducing the editing process time as well as avoiding noise in the analysis of changes and maintaining a better overall quality of the model provided by the higher positional accuracy of the 2010 data.

After finishing the outdating of the reference model to the immediately previous historical survey under analysis, this resulting model was used as a reference model in the preceding survey, repeating the process in all cases until the 1956 DTM was obtained. In Figure 7, the DTMs generated in the unstable area are shown, allowing a visual observation of significant changes between these six dates considered.

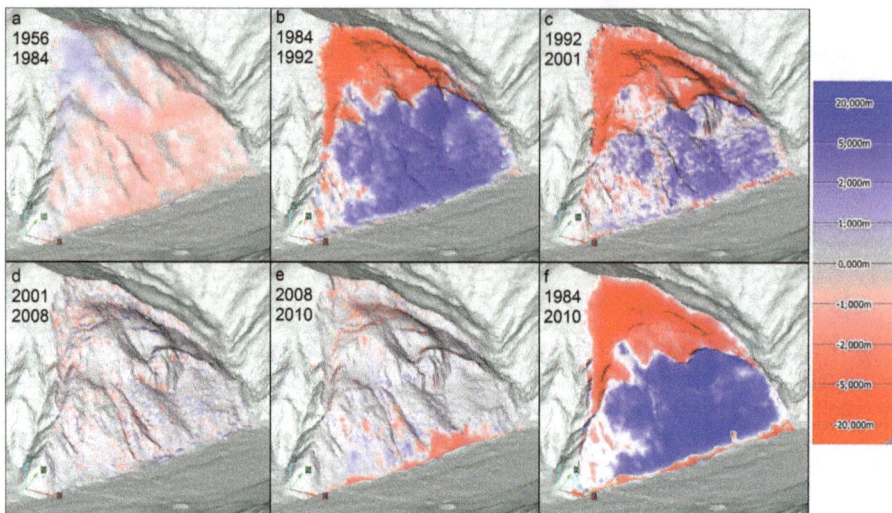

Figure 7. Differential DTMs (Digital Terrain Model). Depletion areas are in red and accumulation areas are in blue: (**a**) 1956–1984; (**b**) 1984–1992; (**c**) 1992–2001; (**d**) 2001–2008; (**e**) 2008–2010; (**f**) 1984–2010.

3.5. Comparison between Models and Calculations

The comparison between models was carried out from different perspectives:

1. Vertical displacements by subtracting DTMs (differential DTMs) (Figure 7). They may be negative or positive, depending on whether the model compared with the reference lies below or above, allowing the identification of areas of mass depletion or accumulation, respectively.
2. Volumetric calculations between models. As in the previous case, the volumes can be negative (depletion areas) or positive (accumulation areas).
3. Analysis of DTMs profiles through the landslides permitting a better observation of terrain displacements (Figure 8). With enough profiles, it is possible to attain an overview of the landslide kinematics, for instance if it is planar or there are rotations around vertical or horizontal axis indicating rotational sliding.

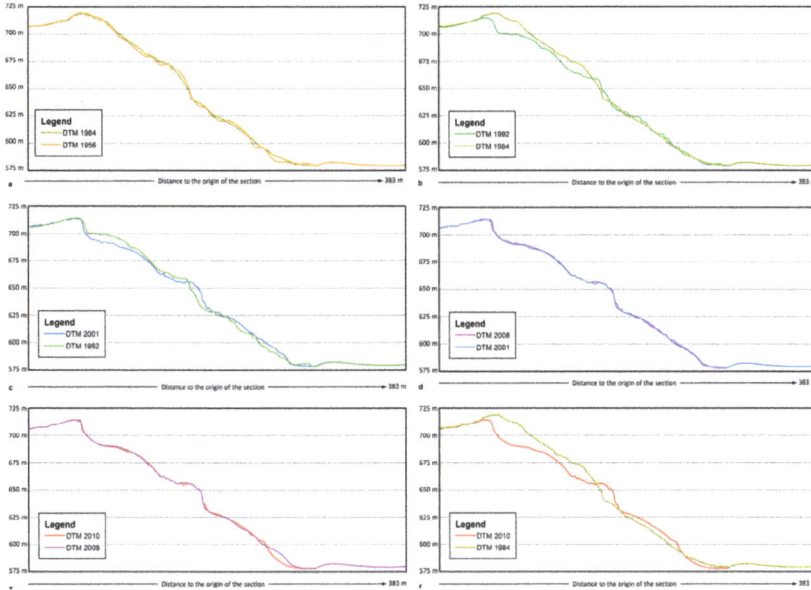

Figure 8. Cross sections or profiles of the DTMs: (**a**) 1956–1984; (**b**) 1984–1992; (**c**) 1992–2001; (**d**) 2001–2008; (**e**) 2008–2010; (**f**) 1984–2010.

These calculations and analyses have been made by means of the Maptek I-Site Studio 6.0 software (Maptek Pty. Ltd., Adelaide, Australia) [69] and ArcGIS 10.2 (Esri, Redlands CA, USA) [70]. The errors and uncertainties—usually expressed as root mean square (RMS) and standard deviation (SD), respectively—of these procedures have to be estimated in order to validate the results of these calculations. In the case of the 2010 flight, the value of standard deviation (0.26 m) is assumed to be vertical uncertainty or the minimum level of detection [52,71,72]. Meanwhile, from previous works [7,23,32] it has been observed that the vertical errors and uncertainties in DEMs obtained by means of matching from photogrammetric oriented flights are about two or three times higher than the orientation residual errors in Z. Therefore the vertical uncertainties of DTMs obtained in this work range from 0.27 to 0.30 m for 2008 and 1992 flights, about 0.60 m for 2001 and 1984 flights, to over 0.70 m for the 1956 flight (Table 4).

Table 4. Estimated uncertainties of DTMs, differential models, and volumetric calculations.

Date	Propagated Error GCPs [1]	Uncertainty in DTMs [1]	Interval	Differential Models [1]	Volumes [2]		
					DEP	ACC	WAS
1956	0.287	0.718	1956–1984	0.958	33,548	22,266	40,265
1984	0.254	0.635	1984–1992	0.702	25,735	17,350	31,037
1992	0.120	0.300	1992–2001	0.660	18,374	15,782	24,222
2001	0.235	0.588	2001–2008	0.647	23,933	23,755	33,720
2008	0.108	0.270	2008–2010	0.375	16,024	14,278	21,462
2010	-	0.260	1956–2010	0.763	26,558	20,600	33,611
-	-	-	1984–2010	0.686	22,717	16,343	27,985

[1] Errors in GCPs and the estimated uncertainty in DTMs (Digital Terrain Model) and differential models are in meters. [2] The estimated uncertainty in volumetric calculations is in m³. Volumes: DEP: Depletion; ACC: Accumulation; WAS: Waste.

Furthermore, the vertical uncertainties of the differential DEMs are estimated as [7,52,71,72]:

$$\text{Uncert. }_{\text{YEAR1-YEAR2}} = (\text{Uncert. }_{\text{YEAR1}}{}^2 + \text{Uncert. }_{\text{YEAR2}}{}^2)^{0.5} \qquad (2)$$

In this case, the uncertainties are lower than 0.40 m for the 2008–2010 period, 0.65–0.70 m for the 2001–2008, 1992–2001, and 1984–1992 periods, and close to 1 m for the 1956–1984 period (Table 4).

In addition, the uncertainties of the volumetric calculations are estimated by multiplying the vertical uncertainties by the areas of depletion and accumulation [52]. The uncertainty of the waste material (difference between depletion and accumulation volumes) is calculated as a propagated error (square root of the sum of squares of depletion and accumulation volumetric uncertainties). These last ones range from 21×10^3 m^3 for the 2008–2010 period, 31–33×10^3 m^3 for the 2001–2008, 1992–2001, and 1984–1992 periods, to about 40×10^3 m^3 for the 1956–1984 period (Table 4).

4. Results

The analysis of orthophotographs and models since 1956 of the area located in the Guadalfeo River, near Almegíjar village, has allowed for the identification of a landslide as well as its characteristics and evolution. The landslide does not appear in the aerial photographs of the 1956 and 1984 flights, so it was triggered after 1984, although located in a broadly unstable zone, with evidence of older activity upslope. Thus in the stereoscopic models of later dates (1992, 2001, 2008, and 2010), the landslide area presents some clear features that include a clean scarp, steps and terraces, minor scarps, and an accumulation zone invading the river channel. The involved mass reaches about 300×10^3 m^3 according to the volumetric calculations, so the landslide can be defined as moderately large. Other features identified through aerial photographs, but also field work, are a large extent of debris flows and the development of pervasive erosion in rills and gullies on the weathered rock-mass surface, even before 1984 and at the start of the main landslide.

The differential models between the 1956 and 2010 surveys (the whole period considered) show two clearly differentiated areas, the depletion area at the upper part of the mass and the accumulation area in the lower part as shown in Figures 7 and 8. As the differential models are calculated by subtracting the older models to the recent ones, negative differences in elevation and volume can be observed in the depleted area, and positive differences are recorded in the accumulation zone. In the Table 5, the proportion of areas of depletion, accumulation, and below the uncertainty is shown. If uncertainty is not taken into account, the areas of depletion and accumulation are practically compensated for (near 50%), although in some periods (1992–2001 and 2008–2010) depletion areas predominate slightly. When the uncertainty areas are introduced, some periods show a high proportion of areas below the uncertainty, such as 1956–1992 and 2001–2008 (less than 10%) or even 2008–2010 (16%). In these cases, the depleted and accumulated areas are compensated for except in the 2008–2010 period where the depleted area predominates. Meanwhile, other periods such as 1984–1992, 1992–2001, and the global periods show a higher proportion of area depleted and accumulated (over 70%), usually compensated for, except in the 1992–2001 period, when depleted area also predominates.

Table 5. Areas resulting in from the comparisons between DTMs.

| Period | Without Uncertainty | | | | With Uncertainty | | | | | |
| | Areas [1] | | Proportion [2] | | Areas [1] | | | Proportion [2] | | |
	DEP	ACC	DEP	ACC	DEP	ACC	UNC	DEP	ACC	UNC
1956–1984	37,846	35,859	51.35	48.65	4738	4048	64,919	6.43	5.49	88.08
198–1992	44,024	29,681	59.73	40.27	33,555	19,950	20,200	45.53	27.07	27.41
199–2001	39,650	34,055	53.80	46.20	28,985	22,764	21,955	39.33	30.89	29.79
200–2008	36,990	36,715	50.19	49.81	5955	5425	62,325	8.08	7.36	84.56
200–2010	38,975	34,730	52.88	47.12	12,232	7448	54,026	16.60	10.10	73.30
195–2010	45,887	27,818	62.26	37.74	37,523	29,523	6659	50.91	40.06	9.03
198–2010	41,509	32,196	56.32	43.68	37,770	28,533	7403	51.24	38.71	10.04

[1] Areas are in m^2. [2] Proportions are in %. Volumes and areas: DEP: Depletion; ACC: Accumulation; UNC: Uncertainty.

Significant vertical displacements are recorded in the landslide area, as shown in Table 6, with mean values higher than an order of magnitude with respect to the estimated uncertainties (Table 4). Thus, in the depleted area, there is an average descent of 8.5 m of the ground surface, with maximum values around 28 m. In addition, in the accumulation area, vertical displacements related to the elevation of the ground surface show an average value of 3.9 m, with maximum peaks of almost 28 m. Considering that the landslide was triggered after 1984, the annual displacement rates reach at least 0.32 m/year and 0.15 m/year in the depleted and accumulation areas, respectively. Consequently, the landslide velocity can be estimated on the order of some few cm/year so the slide can be catalogued as "very slow" if it were considered continuous and uniform in time [73,74].

The analysis of the results of all five time periods considered (Table 6), between the six available flights, reveals that the vertical mean displacements are practically null in the first period of 1956–1984 (0.09 m and 0.08 of terrain elevation in the depletion and accumulation areas), higher in the second period of 1984–1992 (3.70 m of decrease in the depletion area and 3.11 m of elevation in the accumulation area), lower in the third period of 1992–2001 (2.87 m and 1.78 m, respectively), and far lower in the last two periods of 2001–2008 (0.30 and 0.27 m) and 2008–2010 (0.43 and 0.23 m). These partial values are significantly higher than the uncertainties of Table 4, except for the periods of 1956–1984 and 2001–2008 (clearly lower than uncertainties) and the period 2008–2010 (similar to uncertainties). More informative is the comparison of annual rates, with values clearly above average in the period 1984–1992 (0.46 m/year of depletion and 0.39 m/year of accumulation); values near average in the period 1992–2001 (0.32 and 0.20 m/year respectively), and slightly lower in 2008–2010 (0.22 and 0.12 m/year); finally, rate values are clearly lower in the period 2001–2008 (0.04 and 0.04 m/year) and practically null in the period 1956–1984.

Table 6. Vertical displacements resulting from differential DTMs.

Period	Displacements [1]								Rates [2]	
	Depletion Area				Accumulation Area				DEP	ACC
	MIN	MAX	ME	SD	MIN	MAX	ME	SD	ME	ME
1956–1984	−3.39	2.91	−0.09	1.95	−2.83	2.37	−0.08	1.66	0.00	0.00
1984–1992	−12.40	2.57	−3.70	3.94	−2.17	24.75	3.11	4.21	−0.46	0.39
1992–2001	−10.89	2.35	−2.87	2.43	−1.98	19.09	1.78	2.22	−0.32	0.20
2001–2008	−6.14	0.65	−0.30	0.36	−0.83	7.95	0.27	0.41	−0.04	0.04
2008–2010	−5.95	1.02	−0.43	0.34	−1.34	6.14	0.23	0.65	−0.22	0.12
1956–2010	−27.75	3.12	−8.49	6.29	−3.53	27.60	3.94	4.82	−0.16	0.07
1984–2010	−25.67	3.03	−8.41	6.18	−3.23	25.70	3.89	4.74	−0.32	0.15

[1] Displacements are in meters. [2] Displacement rates are in m/year. Rates: DEP: Depletion; ACC: Accumulation. MIN: Minimum; MAX: Maximum; ME: Mean; SD: Standard deviation.

The volume assessment in Table 7 shows a depletion of more than 290×10^3 m^3 of mass in the upper part of the slide, but the accumulated mass in the lower part near the river channel is about 120×10^3 m^3. This implies mass losses of nearly 175×10^3 m^3, probably evacuated by the river flow. These values are slightly modified to 280×10^3 m^3, over 110×10^3 m^3 and 167×10^3 m^3 if a starting date of 1984 is considered. By periods, the values are very low in the first period of 1956–1984 (24×10^3 m^3 of depleted material, 17×10^3 m^3 of accumulated material and 8×10^3 m^3 of waste material), clearly higher in the second period of 1984–1992 (169×10^3, 53×10^3 and 116×10^3 m^3, respectively), lower in the third period of 1992–2001 (86×10^3, 46×10^3 and 40×10^3 m^3), and far lower in the last two periods of 2001–2008 (9.2×10^3, 8.7×10^3 and 0.5×10^3 m^3) and 2008–2010 (18×10^3, 7×10^3 and 11×10^3 m^3). The values corresponding to the periods 1984–1992 and 1992–2001 are clearly higher than the uncertainties in Table 4, but the values of the periods 1956–1984 and 2001–2008 are lower than the uncertainties. The values of the period 2008–2010 are on the same order as the uncertainties, higher in depleted material and lower in accumulated and waste materials.

The annual rates considering the landslide starting after 1984 show values over 10.7×10^3 m^3/year of depletion, near 4.3×10^3 m^3/year of accumulation, and about 6.4×10^3 m^3/year of mass waste. By periods, the annual rates are highest in the interval 1984–1992 (21×10^3 m^3/year of depleted material, almost 7×10^3 m^3/year of accumulated material and 15×10^3 m^3/year of waste material), followed by the intervals of 1992–2001 and 2008–2010, with rates near to those of the whole period (9.2–9.5×10^3, 3.6–5.1×10^3 and 4.4–5.6×10^3 m^3/year, respectively), and much lower in 2001–2008 (1.3×10^3, 1.2×10^3 and 0.1×10^3 m^3/year) and 1956–1992, when the mass losses are practically null.

Table 7. Volumetric calculations between DTMs.

Period	Volumes [1]			Rates [2]		
	DEP	ACC	WAS	DEP	ACC	WAS
1956–1984	−24,503	16,803	−7700	−875	600	−275
1984–1992	−168,806	52,765	−116,041	−21,101	6596	−14,505
1992–2001	−85,784	46,182	−39,602	−9532	5131	−4400
2001–2008	−9252	8705	−547	−1322	1244	−78
2008–2010	−18,389	7124	−11,265	−9195	3562	−5632
1956–2010	−292,333	117,707	−174,626	−5414	2180	−3234
1984–2010	−280,231	112,776	−167,455	−10,778	4338	−6441

[1] Volumes are in m^3; [2] Volume rates are in m^3/year. Volumes and rates: DEP: Depletion; ACC: Accumulation; WAS: Waste.

5. Discussion

As the landslide does not appear in aerial photographs from the 1956 and 1984 flights, the origin of the landslide had to be after 1984, probably in 1989, the rainiest year in this period [75]. However, due to the uncertainty about this date, the calculations presented in Tables 6 and 7 are made assuming the landslide origin in 1984, so the rate values should be considered as minimum values for the period 1984–1992.

The development of recent features, such as a clean scarp, steps and terraces, minor scarps, and an accumulation zone invading the river channel, can be clearly observed in the orthophotographs and models from 1992 to 2010, which evidences significant activity in the last 30 years. As mentioned before, other features identified in the orthoimages of 1956 and 1984 such as debris flows and the development of rills and gullies allow the zone to be described as unstable, with some evidence of older activity upslope.

Furthermore, landslides and these other features evidence a relevant geomorphic activity in the study area, confirmed by large mass losses associated with seasonal "rambla" flash-flooding events. Thus, the recorded negative displacements or volume changes (ground decrease or mass erosion) are larger than the positive changes (ground elevation or mass accumulation), so it can be deduced that the resulting mass losses are due to their evacuation by river flows. The great volume of mass waste is consistent with the significant geomorphic activity and the high indexes of tectonic activity along the southern flank of the Sierra Nevada, where the Almegíjar landslide is set [58]. This currently visible activity is related to the overall tectonic evolution of the region since Late Neogene and through the Holocene [76].

The parameters related to the landslide and determined from the calculation and comparison of models, longitudinal sections, and other evidence (Tables 5–7 and Figures 7 and 8) allow for the assessment of its dimensions and morphology, as well as the identification of trends concerning age, velocity, activity, diachroneity, and intensity following published classifications [63,73,74,77–81].

Thus the affected area is about 75×10^3 m^2, with a balanced distribution of depletion and accumulation areas (near 50%, although with a certain predominance of the depletion areas) that leads to a movement of rockslide type without development of flow mechanism, in which the material can expand downslope. In this sense, according to [60], the landslide, reactivated in the period 2009–2010, shows a strain shortening along the slope with an extension perpendicular to this axis, resulting in a bulging in the lower sector of the displaced mass, following a mechanism described by [61]. The morphology of the landslide scarp and the profiles suggest a planar sliding surface and a mass thickness ranging from 40 to 70 m, partially eroded and affected by debris flow of the weathered rock-mass surface. According to it, the movement can be classified as a rock translational or planar slide [74,77] without evidence of a rotational sliding component in the models and sections. The mobilized (depleted) mass has a volume of about 300×10^3 m^3, which leads us to categorize the landslide as "moderately large" [63,78,80]. Part of the mobilized mass is now displaced downslope, while another part invaded the river channel and was evacuated by the water flow, which has also limited the development of the accumulation areas.

The age is catalogued as "contemporary" [63] and its velocity—in average terms—can be considered as "very slow" [73,74], with vertical displacements of 0.32 and 0.15 m/year in the depleted and in the accumulated area, respectively, over the whole period of 26 years (1984–2010). However, as discussed below, this velocity is not uniform along this period and the landslide has shown periods of different rates of movement, with inactive and active phases, where the velocity ranges from "extremely slow" to "very slow." Thus, more difficult is the assessment of the landslide activity and diachroneity, because the results do not provide a detailed pattern of active periods, which is particularly difficult in these regions where the triggering factors, mainly the rainfall, do not follow a regular or seasonal pattern. However, the analysis of the differential changes between DTMs for the five time periods considered, and some additional data from terrestrial laser scanner (TLS) techniques [60], make it possible to assess certain activity features of interest.

The style of activity [81] is "single," corresponding to the displacement of a well-delimited mass in this single planar slide. On the other hand, the distribution of the activity may be considered "in advance" and also "confined," as the failure plane—shown only at the slide crown of the main scarp—is below the mass at slide toe. Regarding the state of activity [81], the movement can be considered active, although alternating periods of low activity with more active periods during ephemeral floods of the "rambla" [53,58,59].

Furthermore, the results described in the previous section show some periods of activity with displacements rates—measured in the vertical direction—on the order of decimeters by year. Regarding them, the accuracy of the technique has to be taken into account. The accuracy has been established in this work from errors of the flight orientation at check points that allow the estimation of the uncertainty of DTMs and differential DTMs. This uncertainty ranges from 0.36 m in the most favorable case (2008–2010) to 0.95 m in the least favorable (1956–1984), with the other cases between 0.60–0.70 m. Thus the vertical displacements measured in the periods 1984–1992 and 1992–2001, clearly higher than uncertainty, can be considered significant, but those measured in the periods 1956–1984 and 2001–2008 (next to zero values) are considered non-significant. The displacements calculated in the short period of 2008–2010—slightly higher than uncertainty—are at the limit of accuracy of the analysis, although additional data confirm the activity in this period, as will be discussed below. The volumetric accuracy and calculations agree with this, and therefore the depleted, accumulated, and waste volumes in the periods 1984–1992 and 1992–2001 are significant (much larger than the uncertainty), while the other periods (1956–1984, 2001–2008, and 2008–2010) are not significant (lower or similar to the uncertainty).

With this premise, the movement started after 1984 with the maximum rates observed (0.46 m/year of descent in the head or depleted area and 0.39 m/year in the foot or accumulation area), and slowed in the next period of 1992–2001 when lower displacements are observed (0.32 and 0.20 m/year, respectively), until the landslide seemed to stop in the period of 2001–2008, when the measured displacements are not significant. After that, in the last period of 2008–2010 a reactivation of the landslide occurred with displacement rates of about 0.21 and 0.12 m/year. The volumetric calculations confirm this evolution, with volume rates ranging between 10 to 20×10^3 m^3/year in the active periods and values near to 0 in the inactive periods.

At this point, we must mention the complementary results found in this landslide by the combined use of Terrestrial Laser Scanner (TLS) and Global Navigation Satellite Systems (GNSS) between 2008 and 2010 [60], which showed a reactivation in the period March 2009 to June 2010—coinciding with a year of intense rainfall—after an inactive period between September 2008 and March 2009. In the reactivation period a downward movement of 1.20 m at the top was measured, which corresponds to a descent rate of 0.96 m/year; meanwhile, below the middle part of the mass, 1.30 m of advance was established with a displacement rate of 1.04 m/year. These values of the annual rates about 1 m/year express periods of greater activity with a velocity near to that catalogued as "slow" (higher than 1 m/year) between periods of much lower activity in which the velocity is "very slow" to "extremely slow". These data suggest that other vertical displacements and volumes measured between the flights used in this study could have taken place in shorter intervals of time than the periods considered in this work.

Thus, in a more speculative way, if the date of 1989 were considered as the starting point of the landslide—triggered by a period of heavy rainfall [75]—the estimated rates of displacements calculated for the period of 1984–1992 (decimeters by year) would increase to 1.7 m/year in the depleted area and almost 1 m/year in the accumulation area (velocity catalogued as "slow"), on the same order of magnitude as those calculated with TLS measurements [60]. If the period in which the displacement occurred were even shorter, the displacement rates could be even higher, which expresses a large initial impulse for the landslide. The same situation could have occurred in other active periods such as 1992–2001, in relation to the heavy rainfall in the years 1995 to 1998, particularly in the winter of 1996–1997. Again, the displacement rates could increase to values near 1 m/year or higher.

For these landslides of discontinuous activity, an interesting concept is the activity duration or diachroneity [63]. According to the scale established in these works, the degree of diachroneity assigned to the studied landslide is VIII (10 to 100 years). Nevertheless, if only the periods of activity are considered, the degree of diachroneity would be VI (1 month to 1 year).

The triggering mechanism of the landslide was the rainfall, as in other regions near to it [55] and in general in Mediterranean countries [82]. In these areas the movements are triggered occasionally by "rambla" flash floods cutting down through the confined planar slide mass and dragging out the mass accumulation [75,83]. These flood events occur at the beginning of autumn with heavy rain (cold drop phenomenon) or at the beginning of spring after a rainy winter, in which heavy rains coincide with snow melting from the Sierra Nevada range.

Thus, in an approximate way, a relationship between the landslide activity and the rainfalls can be established. The landslide probably was triggered in relation to a rainy event of about 800 mm between November 1989 and January 1990 (more than 400% of the average rainfall for this period). Other equivalent periods occurred between November 1996 and January 1997 with more than 600 mm (more than 300% of the average) [53,55,59,75,83], and in the early months of 2010 with 500 mm between December 2009 and March 2010 (more than 200% of the average rainfall). These events and the general rainfalls of the corresponding years (1989–1990, 1995–1998, and 2009–2011) are related to the higher activity of the periods 1984–1992, 1992–2001, and 2008–2010, respectively. Meanwhile, the lesser activity in the period 2001–2008, when minimum displacements and volumetric changes were recorded, is associated with the absence of relevant precipitation.

This irregular pattern of landslides and rainfall coincides with other evidence such as field observations and landslide inventories along the Betic Cordillera since 1979 [54–57], particularly in the vicinity of this landslide [53,58,59,75,83]. The analysis of the available regional climatic data, reliable in the region since 1940, shows some intense rainy periods similar to those recorded in 1989, 1997, and 2010 [54], triggering slope failures in the study zone. These rainfall patterns, spanning periods of two or three years, register intense peaks in some months. First, we can see five periods exceeding 400 mm in three months, when the mean annual rainfall is 620 mm/y, and also another four periods of accumulated rainfall of more than 300 mm. Thus, the return period for rainfall triggering landslides in the region is around 15 years for the more intense events and eight years if less intense events are taken into account. Other analyses, based on average annual rainfall, show return periods of 18 years for annual precipitation higher than 950 mm, and five years for precipitation higher than 750 mm, these reaching the threshold of new landslides in the region [84]. Finally, recent studies based on the analysis of the anomalies in antecedent cumulative rainfall establish return periods of 12.4 years for these anomalies [85], a value similar to the previous studies.

Comparing these data with climatic studies on the origin of the precipitation in southern Spain, the intense rainfall in the region appears to be related to a negative value of the North Atlantic Oscillation index (NAO index, [86]). Positive values, which are more frequent, are related to more intense precipitation in Northern Europe and drier periods with rainfall in the Mediterranean area. As the periodicity of winter values of this NAO index is eight years [87], this can be seen as coincident with the return periods identified.

6. Conclusions

In conclusion, the real usefulness of LiDAR and digital photogrammetry techniques, as well as the use of differential DEMs, is demonstrated in landslide studies, particularly concerning the determination of the temporal evolution of landslides. The accuracy and consistency of these techniques enables the detection of very detailed slope features and their changes. In this work a new methodology has been developed, based on the use of a stereo photogrammetric workstation for outdating DTMs of more recent flights by editing them only in unstable zones (with clear changes between DTMs), over the stereoscopic models of previous flights. This permits us to reduce the time spent editing DTMs and especially to avoid noise in the differential model calculations.

Concerning the results, the methodology has enabled the detection and quantification of the extent of depletion and accumulation areas of the landslide, with average vertical displacement of about 8.5 m in the depletion area and 4 m in the accumulation area, attaining maximum displacements of almost 30 m in both areas. These displacements can be considered significant because they are an order of magnitude higher than the estimated uncertainty. Also, significant volumetric changes of about 300×10^3 m^3 in depletion area and 120×10^3 m^3 in accumulation areas were quantified, with some 175×10^3 m^3 of materials evacuated afterwards by the river flow.

The displacement and volumetric change rates are not constant over time, attaining maximum values in the initial period 1984–1992 (displacements of about 40 cm/year) and more moderate in the periods 1992–2001 and 2008–2010 (displacements of 15–30 cm/year), during seasons of heavy rainfalls, particularly in 1997 and 2010. The lowest rates were recorded in the period 2001–2008, when the displacement hardly reached 4 cm/year in a long dry period and they can be considered non-significant regarding the accuracy of the technique used. Other observations and data from TLS lead to estimates of displacement rates of higher than 1 m/year in some recent reactivation (2009–2010), similar to that found during the landslide-triggering phase. This difference in slide activity features diachronic slope movements alternating periods of variable activity depending on the influence of triggering factors. From these data, an "extremely to very slow and slow" slide velocity is calculated.

Nevertheless, current research could provide more detailed results on the slide displacements and evolution by using more flights, mainly available in the last decade, and incorporating TLS and UAV photogrammetric data, as well as displacement wireless sensor networks (WSN) and local rainfall data, for easier forecasts of future slide displacements. On the other hand, other techniques such as the determination of displacement vectors in significant slope points or the calculation of absolute distances provide more precise information on the slide kinematics. Finally, regional quantitative studies of other landslides are also needed to draw better quality landslide-hazard maps.

Acknowledgments: This research was funded by the projects P06-RNM-02125 and RNM-06862 (ISTEGEO) funded by the Andalusian Research Plan, projects CGL2008-04854 and TIN2009-09939 funded by the Ministry of Science and Innovation of Spain and Research Groups TEP-213, and RNM 221 of the Andalusian Research Plan.

Author Contributions: Tomás Fernández, José Luis Pérez, Javier Cardenal and Clemente Irigaray conceived and organized the research activity. José Luis Pérez, Javier Cardenal, and Carlos Colomo processed the photogrammetric data, generated the DTMs, and made the calculations (differential models and volumes). Tomás Fernández, José Antonio Palenzuela and Clemente Irigaray made the GIS analysis and Tomás Fernández interpreted the resulting data. Jorge Delgado and José Chacón supervised the research activity. All authors contributed to the writing of this manuscript.

Conflicts of Interest: The authors declare no conflicts of interest.

References

1. Fernández, T.; Pérez, J.L.; Colomo, C.; Mata, E.; Delgado, J.; Cardenal, J.; Irigaray, C.; Chacón, J. Digital photogrammetry and LiDAR techniques to study the evolution of a landslide. In Proceedings of the 8th International Conference on Geo-information for Disaster Management: Best Practices, Enschede, The Netherlands, 13–14 December 2012; Zlatanova, S., Peters, R., Fendel, E.M., Eds.; Universiteit Twente: Enschede, The Netherlands, 2012; pp. 95–104.
2. Mantovani, F.; Soeters, R.; van Westen, C.J. Remote Sensing Techniques for Landslide Studies and Hazard Zonation in Europe. *Geomorphology* **1996**, *15*, 213–225. [CrossRef]
3. Chacón, J.; Irigaray, C.; Fernández, T.; El Hamdouni, R. Engineering geology maps: Landslides and Geographical Information Systems (GIS). *Bull. Eng. Geol. Environ.* **2006**, *65*, 341–411. [CrossRef]
4. Metternicht, G.; Hurni, L.; Gogu, R. Remote sensing of landslides: An analysis of the potential contribution to geo-spatial systems for hazard assessment in mountainous environments. *Remote Sens. Environ.* **2005**, *98*, 284–303. [CrossRef]
5. Tofani, V.; Hong, Y.; Singhroy, V. Introduction: Remote Sensing Techniques for Landslide Mapping and Monitoring. In *Landslide Science for a Safer Geoenvironment*; Sassa, K., Canuti, P., Yin, Y., Eds.; Springer: Cham, Switzerland, 2014; Part III; pp. 301–303.
6. Hapke, C.J. Estimation of Regional Material Yield from Coastal Landslides Based on Historical Digital Terrain Modelling. *Earth Surf. Process. Landf.* **2005**, *30*, 679–697. [CrossRef]
7. Prokešová, R.; Kardoš, M.; Medved'ová, A. Landslide dynamics from high-resolution aerial photographs: A case study from the Western Carpathians, Slovakia. *Geomorphology* **2010**, *115*, 90–101. [CrossRef]
8. Chandler, J.H.; Brunsden, D. Steady state behaviour of the Black Ven Mudslide: The application of archival analytical photogrammetry to studies of landform change. *Earth Surf. Process. Landf.* **1995**, *20*, 255–275. [CrossRef]
9. Gentili, G.; Giusti, E.; Pizzaferri, G. Photogrammetric Techniques for the Investigation of the Corniglio Landslide. In *Applied Geomorphology*; Allison, R.J., Ed.; John Wiley & Sons: Chichester, UK, 2002; pp. 39–48.
10. Casson, B.; Delacourt, C.; Allemand, P. Contribution of multi-temporal remote sensing images to characterize landslide slip surface: Application to the La Clapière landslide (France). *Nat. Hazards Earth Syst. Sci.* **2005**, *5*, 425–437. [CrossRef]
11. Casson, B.; Delacourt, C.; Baratoux, D.; Allemand, P. Seventeen Years of the "La Clapière" Landslide Evolution Analysed from Ortho-Rectified Aerial Photographs. *Eng. Geol.* **2003**, *68*, 123–139. [CrossRef]
12. Van Westen, C.J.; Getahun, F.L. Analyzing the evolution of the Tessina landslide using aerial photographs and digital elevation models. *Geomorphology* **2003**, *54*, 77–89. [CrossRef]
13. Walstra, J.; Chandler, J.H.; Dixon, N.; Dijkstra, T.A. Time for Change—Quantifying Landslide Evolution Using Historical Aerial Photographs and Modern Photogrammetric Methods. In *The International Archives of the Photogrammetry, Remote Sensing and Spatial Information Sciences, Proceedings of the XXth ISPRS Congress, Istanbul, Turkey, 12–23 July 2004*; Altan, O., Ed.; Copernicus Publications: Gottingen, Germany, 2004; Volume XXXV, Part B4; pp. 475–480.
14. Walstra, J.; Chandler, J.H.; Dixon, N.; Dijkstra, T.A. Extracting Landslide Movements from Historical Aerial Photographs. In *Landslides: Evaluation and Stabilization*; Lacerda, W., Erlich, M., Fontoura, S.A.B., Sayao, A.S.F., Eds.; Taylor & Francis: London, UK, 2004; pp. 843–850.
15. Walstra, J.; Chandler, J.H.; Dixon, N.; Dijkstra, T.A. Aerial Photography and Digital Photogrammetry for Landslide Monitoring. *Geol. Soc. (Lond.) Spec. Publ.* **2007**, *283*, 53–63. [CrossRef]
16. Walstra, J.; Dixon, N.; Chandler, J.H. Historical aerial photographs for landslide assessment: Two case histories. *Q. J. Eng. Geol. Hydrogeol.* **2007**, *40*, 315–332. [CrossRef]
17. Baldi, P.; Fabris, M.; Monticelli, R. Monitoring the morphological evolution of the Sciara Del Fuoco during the 2002–2003 Stromboli eruption using multi-temporal photogrammetry. *ISPRS J. Photogramm. Remote Sens.* **2005**, *59*, 199–211. [CrossRef]

18. Baldi, P.; Cenni, N.; Fabris, M.; Zanutta, A. Kinematics of a landslide derived from archival photogrammetry and GPS data. *Geomorphology* **2008**, *102*, 435–444. [CrossRef]
19. Cardenal, J.; Delgado, J.; Mata, E.; González-Díez, A.; Remondo, J.; Díaz de Terán, J.R.; Francés, E.; Salas, L.; Bonachea, J.; Olague, I.; et al. The use of digital photogrammetry techniques in landslide instability. In *Geodetic Deformation Monitoring: From Geophysical to Geodetic Roles*; Gil-Cruz, J., Sanso, F., Eds.; IAG Springer Series: New York, NY, USA; Heidelberg, Germany, 2006; pp. 259–264.
20. Kasperski, J.; Delacourt, C.; Allemand, P.; Potherat, P. Evolution of the Sedrun landslide (Graubünden, Switzerland) with ortho-rectified air images. *Bull. Eng. Geol. Environ.* **2010**, *69*, 421–430. [CrossRef]
21. Fabris, M.; Menin, A.; Achilli, V. Landslide displacement estimation by archival digital photogrammetry. *Ital. J. Remote Sens.* **2011**, *43*, 23–30. [CrossRef]
22. González-Díez, A.; Fernández-Maroto, G.; Doughty, M.W.; Díaz de Terán, J.R.; Bruschi, V.; Cardenal, J.; Pérez, J.L.; Mata, E.; Delgado, J. Development of a methodological approach for the accurate measurement of slope changes due to landslides, using digital photogrammetry. *Landslides* **2014**, *11*, 615–628. [CrossRef]
23. Dewitte, O.; Jasselette, J.C.; Cornet, Y.; Van Den Eeckhaut, M.; Collignon, A.; Poesen, J.; Demoulin, A. Decadal-scale analysis of ground movements in old landslides in western Belgium. *Eng. Geol.* **2008**, *99*, 11–22. [CrossRef]
24. Corsini, A.; Borgatti, L.; Cervi, F.; Dahne, A.; Ronchetti, F.; Sterzai, P. Estimating mass-wasting processes in active earth slides—Earth flows with time-series of High-Resolution DEMs from photogrammetry and airborne LiDAR. *Nat. Hazards Earth Syst. Sci.* **2009**, *9*, 433–439. [CrossRef]
25. Mora, P.; Baldi, P.; Casula, G.; Fabris, M.; Ghirotti, M.; Mazzini, E.; Pesci, A. Global Positioning Systems and Digital Photogrammetry for the Monitoring of Mass Movements: Application to the Ca' Di Malta Landslide (Nothern Apennines, Italy). *Eng. Geol.* **2003**, *68*, 103–121. [CrossRef]
26. Brückl, E.; Brunner, F.K.; Kraus, K. Kinematics of a deep-seated landslide derived from photogrammetric, GPS and geophysical data. *Eng. Geol.* **2006**, *88*, 149–159. [CrossRef]
27. De Bari, C.; Lapenna, V.; Perrone, A.; Puglisi, C.; Sdao, F. Digital photogrammetric analysis and electrical resistivity tomography for investigating the Picerno landslide (Basilicata region, southern Italy). *Geomorphology* **2011**, *133*, 34–46. [CrossRef]
28. Yeh, M.L.; Hsiao, Y.C.; Chen, Y.H.; Chung, J.C. A study on unmanned aerial vehicle applied to acquire terrain information of landslide. In Proceedings of the 32 Asian Conference Remote Sensing, Taipei, Taiwan, 3–7 October 2011; Volume 3, pp. 2210–2215.
29. Niethammer, U.; James, M.R.; Rothmund, S.; Travelletti, J.; Joswig, M. UAV-based remote sensing of the Super-Sauze landslide: Evaluation and results. *Eng. Geol.* **2012**, *128*, 2–11. [CrossRef]
30. Stumpf, A.; Malet, J.P.; Kerle, N.; Niethammer, U.; Rothmund, S. Image-based mapping of surface fissures for the investigation of landslide dynamics. *Geomorphology* **2013**, *186*, 12–27. [CrossRef]
31. Turner, D.; Lucieer, A.; de Jong, S.M. Time Series Analysis of Landslide Dynamics using an Unmanned Aerial Vehicle (UAV). *Remote Sens.* **2015**, *7*, 1736–1757. [CrossRef]
32. Fernández, T.; Pérez, J.L.; Cardenal, F.J.; Gómez, J.M.; Colomo, C.; Delgado, J. Analysis of landslide evolution affecting olive groves using UAV and photogrammetric techniques. *Remote Sens.* **2016**, *8*, 837.
33. Peterman, V. Landslide activity monitoring with the help of unmanned aerial vehicle. In *The International Archives of the Photogrammetry Remote Sensing and Spatial Information Sciences, Proceedings of the International Conference on Unmanned Aerial Vehicles in Geomatics, Toronto, Canada, 30 August–2 September 2015*; Armenakis, E., Ed.; Copernicus Publications: Gottingen, Germany, 2015; Volume XL-1/W4, pp. 215–218.
34. Vrublova, D.; Kapica, R.; Jirankova, E.; Strus, A. Documentation of landslides and inaccessible parts of a mine using an unmanned UAV system and methods of digital terrestrial photogrammetry. *GeoSci. Eng.* **2015**, *LXI*, 8–19.
35. Yang, Z.; Lan, H.; Liu, H.; Li, L.; Wu, Y.; Meng, Y.; Xu, L. Post-earthquake rainfall-triggered slope stability analysis in the Lushan area. *J. Mt. Sci.* **2015**, *12*, 232–242. [CrossRef]
36. Shi, B.; Liu, C. UAV for Landslide Mapping and Deformation Analysis. In Proceedings of the International Conference on Intelligent Earth Observing and Applications, Guilin, China, 23–24 October 2015; Zhou, G., Kang, C., Eds.; Volume 9808.
37. Lindner, G.; Schraml, K.; Mansberger, R.; Hübl, J. UAV monitoring and documentation of a large landslide. *Appl. Geomat.* **2016**, *8*, 1–11. [CrossRef]

38. Hsieh, Y.C.; Chan, Y.; Hu, J. Digital Elevation Model Differencing and Error Estimation from Multiple Sources: A Case Study from the Meiyuan Shan Landslide in Taiwan. *Remote Sens.* **2016**, *8*, 199. [CrossRef]

39. Al-Rawabdeh, A.; He, F.; Moussa, A.; El-Sheimy, N.; Habib, A. Using an Unmanned Aerial Vehicle-Based Digital Imaging System to Derive a 3D Point Cloud for Landslide Scarp Recognition. *Remote Sens.* **2016**, *8*, 95. [CrossRef]

40. Walstra, J.; Chandler, J.H.; Dixon, N.; Wackrow, R. Evaluation of the controls affecting the quality of spatial data derived from historical aerial photographs. *Earth Surf. Process. Landf.* **2010**, *36*, 853–863. [CrossRef]

41. Adams, J.C.; Chandler, J.H. Evaluation of LiDAR and Medium Scale Photogrammetry for Detecting Soft-Cliff Coastal Change. *Photogramm. Rec.* **2002**, *17*, 405–418. [CrossRef]

42. Glenn, N.F.; Streutker, D.R.; Chadwick, D.J.; Thackray, G.D.; Dorsch, S.J. Analysis of LiDAR-derived topographic information for characterizing and differentiating landslide morphology and activity. *Geomorphology* **2006**, *73*, 131–148. [CrossRef]

43. Derron, M.H.; Jaboyedoff, M. LIDAR and DEM techniques for landslides monitoring and characterization. *Nat. Hazards Earth Syst. Sci.* **2010**, *10*, 1877–1879. [CrossRef]

44. Palenzuela, J.A.; Marsella, M.; Nardinocchi, C.; Pérez, J.L.; Fernández, T.; Chacón, J.; Irigaray, C. Landslide detection and inventory by integrating LiDAR data in a GIS environment. *Landslides* **2015**, *12*, 1035–1050. [CrossRef]

45. McKean, J.; Roering, J. Objective landslide detection and surface morphology mapping using high resolution airborne laser altimetry. *Geomorphology* **2004**, *57*, 331–351. [CrossRef]

46. Kasai, M.; Ikeda, M.; Asahina, T.; Fujisawa, K. LiDAR-derived DEM evaluation of deep-seated landslides in a steep and rocky region of Japan. *Geomorphology* **2009**, *113*, 57–69. [CrossRef]

47. Burns, W.J.; Coe, J.A.; Kaya, B.S.; Ma, L. Analysis of Elevation Changes Detected from Multi-Temporal LiDAR Surveys in Forested Landslide Terrain in Western Oregon. *Environ. Eng. Geosci.* **2010**, *16*, 315–341. [CrossRef]

48. DeLong, S.B.; Prentice, C.S.; Hilley, G.E.; Ebert, Y. Multitemporal ALSM change detection, sediment delivery, and process mapping at an active earthflow. *Earth Surf. Proc. Land.* **2012**, *37*, 262–272. [CrossRef]

49. Roering, J.J.; Mackey, B.H.; Marshall, J.A.; Sweeney, K.E.; Deligne, N.I.; Booth, A.M.; Handwerger, A.L.; Cerovski-Darriau, C. "You are HERE": Connecting the dots with airborne lidar for geomorphic fieldwork. *Geomorphology* **2013**, *200*, 172–183. [CrossRef]

50. Cavalli, M.; Trevisani, S.; Comiti, F.; Marchi, L. Geomorphometric assessment of spatial sediment connectivity in small Alpine catchments. *Geomorphology* **2013**, *188*, 31–41. [CrossRef]

51. Tarolli, P. High-resolution topography for understanding Earth surface processes: Opportunities and challenges. *Geomorphology* **2014**, *216*, 295–312. [CrossRef]

52. Bossi, G.; Cavalli, M.; Crema, S.; Frigerio, S.; Quan Luna, B.; Mantovani, M.; Marcato, G.; Schenato, L.; Pasuto, A. Multi-temporal LiDAR-DTMs as a tool for modelling a complex landslide: A case study in the Rotolon catchment (eastern Italian Alps). *Nat. Hazards Earth Syst. Sci.* **2015**, *15*, 715–722. [CrossRef]

53. Fernández, T.; Irigaray, C.; El Hamdouni, R.; Chacón, J. Methodology for landslide susceptibility mapping by means of a GIS. Application to the Contraviesa area (Granada, Spain). *Nat. Hazards* **2003**, *30*, 297–308. [CrossRef]

54. Fernández, T.; Ureña, M.A.; Delgado, J.; Cardenal, J.; Irigaray, C.; Chacón, J. Examples of natural risk analysis from SDI. In Proceedings of the International Cartographic Conference (CO-022), Paris, France, 21–27 June 2011.

55. Irigaray, C.; Lamas, F.; El Hamdouni, R.; Fernández, T.; Chacón, J. The importance of the precipitation and the susceptibility of the slopes for the triggering of landslides along the roads. *Nat. Hazards* **2000**, *21*, 65–81. [CrossRef]

56. Irigaray, C.; Fernández, T.; El Hamdouni, R.; Chacón, J. Evaluation and validation of landslide susceptibility maps obtained by a GIS matrix method: Examples from the Betic Cordillera (southern Spain). *Nat. Hazards* **2007**, *41*, 61–79. [CrossRef]

57. Chacón, J. Landslide susceptibility, hazard & risk GIS mapping in the Betic Cordillera (Spain): Areas with limited information about triggering factors. In *Guidelines for Mapping Areas at Risk of Landslides in Europe, Proc. Experts Meeting Joint Research Center (JRC), Ispra, Italy, 23–24 October 2007*; Hervás, J., Ed.; Office for Official Publications of the European Communities: Luxembourg, Luxembourg, 2007; pp. 23–26.

58. El Hamdouni, R.; Irigaray, C.; Fernández, T.; Chacón, J.; Keller, E.A. Assessment of relative active tectonics, southwest border of the Sierra Nevada (Southern Spain). *Geomorphology* **2008**, *96*, 150–173. [CrossRef]

59. Jiménez-Perálvarez, J.D.; Irigaray, C.; El Hamdouni, R.; Chacón, J. Landslide-susceptibility mapping in a semi-arid mountain environment: An example from the southern slopes of Sierra Nevada (Granada, Spain). *Bull. Eng. Geol. Environ.* **2011**, *70*, 265–277. [CrossRef]

60. Palenzuela, J.A.; Irigaray, C.; Jiménez-Perálvarez, J.D.; Chacón, J. Application of Terrestrial Laser Scanner to the Assessment of the Evolution of Diachronic Landslides. In *Landslide Science and Practice*; Margottini, C., Canuti, P., Sassa, K., Eds.; Springer: Berlin/Heidelberg, Germany, 2013; Section 68; pp. 517–523.

61. Agliardi, F.; Crosta, G.; Zanchi, A. Structural constraints on deep-seated slope deformation kinematics. *Eng. Geol.* **2001**, *59*, 83–102. [CrossRef]

62. Aldaya, F.; Díaz de Federico, A.; García-Dueñas, V.; Martínez García, E.; Navarro-Vila, F.; Puga, E. *Mapa Geológico de España 1:50.000: Hoja 1042 Lanjarón*; 2ª serie; IGME: Madrid, Spain, 1981; p. 32.

63. Chacón, J.; Irigaray, C.; El Hamdouni, R.; Jiménez-Perálvarez, J.D. Diachroneity of landslides. In *Geologically Active*; Williams, A.L., Pinches, G.M., Chin, C.Y., McMorran, T.J., Massey, C.I., Eds.; CRC Press/Balkema, Taylor & Francis Group: Leiden, The Netherlands, 2010; pp. 999–1006.

64. Colomo-Jiménez, C.; Pérez-García, J.L.; Fernández-del-Castillo, T.; Gómez-López, J.M.; Mozas-Calvache, A.M. Methodology for orientation and fusion of photogrammetric and LiDAR data for multitemporal studies. In *The International Archives of the Photogrammetry, Remote Sensing and Spatial Information Sciences, Proceedings of the XXIII ISPRS Congress, Prague, Czech Republic, 12–19 July 2016*; Copernicus Publications: Gottingen, Germany, 2016; Volume XLI-B7, pp. 639–645.

65. *Socet Set 5.2*; Bae Systems Plc.: London, UK, 2009.

66. *Terramatch*; TerraSolid Ltd.: Helsinki, Finland, 2012.

67. Oshel, E.; Liddle, D. Photogrammetric applications: Space Application of Photogrammetry. In *Manual of Photogrammetry*, 6th ed.; McGlone, C., Mikhail, E., Bethel, J., Eds.; American Society for Photogrammetry and Remote Sensing (ASPRS): Bethesda, MD, USA, 2013; p. 1183.

68. Cavalli, M.; Tarolli, P. Application of LiDAR technology for rivers analysis. *Ital. J. Eng. Geol. Environ.* **2011**, *1*, 33–44.

69. *Maptek I-Site Studio 6.0*; Maptek Pty. Ltd.: Adelaide, Australia, 2016.

70. *ArcGIS 10.3*; Esri: Redlands, CA, USA, 2013.

71. Brasington, J.; Rumsby, B.T.; McVey, R.A. Monitoring and modelling morphological change in a braided gravel-bed river using high resolution GPS-based survey. *Earth Surf. Proc. Land.* **2000**, *25*, 973–990. [CrossRef]

72. Wheaton, J.M.; Brasington, J.; Darby, S.E.; Sear, D.A. Accounting for uncertainty in DEMs from repeat topographic surveys: Improved sediment budgets. *Earth Surf. Proc. Land.* **2010**, *35*, 136–156. [CrossRef]

73. WP/WLI. A suggested method for describing the rate of movement of a landslide. *Bull. Eng. Geol. Environ.* **1995**, *52*, 75–78.

74. Hungr, O.; Leroueil, S.; Picarelli, L. The Varnes classification of landslide types, an update. *Landslides* **2014**, *11*, 167–194. [CrossRef]

75. Thornes, J.B.; Alcántara-Ayala, I. Modelling mass failure in a Mediterranean mountain environment: Climatic, geological, topographical and erosional controls. *Geomorphology* **1998**, *24*, 87–100. [CrossRef]

76. García, A.F.; Zhu, Z.; Ku, T.L.; Sanz de Galdeano, C.; Chadwick, O.A.; Chacón, J. Tectonically driven landscape development within the eastern Alpujarran Corridor, Betic Cordillera, SE Spain (Almería). *Geomorphology* **2003**, *50*, 83–110. [CrossRef]

77. Varnes, D.J. Slope movement, types and processes. In *Landslides: Analysis and Control*; Schuster, R.L., Krizek, R.J., Eds.; Transportation Research Board Special Report National Academy of Sciences: Washington, DC, USA, 1978; Volume 176, pp. 12–33.

78. Fell, R. Landslide risk assessment and acceptable risk. *Can. Geotech. J.* **1994**, *31*, 261–272. [CrossRef]

79. Cardinali, M.; Reichenbach, P.; Guzzetti, F.; Ardizzone, F.; Antonini, G.; Galli, M.; Cacciano, M.; Castellani, M.; Salvati, P. A geomorphological approach to the estimation of landslide hazards and risks in Umbria, Central Italy. *Nat. Hazards Earth Syst. Sci.* **2002**, *2*, 57–72. [CrossRef]

80. Rodríguez-Ortiz, J.M.; Hinojosa, J.A.; Prieto, C. Regional studies on mass movements in Spain. In Proceedings of the 3th IAEG Congress, Madrid, Spain, 4–8 September 1978; Volume 1, pp. 267–277.

81. WP/WLI. A suggested method for describing the activity of a landslide. *Bull. Eng. Geol. Environ.* **1993**, *47*, 53–57.

82. Guzzeti, F. Landslide hazard assessment and risk evaluation: Limits and prospectives. In Proceedings of the 4th EGS Plinius Conference, Mediterranean Storms, Mallorca, Spain, 2–4 October 2002.

83. Alcántara-Ayala, I.; Thornes, J.B. Structure and hydrology in controlling mass failure in space and time: The case of the Guadalfeo failures. In Proceedings of the 8th International Conference & Field Trip on Landslides, Granada, Spain, 27–28 September 1996; Chacón, J., Irigaray, C., Fernández, T., Eds.; Balkema: Rotterdam, The Netherlands, 1996; pp. 89–96.

84. Jimenez-Perálvarez, J.D. Movimientos de ladera en la vertiente meridional de Sierra Nevada (Granada, España): Identificación, análisis y cartografía de susceptibilidad y peligrosidad mediante SIG. Ph.D. Thesis, University of Granada, Granada, Spain, July 2012.

85. Palenzuela, J.A.; Jiménez-Perálvarez, J.D.; Chacón, J.; Irigaray, C. Assessing critical rainfall thresholds for landslide triggering by generating additional information from a reduced database: An approach with examples from the Betic Cordillera (Spain). *Nat. Hazards* **2016**, *84*, 185–212. [CrossRef]

86. Trigo, R.M.; Pozo, D.; Timothy, C.; Osborn, J.; Castro, Y.; Gámiz, S.; Esteban, M.J. NAO influence on precipitation, river flow and water resources in the Iberian Peninsula. *Int. J. Climatol.* **2004**, *24*, 925–944. [CrossRef]

87. Pozo, D.; Esteban, M.J.; Rodrigo, F.S.; Castro, Y. An analysis of the variability of the North Atlantic Oscillation in the time and frequency domains. *Int. J. Climatol.* **2000**, *20*, 1675–1692. [CrossRef]

geosciences

MDPI

Article

Ground Motion in Areas of Abandoned Mining: Application of the Intermittent SBAS (ISBAS) to the Northumberland and Durham Coalfield, UK

David Gee [1,*], Luke Bateson [2], Andrew Sowter [3], Stephen Grebby [1], Alessandro Novellino [2], Francesca Cigna [2,†], Stuart Marsh [1], Carl Banton [4] and Lee Wyatt [4]

[1] Nottingham Geospatial Institute, University of Nottingham, Triumph Road, Nottingham NG7 2TU, UK; stephen.grebby@nottingham.ac.uk (S.G.); stuart.marsh@nottingham.ac.uk (S.M.)
[2] British Geological Survey, Keyworth, Nottinghamshire NG12 5GG, UK; lbateson@bgs.ac.uk (L.B.); alessn@bgs.ac.uk (A.N.); francesca.cigna@gmail.com (F.C.)
[3] Geomatic Ventures Limited, Nottingham Geospatial Building, Triumph Road, Nottingham NG7 2TU, UK; andrew.sowter@geomaticventures.com
[4] Coal Authority, 200 Lichfield Lane, Mansfield, Nottinghamshire NG18 4RG, UK; CarlBanton@coal.gov.uk (C.B.); LeeWyatt@coal.gov.uk (L.W.)
* Correspondence: david.gee@nottingham.ac.uk
† Present address: Italian Space Agency (ASI), Via del Politecnico snc, 00133 Rome, Italy.

Received: 20 June 2017; Accepted: 5 September 2017; Published: 13 September 2017

Abstract: In this paper, we investigate land motion and groundwater level change phenomena using differential interferometric synthetic aperture radar (DInSAR) over the Northumberland and Durham coalfield in the United Kingdom. The study re-visits earlier research that applied a persistent scatterers interferometry (PSI) technique to ERS (European Remote Sensing) and ENVISAT (Environmental Satellite) data. Here, the Intermittent Small Baseline Subset (ISBAS) DInSAR technique is applied to ERS, ENVISAT and Sentinel-1 SAR datasets covering the late 1990s, the 2000s and the mid-2010s, respectively, to increase spatial coverage, aid the geological interpretation and consider the latest Sentinel-1 data. The ERS data identify surface depressions in proximity to former collieries, while all three data sets ascertain broad areas are experiencing regional scale uplift, often occurring in previously mined areas. Uplift is attributed to increases in pore pressure in the overburden following the cessation of groundwater pumping after mine closure. Rising groundwater levels are found to correlate to ground motion measurements at selected monitoring sites, most notably in the surrounding area of Ashington. The area is divided by an impermeable EW fault; to the south, surface heave was identified as groundwater levels rose in the 1990s, whereas to the north, this phenomenon occurred two decades later in the 2010s. The data emphasize the complexity of the post-mining surface and subsurface environment and highlight the benefit that InSAR, utilizing the ISBAS technique, can provide in its characterization.

Keywords: ground motion; subsidence; coal mining; DInSAR; Intermittent SBAS

1. Introduction

High precision satellite radar interferometry is an effective and well established technique for monitoring earth surface motion [1]. It is a cost-effective method capable of surveying extensive areas to within millimetre precision regardless of weather and illumination conditions [2]. Differential interferometric synthetic aperture radar (DInSAR) utilizes stacks of high resolution synthetic aperture radar (SAR) data acquired over time from spaceborne platforms. DInSAR utilizes shifts in phase to measure changes in surface height in the satellite line-of-sight (LOS) direction and is capable of computing both average velocities (linear) and time-series (non-linear) to characterize ground motions occurring over the temporal epoch of image acquisitions [3].

A variety of persistent scatterer and small baseline interferometric techniques have been used to delineate and quantify coal mining related surface phenomena in Europe [4,5] and farther afield [6,7]. However, it is common for temporal gaps to restrict multi-temporal DInSAR analysis in areas subject to historical or active mining [7]. Temporal gaps are common due to large deformation gradients often associated with mining subsidence [8,9] and decorrelation, the loss of coherence, which is common over coal mining areas [10].

Coherence, a by-product of the interferogram formation process, is a measure of phase correlation between two corresponding signals [11]. In areas of high coherence, targets are highly reflective and alter minimally over time. Conversely, in areas of low reflectance, where targets change over time (i.e., from year to year or from season to season), coherence is lost. In temperate, vegetated, rural settings the land surface can be variable due to both natural processes (e.g., soil shrink-swell) and human influences (e.g., agriculture) [12]. As a result, when differential interferograms are stacked, points for which coherence is maintained throughout the stack are generally abundant in urban locations, but sparse if not non-existent in rural areas. The optimum results produced by DInSAR are, therefore, obtained in urban or rocky areas, where coherence is recurrently high.

The Intermittent Small Baseline Subset (ISBAS) [10,13] is an adapted version of the established low resolution SBAS [14] DInSAR time-series algorithm. It has been designed to improve the density and spatial distribution of survey points to return measurements in vegetated areas where DInSAR processing algorithms habitually struggle. The amendment of SBAS is introduced during unwrapping, after a set of low-resolution, two-pass, multi-looked interferograms with small perpendicular baselines have been produced. While SBAS unwraps targets where coherence is above a given threshold in all stacked interferograms, ISBAS considers targets where coherence is above a given threshold in a subset of the total number of interferograms within the stack. The algorithm can retrieve measurements for targets that are intermittently coherent throughout the period of observations. As a consequence, the density of survey points returned is higher and their distribution not limited to urban centres [15]. The ISBAS method has been successfully validated with ground truth over an area of gas production and geostorage in North Holland, the Netherlands [16]. The results demonstrated that the ISBAS technique can be used with confidence over locations where traditional ground-based survey measurements are not available.

Amongst many other applications, the ISBAS technique has previously been used for the monitoring of ground motion related to coal mining in the UK [10,17,18]. Sowter et al. (2013) [10] implemented the ISBAS technique on ENVISAT (Environmental Satellite) data covering the South Derbyshire and North West Leicestershire coalfields. The ISBAS-derived uplift, observed in agricultural and forested areas as well as urban centres, closely matched the outcrop of middle Coal Measures near Swadlincote, appearing bound by existing fault structures and spatially correlated to underground mining works abandoned in the early 1990s. Bateson et al. (2015) [18] applied ISBAS to ERS (European Remote Sensing) data acquired over the South Wales coalfield which has a complex history of surface movement due to coal mining and post-glacial stress relief [19]. Two discrete areas of uplift were observed in the proximity of the towns of Bargoed and Bedwas, again in mixed urban and rural land cover. In both investigations, the mechanism of uplift was not definitively confirmed but was thought to be related to increases in pore water pressure and/or the re-activation of existing faults.

Similar regional patterns of uplift following coalfield closures have been observed over other regions in the UK [20]. In Northumberland, a Persistent Scatterers Interferometry (PSI) analysis, an alternative DInSAR technique, was conducted on ERS and ENVISAT data covering the Northumberland and Durham coalfield and confirmed the changing patterns of ground motion over time [21,22]. Some correlations between the geology and PS derived ground motion products were made; however, in some cases the lack of PSI measurements in rural areas meant that the correlation between the motions and the geological and mining information was not clear.

Accordingly, the principle aim of this paper is to better delineate ground motion on a regional scale in the Northumberland and Durham coalfield by applying the ISBAS technique to the ERS and ENVISAT data in order to extend coverage into areas of low coherence. This will subsequently enable a more comprehensive geological interpretation of the factors and processes associated with the observed ground motion. In addition, the launch of Sentinel-1 in April 2014 provides the opportunity to bring the study up to date by revealing more recent patterns of motion, therefore facilitating an investigation of ground deformation spanning three decades.

2. Northumberland and Durham Case Study

2.1. Land Cover

The area of interest (AOI) is located in the north east region of England and covers the counties of Northumberland, Tyne and Wear and County Durham (Figure 1). The AOI covers 2000 km^2, 40 km in width and 50 km in length. The north of the AOI extends up to the towns of Ashington and Morpeth and South to the towns of Houghton-le-Spring and Seaham. Agricultural land is most dominant within the AOI, with small pockets of forested and semi natural areas contained within these expanses. Notable urban regions exist along the north east coastline where the cities of Newcastle-upon-Tyne and Sunderland are situated on the banks of the River Tyne and River Wear respectively.

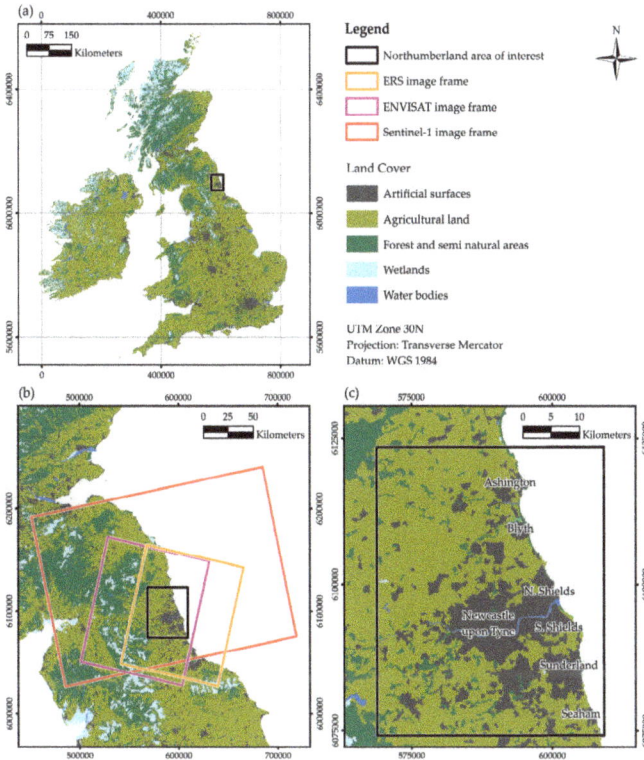

Figure 1. A land cover classification of the United Kingdom and Ireland taken from the 2012 CORINE (Co-ordinated Information on the Environment) Land Cover inventory [23]: (**a**) A map detailing the location of the Northumberland area of interest (AOI) within the United Kingdom; (**b**) Footprints of the ERS, ENVISAT and Sentinel-1 image frames; (**c**) The AOI with major towns and cities labelled. Copyright holder: European Environment Agency (EEA).

2.2. Geological Setting

The geological sequence of interest includes bedrock of Carboniferous strata within the southern sector of the Northumberland Trough: the Yoredale Group (Stainmore and Alston Formation) and the Coal Measures Group (Figure 2). The Northumberland Trough is a major, asymmetrical SW-NE basin bounded to the south by the Stublick and Ninety Fathom faults and by the Cheviot Massif to the north [24]. The Carboniferous rocks dip gently eastwards and are overlaid unconformably by the Permian strata of the basal Yellow Sands Formation and overlying Magnesian limestone formations of the Zechstein Group [25]. The continuity of the outcrops of these units is interrupted by a number of normal faults. The edge of the Zechstein Group outcrop is typically marked by prominent scarp features overlooking the lower lying ground of the Coal Measures Group [22]. Superficial deposits of the Quaternary period, mainly resulting from the last glaciation and subsequent Holocene processes, are present throughout the area and intermittently blanket the bedrock with areas of the Zechstein Group limestone virtually free of cover deposits.

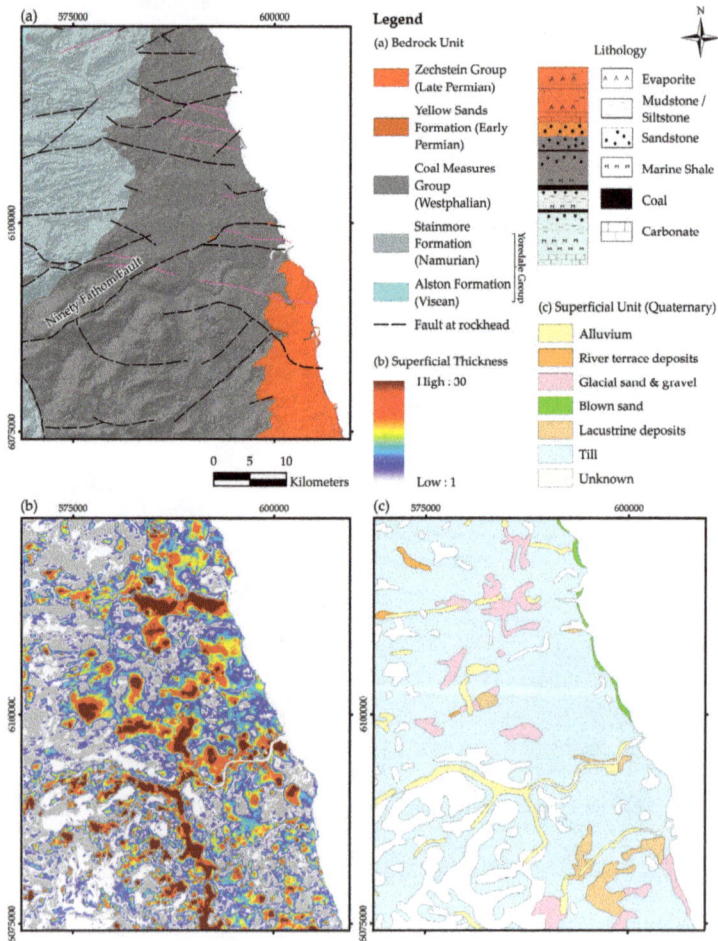

Figure 2. The geology of the AOI: (**a**) Bedrock geology overlain on topography; (**b**) Superficial thickness (interpolated from borehole data) [26]; (**c**) Superficial geology. 'Reproduced with the permission of the British Geological Survey ©NERC. All rights Reserved'.

2.3. Northumberland and Durham Coalfield

The Northumberland and Durham coalfield has a working history dating back to Roman times. Over twenty coal seams have been mined underground and the coalfield has been one of the major sources of opencast (surface-mined) coal in Great Britain. The coal-bearing strata dip gently to the east, with part of the coalfield concealed beneath the Permian strata to the south of the River Tyne. The geological structure of the area has governed the development of the coalfield with faults serving to divide the area into zones of "take". Early mining started in areas most accessible (i.e., inland in the west), where seams were often worked from the outcrop [27]. As mining and pumping methods developed, extraction from greater depths became viable; consequently, mining spread towards the eastern coast where deeper seams are situated, eventually advancing under the North Sea.

The coalfield has been extensively worked by both room and pillar and total extraction techniques. Longwall mining, where coal bearing rocks are machine worked from "longwall" faces at right angles from roadways running parallel to the face, is a method of total extraction and results in the immediate collapse of the roof and overlying rock into the void created behind. The displacement of rock can result in surface subsidence if the ratio of the width of extracted coal measures to the depth of the overburden exceeds a value between 0.1 and 0.5, where the structure and strength of the overburden controls the precise ratio value. Room and pillar mining, where mined material is extracted across a horizontal plane creating horizontal arrays of pillars and rooms, results in between 15% and 90% of the material being extracted. Consequently, load on the pillars increases by up to 90% which can result in spalling, brittle failure or ductile failure. The potential for failure depends on the physical properties of the coal and the geometry of the mine; the collapse of multiple pillars can result in a surface depression, known as areal subsidence. The mechanical aspects of subsidence produced from longwall mining methods are generally better understood compared with room and pillar mining [28].

3. SAR Processing

3.1. ERS and ENVISAT

Archive ERS and ENVISAT C-band (5.6 cm wavelength—5.3 GHz frequency) Stripmap SAR products were obtained and processed separately. The Single Look Complex (SLC) image stacks are acquired in the radar line-of-sight (LOS), the incidence angle ranges between 20.1° and 25.9° from near to far range with respect to the surface normal. Twenty-five ERS descending images spanning nearly a five year period from 24 May 1995 to 30 December 1999 (Appendix A) and twenty-one ENVISAT descending images spanning approximately a six year period from 3 December 2002 to 7 October 2008 (Appendix B) were identified for processing as they imaged the AOI with greatest frequency. The complete image frames were processed on both occasions which covers an area of approximately 100 km × 100 km, at a ground spatial resolution of 25 m in range and 5 m in azimuth [29].

The ERS stack was co-registered with respect to the master acquired on 20 November 1997, while the ENVISAT stack was co-registered to the scene acquired on 11 January 2005. Interferograms were generated with restrictions of 4 years on the temporal baseline and 250 m on the perpendicular orbital baseline. These restrictions are common to small baseline surveys using ERS and ENVISAT data, minimizing both temporal and spatial decorrelation in the interferograms and enhancing the quality of phase. Multilooking, applied to reduce noise and increase coherence, was implemented by a factor of 4 in range and 20 in azimuth to produce pixels of approximately 100 m × 100 m. Pixels deemed coherent within each interferogram displayed a coherence ≥ 0.25.

An ISBAS analysis improves ground coverage by utilizing targets that are not high quality in every interferogram. As detailed by Sowter et al. (2013) [10], the algorithm accepts pixels, termed coherent pixels, which meet a minimum quality standard. Those retained in the analysis displayed a coherence ≥ 0.25 in a minimum number of (m) interferograms, when (m) $\leq$ to the total number of interferograms (N). By unwrapping each interferogram individually the technique accepts intermittently coherent targets throughout the stack. The choice of interferogram threshold (m) is a trade-off between coverage and quality, the value of which has a direct correlation with the standard error of the derived velocities [30].

The ERS analysis produced 102 multi-looked differential interferograms where an interferogram threshold (*m*) of 30 was determined. The temporal and orbital baselines of image pairs are plotted relative to the master in Figure 3a. All velocities are computed with respect to a reference point, it is important that this location is both stable and coherent in every interferogram. A reference point in North Shields (55.00° N, −1.46° E) was chosen based upon the data from the British Geological Survey (BGS) GeoSure natural geohazard datasets and its urban location. The control point is in an area of low susceptibility to geohazards and is surrounded by coherent targets which helps to minimize the propagation of errors during phase unwrapping.

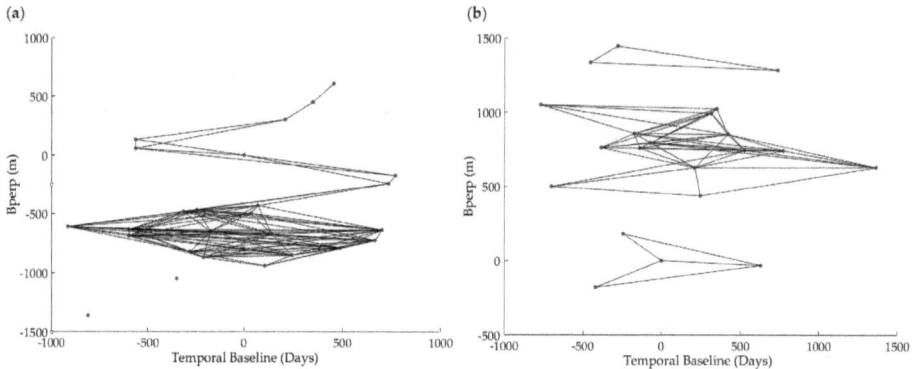

Figure 3. Temporal and perpendicular baselines of image pairs relative to the master. Image acquisitions are represented by points and differential interferograms, satisfying the orbital and temporal requirements, by lines: (**a**) ERS Synthetic Aperture Radar (SAR) (master: 20 November 1997); (**b**) ENVISAT Advanced Synthetic Aperture Radar (ASAR) (master: 11 January 2005).

The ENVISAT analysis produced 60 multi-looked differential interferograms (Figure 3b), from which an interferogram threshold (*m*) of 15 was applied. The reference point was, correspondingly to the ERS analysis, situated in North Shields for the aforementioned reasons.

Topographic phase was removed using a 90 m Shuttle Radar Topography Mission (SRTM) Digital Elevation Model (DEM) [31] and coherent and intermittently coherent pixels within each differential interferogram were unwrapped using a statistical-cost network-flow algorithm [32]. Following the method proposed by Berardino et al. (2002) [14] both linear and non-linear models of deformation were applied to compute average rates of motion and time-series per pixel. Additionally, the standard error for each pixel is computed via a least squares covariance analysis which expresses the goodness-of-fit between the interferogram values and derived linear velocity. Finally, due to the different incidence angles of ERS/ENVISAT (20.1–25.9°) and Sentinel-1 (29.1–46°), LOS velocities were converted into vertical velocities by means of dividing by the cosine of the incidence angle for each pixel.

3.2. Sentinel-1

Thirty-six Sentinel-1 Interferometric Wide Swath SLC images were identified as most suitable for analysis over the AOI (Appendix C). Radar LOS varies from 29.1° to 46.0° across the swath from near to far range, with a single look resolution of 5 m in range and 20 m in azimuth [33]. The full Sentinel-1 scenes are 250 km in width, in this case much of which is located over North Sea. A subset of the image was processed, approximately 90 km × 90 km, contained within IWS1 and IWS2.

The nominal repeat cycle of the Sentinel-1 imagery used is 12 days, less than the 35 day repeat cycles of the ERS and ENVISAT satellites. In the period of approximately 12 months, Sentinel-1 acquired more SLC images of the AOI than ENVISAT in its total operational lifetime. This is of great benefit to a DInSAR analysis as more observations generally implies a lower standard deviation of error, as demonstrated for the area of Doncaster in the UK by Novellino et al. (2017) [34]. The C-band (5.555 cm wavelength—5.405 GHz frequency) level-1 images processed were acquired on an ascending geometry; a stack of descending Sentinel-1 imagery were also considered to try and maintain uniformity with the geometries of the ERS and ENVISAT data, however only 26 images covered the full AOI and therefore generated a less reliable result. The ISBAS processing of the Sentinel-1 data followed the approach defined in Sowter et al. (2016) [15].

Coregistration was performed with respect to a master image acquired on 25 January 2016. Multilooking was implemented by a factor of 22 in range and 5 in azimuth which increased pixel size to approximately 90 m in ground range. A dense network of 520 differential interferograms were generated, with restrictions of 250 m on the perpendicular orbital baseline and one year on the temporal baseline, plotted in Figure 4. Comparing against the networks produced from the ERS and ENVISAT analysis (Figure 3) the improved accuracy of Sentinel-1 orbits is evident, designed to operate within a narrow orbital tube radius of 50 m for the majority of its operational lifetime [35]. Consequently, in concurrence with the reduced revisit time and increased number of acquisitions, there is far more redundancy in the network.

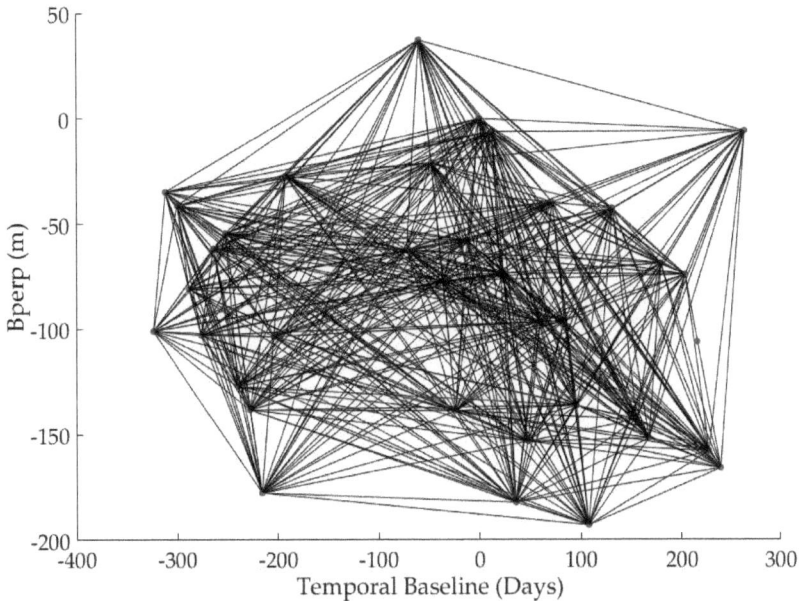

Figure 4. Sentinel-1 Synthetic Aperture Radar (SAR) temporal and perpendicular baselines of image pairs relative to the master acquired on 25 January 2016.

An interferogram threshold of 135 was selected and, following the same approach as for the ERS and ENVISAT processing, the reference point was located in North Shields (55.00° N, −1.46° E), topographic phase was removed using a STRM DEM and SAR orbital parameters, coherent and intermittently coherent pixels were unwrapped to derive linear velocities, standard errors and time-series per pixel which were subsequently projected into the vertical direction.

4. Ground Motion in North East England

4.1. ISBAS Processing Coverage

The analysis covered 99%, 89% and 95% of the land surface area of the full extents processed in the cases of ERS, ENVISAT and Sentinel-1 respectively, providing a broad and complete distribution of velocities over all land cover types (Figure 5). Had velocities been derived for targets where coherence is recurrent in every interferogram, coverage would have only been 7%, 2% and 12% for the respective data sets. Dense coverage was achieved not only in areas of high coherence, but over agricultural land and woodland areas. Standard errors are lowest in urban areas due to increased coherence and consequently more coherent interferograms to derive linear velocities.

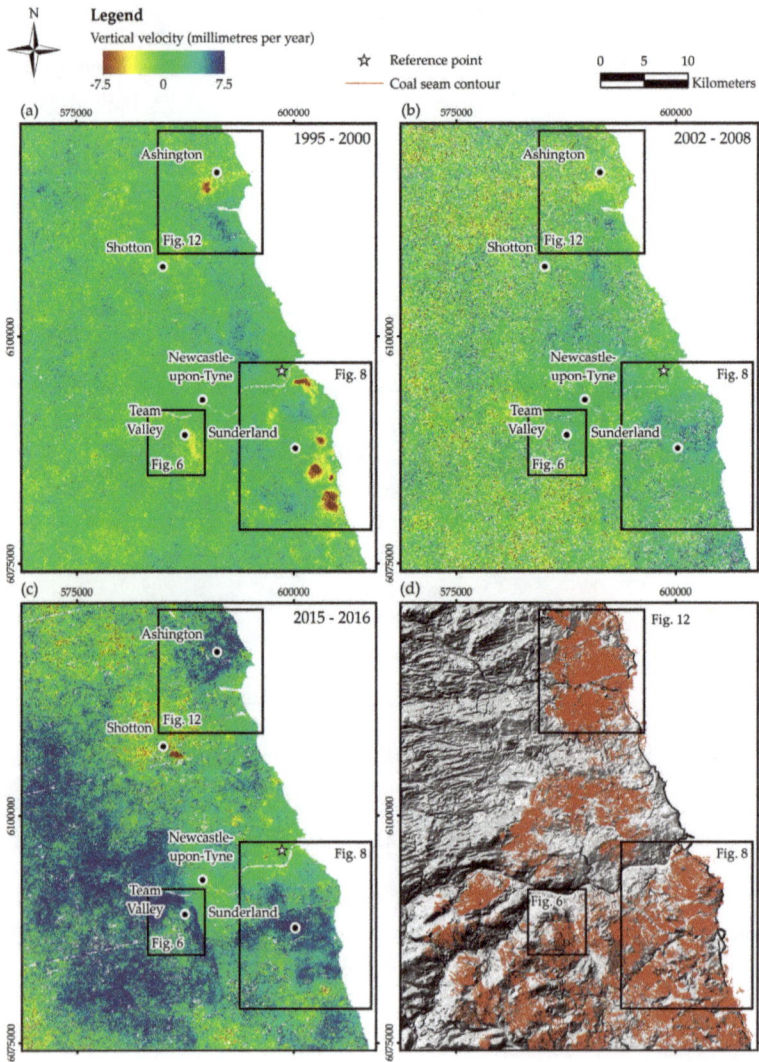

Figure 5. ISBAS average vertical velocities (millimetres per year): (**a**) ERS (**b**) ENVISAT (**c**) Sentinel-1 (**d**) Coal seam contours and topography.

4.2. Linear Velocities

Several notable areas of localised subsidence were identified along the coastline in the ERS analysis, while uplift occurred in the northern half of the AOI and approximately 5 km inland in the south (Figure 5). Uplift occurs over larger regional areas, while subsidence is localised and of a greater magnitude. There has been a significant change in the spatial patterns of motion between the ERS and ENVISAT data sets; areas of localised subsidence present in the ERS data have stabilised and uplift in the north has reduced in velocity. The results achieved in the ERS and ENVISAT analysis compare well to results produced by FugroNPA using PSI [21]. These were supplied for the Terrafirma Product Interpretation Report: Northumberland (ESRIN/Contract No. 4000101274/10/I-AM) [22] commissioned by the Coal Authority and confirm the findings of a distinct change in velocities between data sets. The Sentinel-1 analysis shows that the trend of regional uplift has strengthened. Uplift occurring around Sunderland during the ENVISAT epoch has increased in velocity, while a large regional pattern of uplift has developed inland in the western half of the AOI.

Table 1 summarizes the statistics of the linear velocities and standard errors. It might be expected, since there are approximately 50% more Sentinel-1 acquisitions than ERS/ENVISAT, that standard errors are lowest for Sentinel-1, which is not the case. However, data acquired over shorter epochs reduces the ability to accurately retrieve the underlying deformation signal. The relatively short time period of the Sentinel-1 analysis ($\approx$18 months) is reflected in the amount of noise present in the linear velocities and the standard errors. Atmospheric signal, a prevailing InSAR error source, can be accurately removed via the stacking of numerous acquisitions to retrieve the underlying deformation based upon the assumption that atmospheric delay is random over time. The stratified troposphere is characterized by temporal variations, principally due to precipitation. Interferometric signals, therefore, have a seasonal component which can be aliased by the temporal sampling of SAR acquisitions [36,37]. Furthermore, the TOPS configuration of Sentinel-1 products generates additional difficulties, with respect to stripmap products, and requires a greater amount of acquisitions to accurately sample the atmosphere since constitutive bursts are imaged through different section of the atmosphere [33].

Table 1. A comparison of the statistics of ERS, ENVISAT and Sentinel-1 ISBAS analyses over the Northumberland AOI.

Statistic	ERS SAR	ENVISAT ASAR	Sentinel-1 SAR
Velocity (Millmetres/Year)			
Mean	0.9	0.7	3.5
Minimum	−14.0	−14.5	−16.0
Maximum	10.6	14.6	19.0
Standard Deviation	1.9	2.5	3.4
Standard Error (Millimetres)			
Mean	1.6	2.4	2.2
Minimum	0.5	0.6	0.8
Maximum	4.0	8.0	4.9
Standard Deviation	0.4	0.8	0.7

5. Discussion

As mentioned above, the ground motion results reveal large regional extents of uplift with localized hotspots of subsidence. The following section discusses the link between the observed patterns of ground motion and the geology, history of coal mining and the hydrogeology of the Northumberland region.

5.1. Relationships between Ground Motion and Geology

5.1.1. Bedrock Geology

A number of potential relationships between the velocities and the geology are apparent. However, on the whole, there appears to be no correlation between any of the ERS, ENVISAT or Sentinel-1 velocities and the bedrock geology, with no evidence to indicate that specific geological units are responsible for the observed surface motion.

5.1.2. Superficial Deposits

In general, there is little correlation between any ground motion and the distribution or thickness of superficial deposits, particularly compressible deposits and those with potential for shrink-swell. The exception, however, is in the area of Team Valley, where the land surface changes from subsidence to stability to uplift over the course of the ERS, ENVISAT and Sentinel-1 analyses (Figure 6). Team Valley is a deep buried valley which characterizes the pre-glacial channel of the River Wear. The valley extends downward through Coal Measures to −46 m above ordnance datum and is infilled with an interlensing complex of glacial clays, laminated-clays, silts and sands, overlain by alluvium [38]. Subsidence identified in the ERS data is situated within the deepest parts of the channel, while the glacial till at the margins are stable. Subsidence could be attributed to progressive compaction as the area coincides with an extensive area of industrial development in the Team Valley and Gateshead Metro Centre. Large areas are covered with artificial surfaces sealing the subsurface from rainfall recharge; it is plausible that de-watering of Quarternary sediments in combination with loading, albeit from comparatively light-weight structures, could be significant in creating subsidence.

The ENVISAT data reveals that areas experiencing subsidence in the 1990s, have stabilized, with a small degree of uplift identified further north in the valley. Sentinel-1 velocities indicate this trend has strengthened with strong uplift in the Team Valley. The pumping of mine water within Team Valley could have significant effects on the land surface as a known programme of ground water control is present in this part of the former coalfield.

5.1.3. Geological Structure

When comparing derived velocities to the geological structure, a number of cases emerge for which patterns of motion appear to be associated with major geological faults (Figures 7 and 8). In such cases there is no indication that the faults are directly responsible for the motion, but influence motion by accommodating or constraining it. The constraining of motion appears to be related to both regional patterns of uplift, as well as localised subsidence. Examples of such a relationship in the surrounding areas of Sunderland and Ashington are discussed in more detail in Section 5.4.

5.2. Relationships between Ground Motion and Coal Mining

Subsidence is an inevitable consequence of mining activity [28]. It has been approximated that 4800 million tonnes of coal has been extracted from UK mines, consequently leaving 1000 million cubic metres of voids [39]. Five mines in the AOI were active at some point during the study period; four opencast and one deep mine, Ellington. Ellington was closed in 2005 [40] but there is no spatially correlated subsidence associated with this mine as would be expected since the workings are located under the North Sea (Figure 5). One localized hotspot of deformation is revealed in the Sentinel-1 analysis, corresponding to Shotton opencast coal mine, which was operational throughout the period of Sentinel-1 SAR acquisitions (Figure 5c). A number of potential relationships between ground motion and coal mining were examined including: mine entries; coal seam contours and underground workings.

Within the AOI there are a total 7811 mines entries, the majority of which exist in the Pennine Middle Coal Measures Formation with far fewer located in the Zechstein Formation where Coal Measures are deepest (Figure 2). Areas of subsidence present in the ERS analysis are in proximity to entries to former collieries, for example at Westoe, Wearmouth, Ryhope in the Sunderland area (Figure 8). However, it is challenging to make correlations with this data since there are many mine entries within the AOI where subsidence is not present.

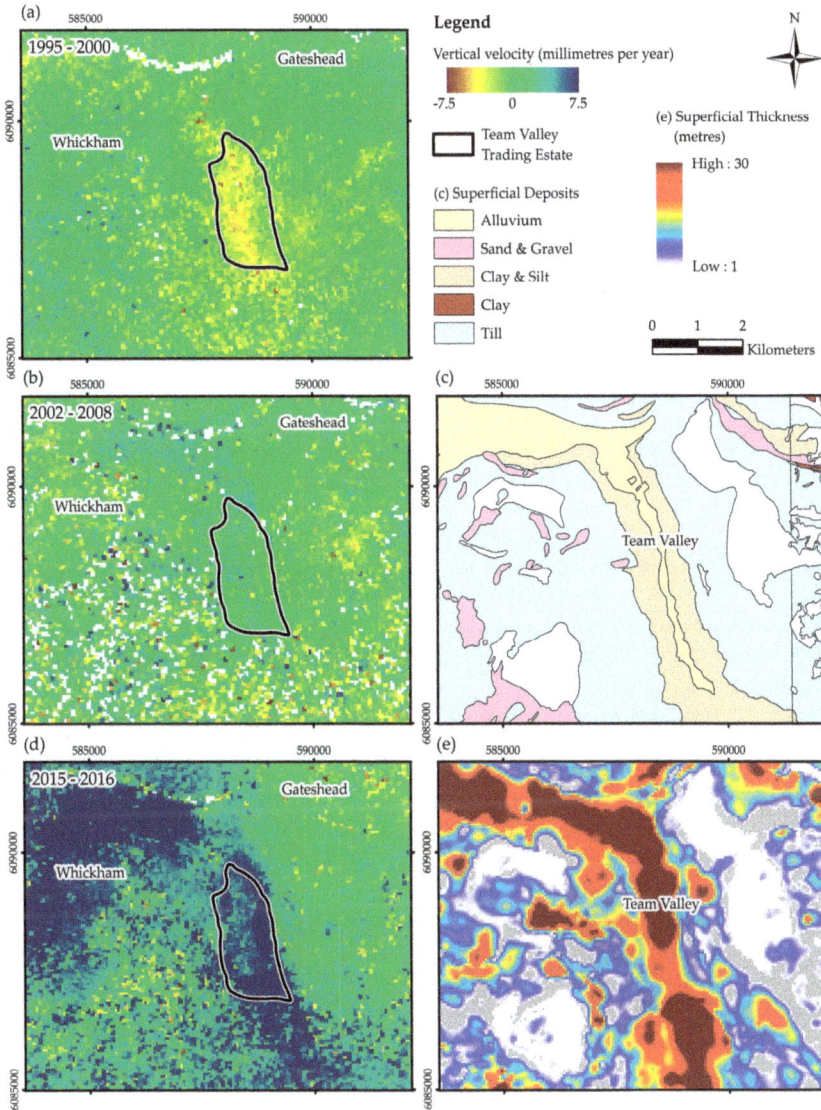

Figure 6. (**a**) ERS ISBAS vertical velocities (millimetres per year); (**b**) ENVISAT ISBAS vertical velocities (millimetres per year); (**c**) Superficial geology; (**d**) Sentinel-1 ISBAS vertical velocities (millimetres per year); (**e**) Superficial thickness. 'Reproduced with the permission of the British Geological Survey ©NERC. All rights Reserved'.

There is a relationship between areas of motion and seam contours at many locations in all three data sets (Figure 5). As faults often delimit Coal Measures it is perceptible that areas of motion are delimited by these boundaries. An example of such a relationship occurs in the surrounding area of Ashington, with evidence to suggest uplift has been caused by former mining activities (Figure 5c,d). The majority of mining and resultant ground motion would have been completed long before the satellite data was acquired, therefore subsidence associated with surface hazards or pillar collapse are localized. No relationships were observed when analysing any ground motion in conjunction with coal seams categorised by depth or working dates.

There is no distinct relationship between underground workings and ground motion in any of the SAR observations. Nonetheless, establishing relationships with these data is challenging because the majority of Coal Measures in the AOI have been worked extensively, at multiple depths by both pillar and stall and total extraction techniques. Coal mine abandonment plans were not a requisite in the United Kingdom until 1859, hence not all mining has been documented. Additionally, old mine plans are often inaccurate due to poor surveying techniques and pillar robbing [41].

5.3. Relationships between Ground Motion and Ground Water Levels

Groundwater rebound is the reasoned cause of regional-scale surface uplift [42,43]. Surface heave phenomena has previously been identified by geodetic levelling [44–46] and radar interferometry [10,13,17,18,20–22,47–51] above a number of abandoned coalfields in different geographies. Deep underground coal extraction requires the pumping of groundwater to maintain safe working conditions within the mine [52]. Systematic pumping ordinarily ceases following the abandonment of underground coal mines and groundwater levels begin to rebound. Monitoring by the Coal Authority in the Durham Coalfield clearly identifies mine water recovery, although part of the coalfield is still controlled by pumping. As the rock matrix reverts back to previously saturated conditions, pore water pressure increases and consequently the rock expands which can produce an expression at the surface [43]. Surface elevation is thought to only represent several percent of any previously identified subsidence during active mining and to occur gradually over a larger area than previous subsidence basins [53].

Broadly, coalfields can become associated with a number of stability and contamination issues following abandonment, due to prominent changes in topography and hydrogeological regime [54]. The rise of mine water and flooding of abandoned workings can have some detrimental consequences including: residual subsidence; the reactivation of existing faults; mine gas migration; slope instability; and influence the geotechnical properties of the ground [55]. Individual coalfields may be subject to some or none of the identified issues. Subsidence and slope instability have not been associated with mine water recovery in British coalfields.

Historical indications demonstrate that there remains a risk of ground instability long after pillar and stall and longwall workings have been abandoned. The delayed onset of motion can be initiated by a number of factors including creep of the overburden, pillar deterioration, collapse of the roof and failure of the floor as a result of seatearth softening. There is potentially no time limit for pillar and stall instability. Furthermore, deeper mine activity can trigger events in shallower, often unrecorded, mines [28]. The risk of surface instability is substantially higher during the transitory phase of groundwater rise, due to the redistribution of stress within the rock matrix. Risk is reduced once a new hydraulic and hydrogeological equilibrium is reached, following saturation to historical levels [53]. In most cases in the UK, the pre mining hydrogeology is not known but it is assumed rebound is to the pre-mining levels.

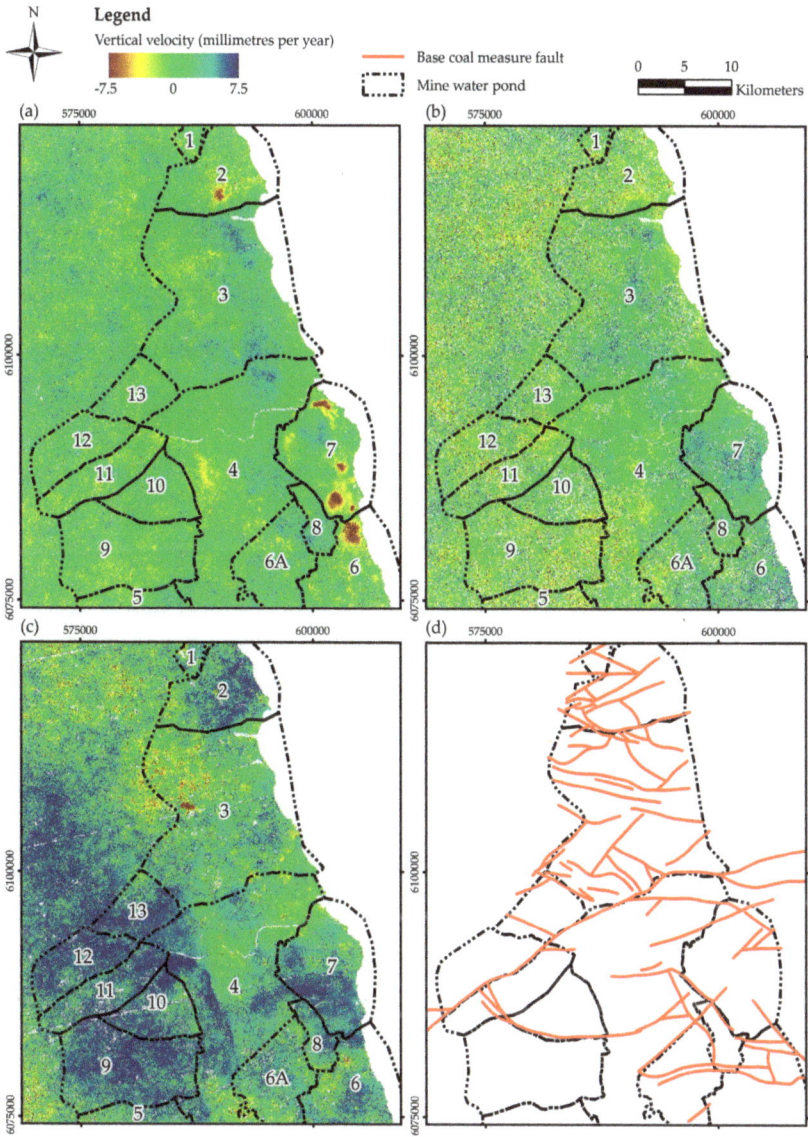

Figure 7. ISBAS vertical velocities (millimetres per year) with minewater ponds overlaid: (**a**) ERS (**b**) ENVISAT (**c**) Sentinel-1; (**d**) Minewater ponds and base coal measure faults.

Rising groundwater creates a complex subsurface environment. Water ingress into former workings does provide some support pressure inside the mined voids. Ingress is aided by the disturbance, fracturing and disintegration of roof layers and overlying rock mass during mine operation, consequently permeability and porosity are increased [44,53]. However, Coal Measures rocks can lose significant strength on water saturation as a result of mineralogical changes to the clays within the rock matrix, stress corrosion or water absorption [56] and has been associated with pillar failure in within former mine workings [57,58]. Additionally, the chance of failure can increase as the saturation of soil within the overburden increases vertical stress on the strata [44].

Fault reactivation is a further consequence of rising minewater [59]. The incidence of fluids within fault zones have been shown to induce seismicity [60,61]. An increase in fluid pore pressure reduces fault plane shear strength. Reactivation can occur if the reduction in shear strength counterpoises the normal stress acting across the fault [62]. Since Coal Measures sandstones in the Durham coalfield possess very little intergranular permeability, groundwater movement is principally through fractures in the sandstone [63]. Mining has extensively disrupted and complicated the hydrogeology and increased permeability. Rising groundwater levels have previously been associated with incidences of fault reactivation in the Durham Coalfield [64].

In the area of study, rather than abruptly turn off all pumps in the abandoned coalfield, the Coal Authority have continued to pump minewater from a number of sites as part of a strategy to control the rise of water levels. Since mining was operational there has been a reduction in the number of operational pumping sites and, consequently, groundwater levels in some eastern parts of the coalfield have recovered to levels close to the base of the Permian rocks. However, in more recent years ($\approx$10 years) the number of operational sites may have increased; active management of groundwater is important so to prevent potential contamination of important aquifers within these rocks. For example, it is known that approximately 4 km south of the study area that groundwater pumping began at a new facility at the former Horden Colliery in July 2004 [65].

Newcastle University, the Environment Agency and Coal Authority have conducted previous work over the AOI to better understand the hydrogeological regime and the effects of altering pumping regimes following mine closure [21,22,66–70]. Modelling studies have previously attempted to predict minewater rebound. The coalfield is divided into "ponds", considered single hydraulic units which homogenously rise or fall principally depending on ground water pumping. Some of these ponds are connected (hydraulically) to other ponds, whilst some are not. Some of the connections between ponds are unclear and can change over time. The connections can be at different depths, hence some ponds can become joined or isolated with groundwater level changes. From the information made available, the location of these ponds appears to be controlled in many cases by the geological structure, specifically by geological faults (Figure 7). The ponds, which cover a greater area than the AOI, are thought to refill in increasing numerical order defined in Figure 7.

The filling of ponds in numerical order is largely supported by the Sentinel-1 data (Figure 7c). The rates of uplift have increased over time in ponds 4–13. Pond "2" does not support this premise, experiencing strong uplift in the Sentinel-1 analysis; however, it does support the proposition of mine water ponds moving homogenously, with the extents of motion defined by the pond. One potential contributory factor could be the accuracy of the divisions between ponds. For example, pond "4", which extends from Newcastle to Durham, is considered to be sub-divided into further blocks. There are three major pumping stations within this block, the relationships between these pumping stations are complex and not well understood. Surface motion in the Sentinel-1 data supports the premise that pond "4" might be further divided into moving blocks. The results show InSAR can provide new information to help determine these boundaries.

For further scrutiny, comparisons are made in the following section between groundwater levels and ground motion at specific monitoring sites.

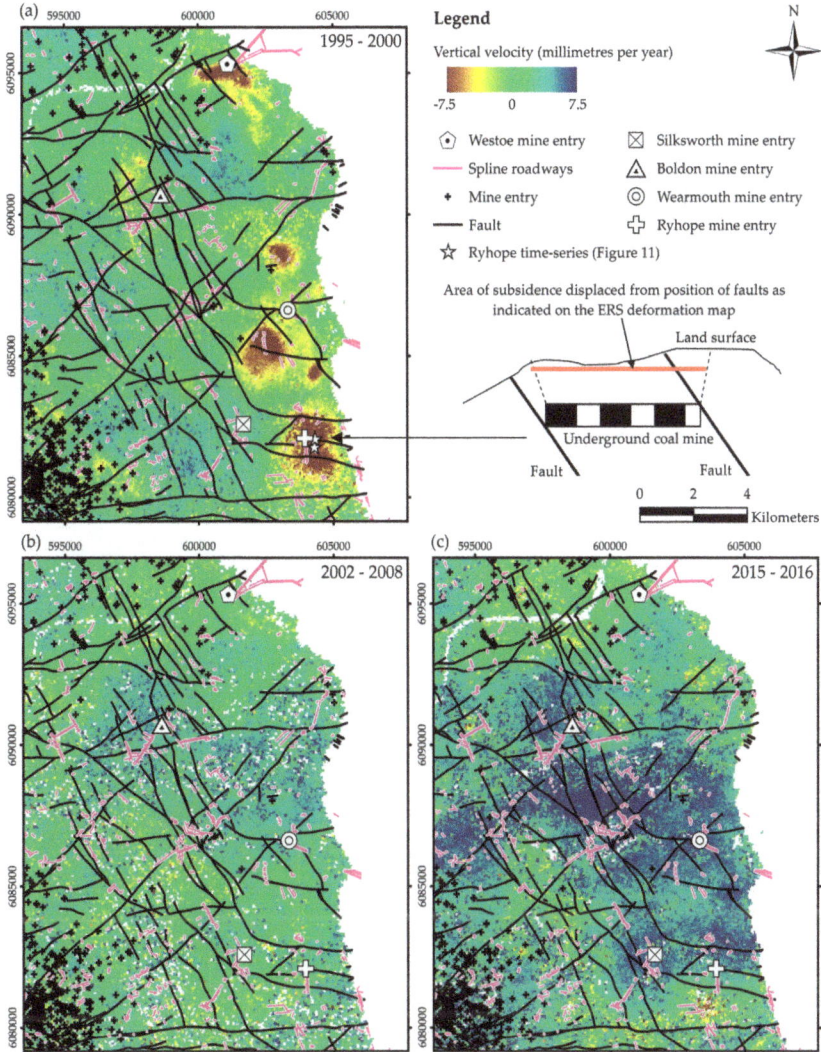

Figure 8. ISBAS vertical velocities (millimetres per year) in the surround area of Sunderland with faults, mine entries and spline roadways overlaid: (**a**) ERS (**b**) ENVISAT (**c**) Sentinel-1.

5.4. Case Studies

5.4.1. Sunderland

It was not until 1788, as exploration methods reached greater depths, that serious prospecting began in the Durham coastal limestone. Wearmouth colliery struck coal in 1834 at 480 m below sea level, making it the deepest coal mine in the world at the time. The Wearmouth Coal Company subsequently undermined most of the Sunderland area and in 1912 were calculating how much they could extract from underneath churches, hospitals, schools and St George's square before subsidence claims offset profits. In 1958, Wearmouth was identified as a 'super pit' from which 500 million tonnes of coal were extracted from under the North Sea, until the colliery was closed in 1993 [71]. Other smaller mines, just outside Sunderland, were later founded such as Ryhope (1859–1966), Boldon (1869–1982), Silksworth (1873–1971) and Westoe (1909–1993) [40].

The ERS analysis indicates a complex environment, identifying hotspots of deformation as well as areas of uplift (Figure 8). The most prominent subsidence identified in the ERS analysis spatially correlates to the collieries at Westoe, Wearmouth and Ryhope. Westoe and Wearmouth collieries shut in 1993, shortly before the period of ERS imagery (1995–2000), while, curiously, Ryhope closed in 1966, nearly four decades before the ERS analysis. The later ENVISAT data indicates a reversal in the velocities, with uplift present in some areas where subsidence was previously identified. By 2015–2016, the Sentinel-1 data reveal that uplift has both strengthened and spread out over a greater area.

Neighbouring time-series and linear velocities from ENVISAT and Sentinel-1 data correlate with rising groundwater monitored at Westoe, Wearmouth and Boldon, as shown in Figures 9 and 10. The ENVISAT time series are the noisiest due to the reduced amount of data acquisitions. There is a lack of correlation at Westoe and Wearmouth during the ERS time period where localised subsidence is identified; no groundwater data is available at Ryhope. From data provided by the Coal Authority and Harrison et al. (1989) [72] it is known that groundwater levels were rising sharply, through former workings, at Westoe and Wearmouth in the 1990s (Figures 9 and 10). Levels rose approximately 240 m at Westoe during this decade and approximately 315 m at Wearmouth between 1989 and 2002. It would be expected that if there is a collapse, for example caused by pillar failure, that strong localised subsidence will dominate the deformation signal over any moderate uplift caused by a rise in groundwater.

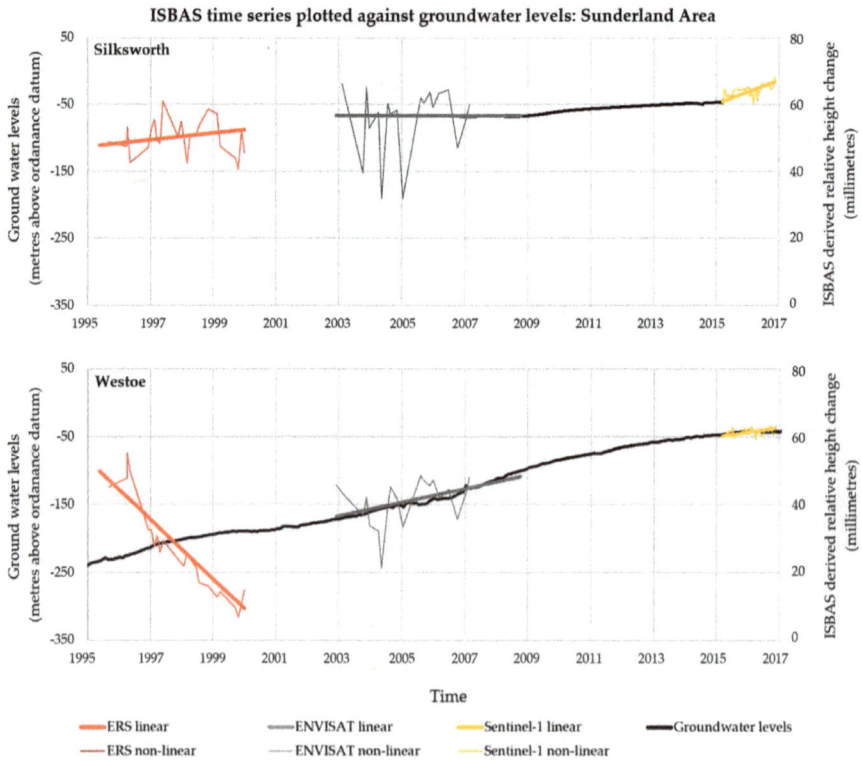

Figure 9. Groundwater levels plotted against neighbouring ISBAS time-series and linear velocities (derived separately) at Silksworth and Westoe, the locations of which are marked on Figure 8.

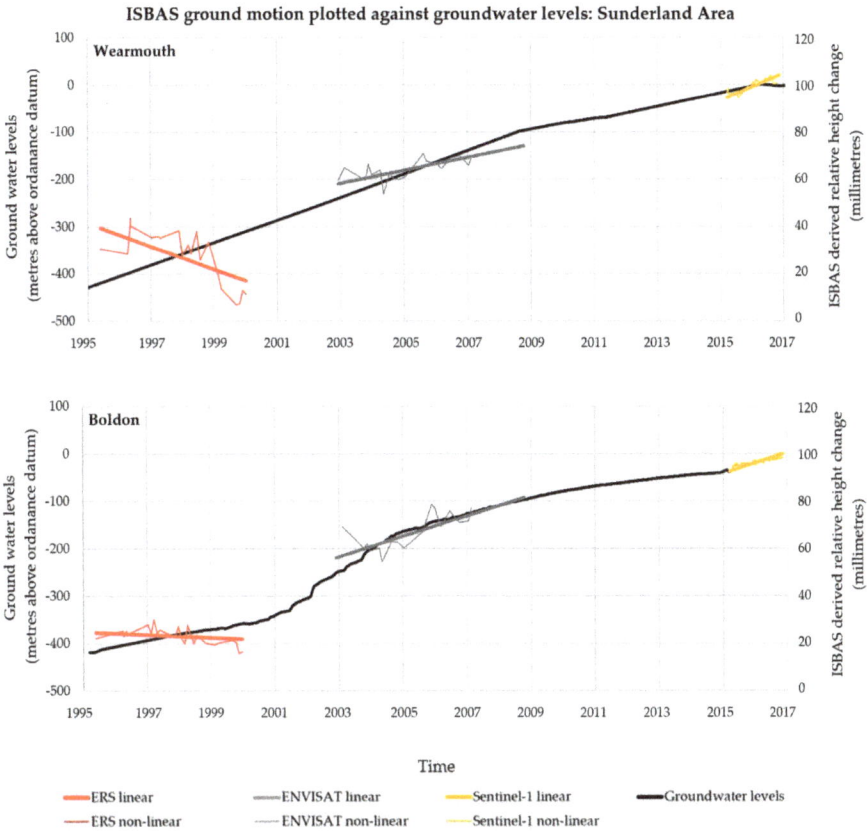

Figure 10. Groundwater levels plotted against neighbouring ISBAS time-series and linear velocities (derived separately) at Wearmouth and Boldon, the locations of which are marked on Figure 8.

Between 1998 and 1999, 20 earth tremors in the Ryhope area were reported over a 12 month period, with local residents from 14 different streets reporting vibrations [19]. Following an inconclusive investigation into the cause, four possible rationale were put forward: neotectonic processes; collapse of sub-surface cavities in the Magnesian limestone that overlie the Coal Measures; residual mining subsidence or rising mine water; the displacement of mine gases along the fault zone [73]. ERS time-series at Ryhope indicate an acceleration in the subsidence from July 1998, shortly before the period when the tremors were felt as shown in Figure 11.

A geological justification for the apparent offset of subsidence to faults thought to be controlling the motion is given in Figure 8. Subsidence at Ryhope, for example, appears bound by two east-west trending faults, albeit displaced slightly to the south with respect to the faults. The apparent offset of the subsidence to the bounding faults is likely due to the dip of the faults, with the assumed southward dipping fault planes intersecting the rockhead (or ground surface) at a more northerly point than that at which it confines the coal seam being mined at depth.

ERS time-series at Ryhope

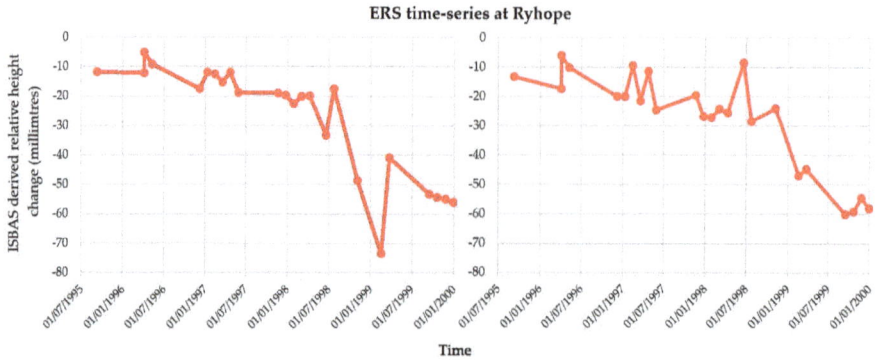

Figure 11. Selected ERS ISBAS time-series in the Ryhope subsidence depression, marked on Figure 8.

5.4.2. Ashington

The Ashington area has been extensively mined since the Bothal mining shaft was sunk in 1867, situated just over 3 km to the west of the Ashington shafts. Collieries quickly developed in the Ashington area remaining active until the rapid decline of the UK coal industry in the 1970s and 1980s, including: Ashington (1867–1986); Ellington (1905–2005); Woodhorn (1894–1981); Bates (1932–1986) and Bedlington (1838–1971) [40].

The land surface around Ashington is highly dynamic, where patterns of motion change significantly over the sequence of satellite acquisitions (Figure 12). During the ERS period of observations a localised hotspot of deformation spatially correlates with Ashington colliery, where seven mine entry shafts are connected to areas of underground workings via spline roadways. Groundwater levels at Ashington rose 30 m through former workings during this period of collapse. Further south, a large area of uplift is present, bound to the north by the ENE to WSW Stakeford fault, where differential motion is present over the fault. The southern tip of this area of uplift also appears delimited by faults. Subsidence occurring over Ashington has reduced during the ENVISAT period of observations, appearing less spatially correlated and distinct. To the south uplift is still present but the size and velocity has reduced. The Sentinel-1 analysis indicates the region to the north of the Stakeford fault, is homogenously uplifting. Similarly to the temporal evolution of motion occurring around Sunderland, uplift in the Sentinel-1 analysis occurs over a broader area than previously identified subsidence. Motion is constrained by faults, most notably to the south by the Stakeford fault, and uplift correlates strongly with mined Coal Measures. South of the Stakeford fault, uplift has ceased, with the ground surface reaching approximate equilibrium.

Neighbouring time-series and linear velocities were compared against groundwater levels at former collieries to the south of the Stakeford fault, at Bates and Bedlington, and to the north, at Ellington and Woodhorn (Figure 13). These areas belong to separate mining blocks which the Stakeford fault delimits; there are no known connections through this boundary. At Bates and Bedlington, rising groundwater during the 1990s correlates with upward surface motion, a relationship which continues as groundwater levels return to the level of the ordnance datum in the early 2000s and the surface reaches equilibrium.

Subsidence had previously been identified at Bedlington, which occurred between 1990 and 1994 and produced a subsidence trough with a maximum settlement of 260 mm at the centre [74]. This resulted in the demolishing of Cloverdale terrace due to subsidence damage. Borehole investigations took place in 1994 and 1997; less than half of those drilled that were expected to encounter coal did. The boreholes identified much evidence of roof collapse and suggested that the pillars had deteriorated badly with time or had been reworked on retreat. It is thought areal subsidence was caused by small scale pillar failure, here, and at a similar nearby case study [75]. Mining at Bedlington ceased in 1971, long before the identified motion. From the groundwater levels provided by the Coal Authority and those published in Harrison et al. (1989) [72] it is known that groundwater levels rose approximately 150 m during the course of the 1990s; uplift is present during the ERS analysis in the latter half of the 1990s (Figure 12a).

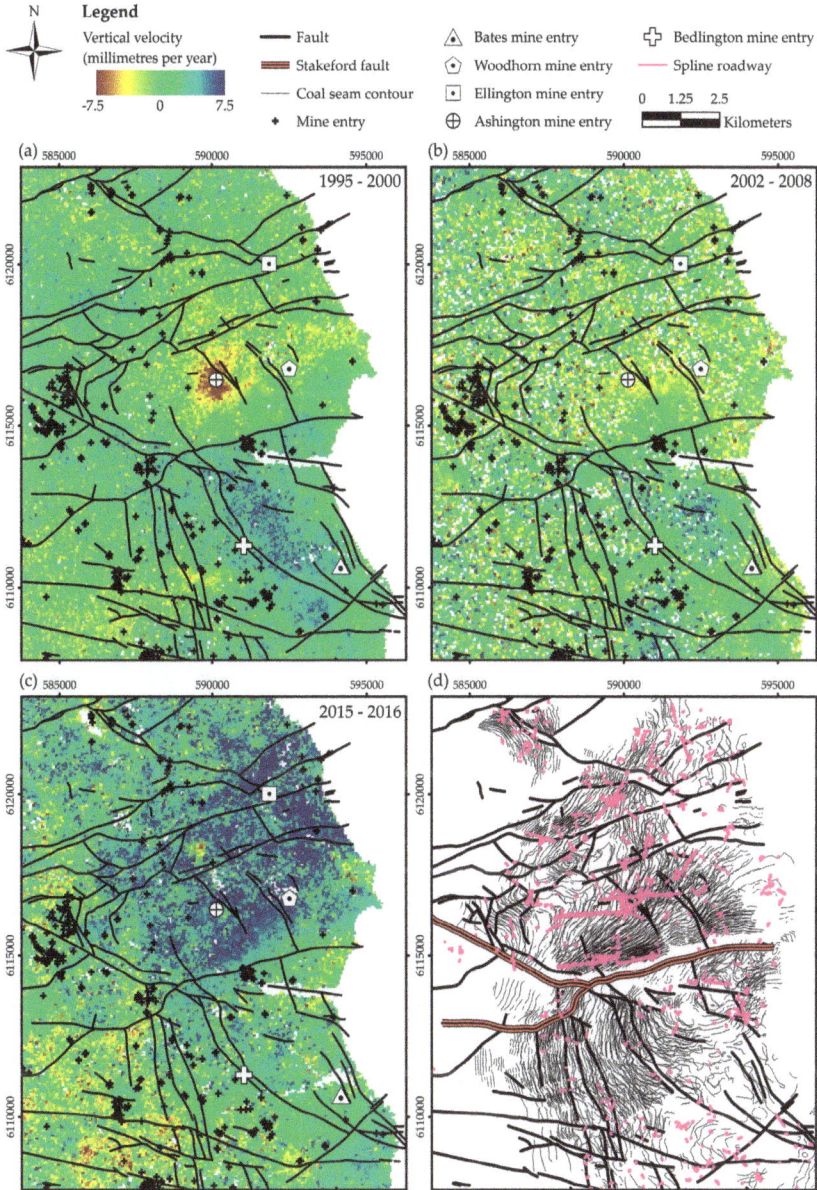

Figure 12. ISBAS vertical velocities (millimetres per year) in the surrounding area of Ashington with faults and mine entries overlaid: (**a**) ERS; (**b**) ENVISAT; (**c**) Sentinel-1; (**d**) Coal seam contours, faults and spline roadways.

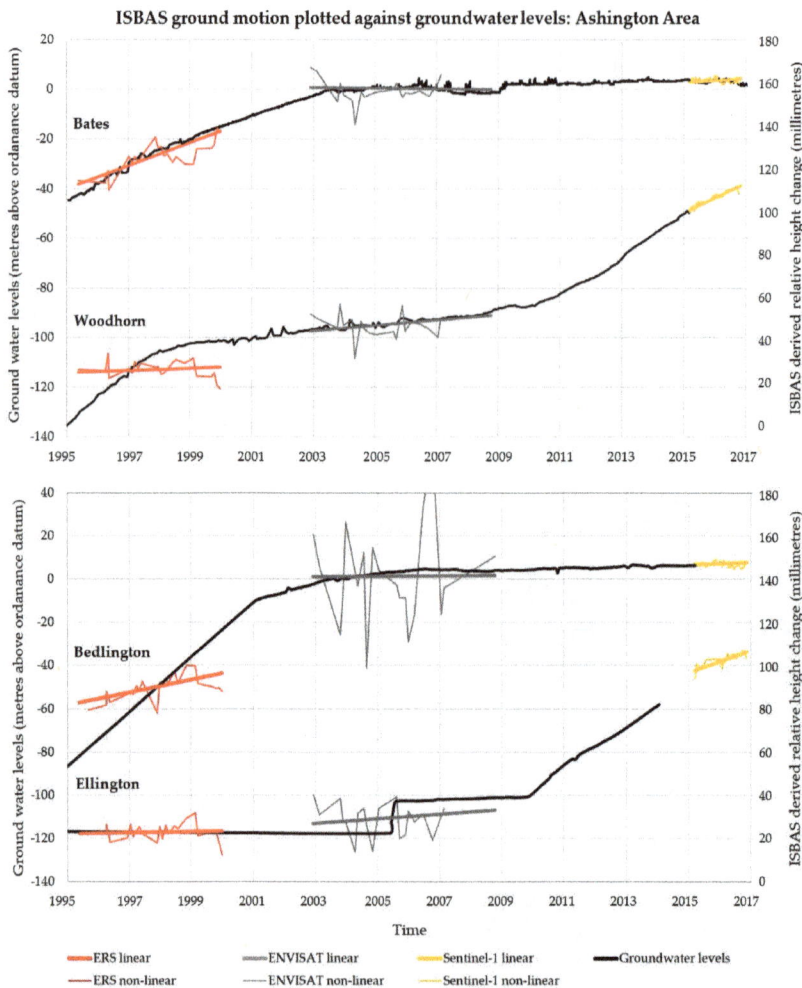

Figure 13. Groundwater levels plotted against neighbouring ISBAS time-series and linear velocities (derived separately) at Bates, Woodhorn, Bedlington and Ellington, the locations of which are marked on Figure 12.

At Ellington and Woodhorn the relationship between rising groundwater and surface velocities continues as strong uplift is identified in the Sentinel-1 data (Figure 13). The rate of groundwater recovery increases after approximately 2010, which is reflected in surface uplift in the Sentinel-1 analysis. Although groundwater data at these locations do not continue over the period of Sentinel-1 observations, on the assumption that levels continue to rebound as they have over the previous five years, a correlation between rising groundwater and the surface is present. Relative stability is identified at Ellington and Woodhorn during the ERS and ENVISAT observations.

Curiously, groundwater levels were rising at Woodhorn during the ERS time period, however the surface was relatively stable. It is, therefore, too simplistic to suggest surface uplift will occur in the surrounding locations where groundwater levels are rising. The volume of void space present in the coalfield could be one plausible explanation for the where there is no correlation. A rise in groundwater might only influence the surface when it is confined within the mine, i.e., once the void space in the mine has been filled, groundwater has reached the roof and is rising through the overburden. The depth to groundwater could also be a consideration; for example, it might be expected that a rise of 10 m in groundwater has less of an impact on the surface at a depth of −400 m AOD, than it will at −30 m AOD. Further research might also consider the potential effects aquifer groundwater levels could have on surface motion.

It should be noted that making a direct correlation between groundwater levels and deformation time-series here is challenging; at the majority of the sites investigated in this study groundwater measurements do not cover the epoch of InSAR acquisitions with sufficient frequency, particularly for ERS and Sentinel-1. For example, at Bedlington there is over a ten year gap (1989 and 2001) between groundwater measurements and only three out of eight sites investigated cover the entire Sentinel-1 period. At Bates, there is sufficient data to cover the all three InSAR acquisition periods; surface deformation is plotted against groundwater levels for the ERS time period (1995–2000) in Figure 14. Here, the coefficient of determination demonstrates that ≈65% of surface deformation is predicted by a rise in groundwater and that under these geological conditions a rise of 10m in groundwater induces surface uplift of 8mm.

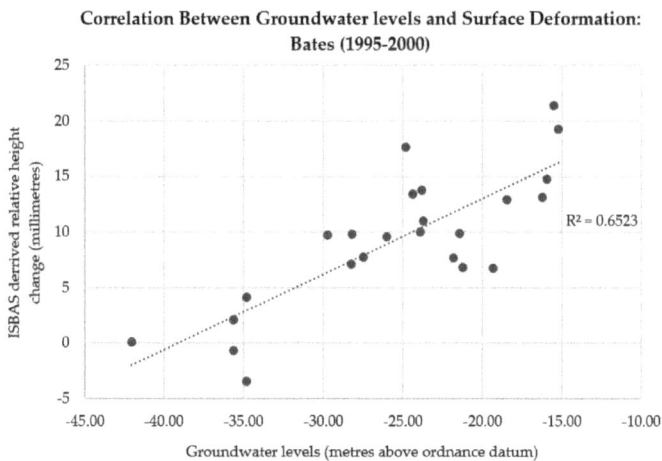

Figure 14. A linear regression between groundwater levels and surface deformation at Bates (Figure 12) for the ERS time period.

6. Conclusions

This study was undertaken to better delineate ground motion in the Northumberland and Durham coalfield over a 20-year time period. The ISBAS algorithm employed provides ground motion measurements that are consistent with those previously obtained using persistent scatterers interferometry on ERS and ENVISAT data, although the ISBAS approach provides a more complete spatial coverage of measurements.

The results show that unexpected amounts of ground motion are present in areas of former mining, occurring long after coal extraction has ceased. Localized subsidence has been identified over former mines during the late 1990s, with subsidence hotspots identified in proximity to former collieries. Regional patterns of uplift are present in all three data sets occurring over a larger spatial extent than any previously observed subsidence, often correlating with mined Coal Measures. An association was ascertained in proximity to former collieries that rising groundwater can be significant in influencing surface motion. The degree of influence rising mine water has on the surface requires further scrutiny in different geographies.

This is important for the post-mining environment, and demonstrates that InSAR can provide a cost-effective regional outlook for making informed decisions to mitigate detrimental environmental effects in abandoned coalfields. The data shows how multifaceted post-closure ground deformation regimes can be with a complex interaction between the local stress field, regional groundwater recovery and accommodation and delimiting of motion by faults.

Acknowledgments: The work conducted here has been funded by the GeoEnergy Research Centre (GERC) and Geomatic Ventures Limited. The authors would like to acknowledge ESA for the provision of ERS, ENVISAT and Sentinel-1 SAR data; the Coal Authority for mining and hydrogeological data; the British Geological Survey for geological data and Chris Satterley for his role in the co-ordination of the study. Alessandro Novellino, Luke Bateson and Francesca Cigna publish with the permission of the Executive Director of BGS.

Author Contributions: David Gee, Andrew Sowter, Alessandro Novellino, Luke Bateson, Stephen Grebby Stuart Marsh and Chris Satterley conceived and designed the experiments; David Gee performed the experiments; all the authors analysed the data; David Gee, Luke Bateson and Stephen Grebby wrote the paper.

Conflicts of Interest: The authors declare no conflict of interest.

Appendix A

ERS SAR image dates and perpendicular baselines $B_\perp$ (m) in reference to the master image on 20 November 1997.

Image Date	$B_\perp$ (m)	Image Date	$B_\perp$ (m)
20 November 1997	0	29 January 1998	−431
24 May 1995	−610	5 March 1998	−943
7 September 1995	−1361	9 April 1998	−674
3 April 1996	−634	18 June 1998	302
4 April 1996	−689	23 July 1998	−852
8 May 1996	133	5 November 1998	449
9 May 1996	58	18 February 1999	606
9 January 1997	−483	25 March 1999	−790
13 February 1997	−821	16 September 1999	−727
20 March 1997	−470	21 October 1999	−639
24 April 1997	−870	25 November 1999	−240
29 May 1997	−649	30 December 1999	−172
25 December 1997	−499		

Appendix B

ENVISAT ASAR image dates and perpendicular baselines $B_\perp$ (m) in reference to the master image on 11 January 2005.

Image Date	$B_\perp$ (m)	Image Date	$B_\perp$ (m)
11 January 2005	0	9 August 2005	627
3 December 2002	1050	13 September 2005	437
11 February 2003	500	22 November 2005	991
14 October 2003	1336	27 December 2005	1023
18 November 2003	−179	7 March 2006	850
23 December 2003	762	20 June 2006	744
6 April 2004	1446	3 October 2006	−33
11 May 2004	179	16 January 2007	1284
20 July 2004	857	20 February 2007	741
24 August 2004	759	7 October 2008	629
2 November 2004	794		

Appendix C

Sentinel-1 IW SAR image dates and perpendicular baselines $B_\perp$ (m) at scene centre in reference to the master image on 25 January 2006.

Image Date	$B_\perp$ (m)	Image Date	$B_\perp$ (m)
25 January 2016	0	13 January 2016	−58
7 March 2015	−101	6 February 2016	−6
19 March 2015	−35	18 February 2016	−73
31 March 2015	−41	1 March 2016	−182
12 April 2015	−81	13 March 2016	−153
24 April 2015	−103	25 March 2016	−97
6 May 2015	−62	6 April 2016	−40
18 May 2015	−55	18 April 2016	−96
30 May 2015	−127	30 April 2016	−136
11 June 2015	−138	12 May 2016	−193
23 June 2015	−178	5 June 2016	−43
5 July 2015	−104	29 June 2016	−143
17 July 2015	−27	11 July 2016	−152
14 November 2015	−62	23 July 2016	−69
26 November 2015	37	16 August 2016	−74
8 December 2015	−21	9 September 2016	−158
20 December 2015	−77	21 September 2016	−166
1 January 2016	−139	15 October 2016	−6

References

1. Rosen, P.A.; Hensley, S.; Joughin, I.R.; Li, F.K.; Madsen, S.N.; Rodriguez, E.; Goldstein, R.M. Synthetic aperture radar interferometry. *Proc. IEEE* **2000**, *88*, 333–382. [CrossRef]
2. Massonnet, D.; Feigl, K.L. Radar interferometry and its application to changes in the Earth's surface. *Rev. Geophys.* **1998**, *36*, 441–500. [CrossRef]
3. Hooper, A.; Bekaert, D.; Spaans, K.; Arıkan, M. Recent advances in SAR interferometry time series analysis for measuring crustal deformation. *Tectonophysics* **2012**, *514*, 1–3. [CrossRef]
4. Colesanti, C.; Mouelic, S.L.; Bennani, M.; Raucoules, D.; Carnec, C.; Ferretti, A. Detection of mining related ground instabilities using the Permanent Scatterers technique—A case study in the east of France. *Int. J. Remote Sens.* **2005**, *26*, 201–207. [CrossRef]
5. Herrera, G.; Tomás, R.; López-Sánchez, J.M.; Delgado, J.; Mallorqui, J.J.; Duque, S.; Mulas, J. Advanced DInSAR analysis on mining areas: La Union case study (Murcia, SE Spain). *Eng. Geol.* **2007**, *90*, 148–159. [CrossRef]
6. Du, Z.; Ge, L.; Li, X.; Ng, A.H. Subsidence monitoring over the Southern Coalfield, Australia using both L-Band and C-Band SAR time series analysis. *Remote Sens.* **2016**, *8*, 543. [CrossRef]
7. Yang, Z.; Li, Z.; Zhu, J.; Yi, H.; Hu, J.; Feng, G. Deriving dynamic subsidence of coal mining areas using InSAR and logistic model. *Remote Sens.* **2017**, *9*, 125. [CrossRef]
8. Kratzsch, H. *Mining Subsidence Engineering*; Springer: Berlin/Heidelberg, Germany, 1983.
9. Peng, S.S.; Ma, W.; Zhong, W. *Surface Subsidence Engineering*; Society for Mining, Metallurgy, and Exploration: Littleton, CO, USA, 1992.
10. Sowter, A.; Bateson, L.; Strange, P.; Ambrose, K.; Syafiudin, M.F. DInSAR estimation of land motion using intermittent coherence with application to the South Derbyshire and Leicestershire coalfields. *Remote Sens. Lett.* **2013**, *4*, 979–987. [CrossRef]
11. Zebker, H.A.; Villasenor, J. Decorrelation in interferometric radar echoes. *IEEE Trans. Geosci. Remote Sens.* **1992**, *30*, 950–959. [CrossRef]
12. Ferretti, A.; Colesanti, C.; Perissin, D.; Prati, C.; Rocca, F. Evaluating the effect of the observation time on the distribution of SAR permanent scatterers. In Proceedings of the FRINGE 2003 Workshop, Frascati, Italy, 1–5 December 2003; pp. 1–5.

13. Cigna, F.; Sowter, A. The relationship between intermittent coherence and precision of ISBAS InSAR ground motion velocities: ERS-1/2 case studies in the UK. *Remote Sens. Environ.* **2017**. [CrossRef]

14. Berardino, P.; Fornaro, G.; Lanari, R.; Sansosti, E. A new algorithm for surface deformation monitoring based on small baseline differential SAR interferograms. *IEEE Trans. Geosci. Remote Sens.* **2002**, *40*, 2375–2383. [CrossRef]

15. Sowter, A.; Amat, M.B.; Cigna, F.; Marsh, S.; Athab, A.; Alshammari, L. Mexico City land subsidence in 2014–2015 with Sentinel-1 IW TOPS: Results using the Intermittent SBAS (ISBAS) technique. *Int. J. Appl. Earth Obs. Geoinf.* **2016**, *52*, 230–242. [CrossRef]

16. Gee, D.; Sowter, A.; Novellino, A.; Marsh, S.; Gluyas, J. Monitoring land motion due to natural gas extraction: Validation of the Intermittent SBAS (ISBAS) DInSAR algorithm over gas fields of North Holland, the Netherlands. *Mar. Pet. Geol.* **2016**, *77*, 1338–1354. [CrossRef]

17. Novellino, A.; Athab, A.D.; bin Che Amat, M.A.; Syafiudin, M.F.; Sowter, A.; Marsh, S.; Cigna, F.; Bateson, L. *Intermittent SBAS Ground Motion Analysis in Low Seismicity Areas: Case Studies in the Lancashire and Staffordshire Coalfields, UK, Seismology from SPACE: Geodetic Observations and Early Warning of Earthquakes*; Royal Astronomical Society, Burlington House: London, UK, 2014.

18. Bateson, L.; Cigna, F.; Boon, D.; Sowter, A. The application of the Intermittent SBAS (ISBAS) InSAR method to the South Wales Coalfield, UK. *Int. J. Appl. Earth Obs. Geoinf.* **2015**, *34*, 249–257. [CrossRef]

19. Donnelly, L.J. A review of coal mining induced fault reactivation in Great Britain. *Q. J. Eng. Geol. Hydrogeol.* **2006**, *39*, 5–50. [CrossRef]

20. Culshaw, M.G.; Tragheim, D.; Bateson, L.; Donnelly, L.J. Measurement of ground movements in Stoke-on-Trent (UK) using radar interferometry. In Proceedings of the 10th Congress of the International Association for Engineering Geology and the Environment, IAEG2006, Nottingham, UK, 6–10 September 2006; Geological Society: London, UK, 2006; pp. 1–10.

21. Banton, C.; Bateson, L.; Mccormack, H.; Holley, R.; Watson, I.; Burren, R.; Lawrence, D.; Cigna, F. Monitoring post-closure large scale surface deformation in mining areas. In Proceedings of the Mine Closure 2013, Eighth International Conference on Mine Closure 2013, Australian Centre for Geomechanics, Perth Eden Project, Cornwall, UK, 18–20 September 2013.

22. Bateson, L.; Lawrence, D. Terrafirma Product: Interpretation Report. V1. 1: Northumberland. Available online: http://nora.nerc.ac.uk/21035/1/OR12054.pdf (accessed on 8 September 2017).

23. Copernicus Land Monitoring Service. Available online: http://land.copernicus.eu/pan-european/corine-land-cover/clc-2012/view (accessed on 15 May 2017).

24. Clarke, S.M. *The Geology of NY76NW (S), Cawfields, Northumberland*; British Geological Survey Open Report, OR/07/034; British Geological Survey: Nottingham, UK, 2007; p. 28.

25. Stone, P.; Millward, D.; Young, B.; Merritt, J.W.; Clarke, S.M.; McCormac, M.; Lawrence, D.J.D. *British Regional Geology: Northern England*, 5th ed.; British Geological Survey: Nottingham, UK, 2010.

26. Lawley, R; Garcia-Bajo, M. *The National Superficial Deposit Thickness Model (Version 5)*; British Geological Survey Internal Report, OR/09/049; British Geological Survey: Nottingham, UK, 2009; p. 18.

27. Smailes, A.E. The development of the Northumberland and Durham coalfield. *Scott. Geogr. Mag.* **1935**, *51*, 201–214. [CrossRef]

28. Clarke, B.G.; Welford, M.; Hughes, D.B. The threat of abandoned mines on the stability of urban areas. In Proceedings of the 10th Congress of the International Association for Engineering Geology and the Environment, IAEG2006, Nottingham, UK, 6–10 September 2006; Geological Society: London, UK, 2006.

29. European Space Agency: SAR Image Mode. Available online: https://earth.esa.int/web/guest/missions/esa-operational-eo-missions/ers/instruments/sar/design (accessed on 15 May 2017).

30. Cigna, F.; Rawlins, B.G.; Jordan, C.J.; Sowter, A.; Evans, C. Intermittent Small Baseline Subset (ISBAS) InSAR of rural and vegetated terrain: A new method to monitor land motion applied to peatlands in Wales, UK. In Proceedings of the EGU General Assembly, Vienna, Austria, 27 April–2 May 2014.

31. Farr, T.G.; Rosen, P.A.; Caro, E.; Crippen, R.; Duren, R.; Hensley, S.; Kobrick, M.; Paller, M.; Rodriguez, E.; Roth, L.; et al. The shuttle radar topography mission. *Rev. Geophys.* **2007**, *45*. [CrossRef]

32. Chen, C.W.; Zebker, H.A. Two-dimensional phase unwrapping with use of statistical models for cost functions in nonlinear optimization. *J. Opt. Soc. Am. A Opt. Image Sci. Vis.* **2001**, *18*, 338–351. [CrossRef] [PubMed]

33. European Space Agency: Interferometric Wide Swath. Available online: https://sentinel.esa.int/web/sentinel/user-guides/sentinel-1-sar/acquisition-modes/interferometric-wide-swath (accessed on 15 May 2017).
34. Novellino, A.; Cigna, F.; Brahmi, M.; Sowter, A.; Bateson, L.; Marsh, S. Assessing the feasibility of a national InSAR ground deformation map of Great Britain with Sentinel-1. *Geosciences* **2017**, *7*, 19. [CrossRef]
35. Davidson, G.; Mantle, V.; Rabus, B.; Williams, D.; Geudtner, D. Implementation of TOPS mode on RADARSAT-2 in support of the Sentinel-1 mission. In Proceedings of the Living Planet Symposium, Edinburgh, UK, 9–13 Septmber 2013; pp. 1–22.
36. Jolivet, R.; Grandin, R.; Lasserre, C.; Doin, M.P.; Peltzer, G. Systematic InSAR tropospheric phase delay corrections from global meteorological reanalysis data. *Geophys. Res. Lett.* **2011**, *38*. [CrossRef]
37. González, P.J.; Fernandez, J. Error estimation in multitemporal InSAR deformation time series, with application to Lanzarote, Canary Islands. *J. Geophys. Res. Solid Earth* **2011**, *116*. [CrossRef]
38. Mills, D.A.; Holliday, D.W. *Geology of the District around Newcastle upon Tyne, Gateshead and Consett: Memoir for 1:50,000 Geological Sheet 20 (England and Wales)*; British Geological Survey: Nottingham, UK, 1998; p. 20.
39. Norton, P.J. Mine closure and associated hydrological effects on the environment: Some case studies. In *Minerals, Metals and the Environment II. Institute of Mining and Metallurgy*; Elsevier Applied Science: London, UK, 1996; pp. 263–270.
40. Durham Mining Museum: Collieries. Available online: http://www.dmm.org.uk/colliery/ (accessed on 15 May 2017).
41. Bell, F.G.; Genske, D.D. The influence of subsidence attributable to coal mining on the environment, development and restoration; some examples from Western Europe and South Africa. *Environ. Eng. Geosci.* **2001**, *7*, 81–99. [CrossRef]
42. Pôttgens, J.J. Uplift as a result of rising mine waters (in German). The Development Science and Art of Minerals Surveying. In Proceedings of the 6th International Congress of the International Society for Mine Surveying, Harrogate, UK, 9–13 September 1985; Volume 2, pp. 928–938.
43. Donnelly, L.J. A review of international cases of fault reactivation during mining subsidence and fluid abstraction. *Q. J. Eng. Geol. Hydrogeol.* **2009**, *42*, 73–94. [CrossRef]
44. Bekendam, R.F.; Pottgens, J.J. Ground movements over the coal mines of southern Limburg, The Netherlands, and their relation to rising mine waters. *IAHS Publ.-Ser. Proc. Rep.-Int. Assoc. Hydrol. Sci.* **1995**, *234*, 3–12.
45. Chen, C.T.; Hu, J.C.; Lu, C.Y.; Lee, J.C.; Chan, Y.C. Thirty-year land elevation change from subsidence to uplift following the termination of groundwater pumping and its geological implications in the Metropolitan Taipei Basin, Northern Taiwan. *Eng. Geol.* **2007**, *95*, 30–47. [CrossRef]
46. Devleeschouwer, X.; Declercq, P.Y.; Flamion, B.; Brixko, J.; Timmermans, A.; Vanneste, J. Uplift revealed by radar interferometry around Liège (Belgium): A relation with rising mining groundwater. In Proceedings of the Post-Mining Symposium, Nancy, France, 6–8 February 2008; pp. 6–8.
47. Cuenca, M.C.; Hooper, A.J.; Hanssen, R.F. Surface deformation induced by water influx in the abandoned coal mines in Limburg, The Netherlands observed by satellite radar interferometry. *J. Appl. Geophy.* **2013**, *88*, 1–11. [CrossRef]
48. Samsonov, S.; d'Oreye, N.; Smets, B. Ground deformation associated with post-mining activity at the French–German border revealed by novel InSAR time series method. *Int. J. Appl. Earth Obs. Geoinf.* **2013**, *23*, 142–154. [CrossRef]
49. Graniczny, M.; Colombo, D.; Kowalski, Z.; Przyłucka, M.; Zdanowski, A. New results on ground deformation in the Upper Silesian Coal Basin (southern Poland) obtained during the DORIS Project (EU-FP 7). *Pure Appl. Geophys.* **2015**, *172*, 3029–3042. [CrossRef]
50. Przyłucka, M.; Herrera, G.; Graniczny, M.; Colombo, D.; Béjar-Pizarro, M. Combination of conventional and advanced DInSAR to monitor very fast mining subsidence with TerraSAR-X Data: Bytom City (Poland). *Remote Sens.* **2015**, *7*, 5300–5328. [CrossRef]
51. Vervoort, A.; Declercq, P.Y. Surface movement above old coal longwalls after mine closure. *Int. J. Min. Sci. Technol.* **2017**, in press. [CrossRef]
52. Younger, P.L. Coalfield closure and the water environment in Europe. *Min. Technol.* **2002**, *111*, 201–209. [CrossRef]

53. Wojtkowiak, F.; Couillet, J.C.; Daupley, X.; Tauziede, C. Geotechnical and environmental impacts on the surface of the water rising in French underground coal mines after closure. In Proceedings of the 7th International Mine Water Association Symposium (IMWA 2000), Ustron, Poland, 11–15 September 2000; pp. 180–194.

54. Smith, F.W.; Underwood, B. Mine closure: The environmental challenge. *Min. Technol.* **2000**, *109*, 202–209. [CrossRef]

55. Donnelly, L. Investigation of Geological Hazards & Mining Risks, Gallowgate, Newcastle-upon-Tyne. In Proceedings of the 10th IAEG International Congress, (IAEG 2006), Nottingham, UK, 6–10 September 2006.

56. Poulsen, B.A.; Shen, B.; Williams, D.J.; Huddlestone-Holmes, C.; Erarslan, N.; Qin, J. Strength reduction on saturation of coal and coal measures rocks with implications for coal pillar strength. *Int. J. Rock Mech. Min. Sci.* **2014**, *71*, 41–52. [CrossRef]

57. Knott, D.L. Assessment of potential subsidence impacts from coal mining using test borings, mine maps and empirical methods. In Proceedings of the 2006 Interstate Technical Group on Abandoned Underground Mines Meeting, Rochester, NY, USA, 14–16 June 2006; pp. 1–40.

58. Castellanza, R.; Gerolymatou, E.; Nova, R. An attempt to predict the failure time of abandoned mine pillars. *Rock Mech. Rock Eng.* **2008**, *41*, 377. [CrossRef]

59. Smith, J.; Colls, J.J. Groundwater rebound in the Leicestershire Coalfield. *Water Environ. J.* **1996**, *10*, 280–289. [CrossRef]

60. Raleigh, C.B.; Healy, J.H.; Bredehoeft, J.D. An experiment in earthquake control at Rangely, Colorado. *Science* **1976**, *191*, 1230–1237. [CrossRef] [PubMed]

61. Ingebritsen, S.E.; Sanford, W.E. *Groundwater in Geologic Processes*; Cambridge University Press: Cambridge, UK, 1999.

62. Donnelly, L.J. Reactivation of geological faults during mining subsidence from 1859 to 2000 and beyond. *Min. Technol.* **2000**, *109*, 179–190. [CrossRef]

63. Younger, P.L. Hydrogeochemistry of minewaters flowing from abandoned coal workings in County Durham. *Q. J. Eng. Geol. Hydrogeol.* **1995**, *28* (Suppl. 2), S101–S113. [CrossRef]

64. Yu, M.H.; Jefferson, I.F.; Culshaw, M.G. Fault reactivation, an example of environmental impacts of groundwater rising on urban area due to previous mining activities. In Proceedings of the 11th Congress of the International Society for Rock Mechanics, Lisbon, Portugal, 9–13 July 2007.

65. Environment Agency. *Personal Communication*; Environment Agency: Bristol, UK, 2012.

66. Sherwood, J.M.; Younger, P.L. Modelling groundwater rebound after coalfield closure: An example from County Durham, UK. In Proceedings of the 5th International Mine Water Congress, University of Nottingham and IMWA, Nottingham, UK, 18–23 September 1994; pp. 767–777.

67. Younger, P.L.; Adams, R. *Predicting Mine Water Rebound: Research & Development Technical Report W179*; Environment Agency: Bristol, UK, 1999. [CrossRef]

68. Adams, R.; Younger, P.L. A strategy for modeling ground water rebound in abandoned deep mine systems. *Ground Water* **2001**, *39*, 249–261. [CrossRef] [PubMed]

69. Kortas, L.; Younger, P.L. Using the GRAM model to reconstruct the important factors in historic groundwater rebound in part of the Durham Coalfield, UK. *Mine Water Environ.* **2007**, *26*, 60–69. [CrossRef]

70. Adams, R. A Review of mine water rebound predictions from the VSS–NET model. *Mine Water Environ.* **2014**, *33*, 384–388. [CrossRef]

71. Victoria County History: Coal-Mining. Available online: https://www.victoriacountyhistory.ac.uk/sites/default/files/work-in-progress/coal-mining.pdf (accessed on 15 May 2017).

72. Harrison, R.; Scott, W.B.; Smith, T. A note on the distribution, levels and temperatures of minewaters in the Northumberland and Durham coalfield. *Q. J. Eng. Geol. Hydrogeol.* **1989**, *22*, 355–358. [CrossRef]

73. IMC Consulting Engineers Ltd. *Report on Earth Tremors at Ryhope*; The Coal Authority: Mansfield, UK, 1999.

74. Sizer, K.E.; Gill, M. Pillar failure in shallow coal mines—A recent case history. *Min. Technol.* **2000**, *109*, 146–152. [CrossRef]

75. Coal Authority. *Personal Communication*; Coal Authority: Mansfield, UK, 2017.

geosciences

MDPI

Article

Ground Stability Monitoring of Undermined and Landslide Prone Areas by Means of Sentinel-1 Multi-Temporal InSAR, Case Study from Slovakia

Richard Czikhardt [1,*], Juraj Papco [1], Matus Bakon [2], Pavel Liscak [3], Peter Ondrejka [3] and Marian Zlocha [3]

[1]   Department of Theoretical Geodesy, Faculty of Civil Engineering, Slovak University of Technology, Radlinskeho 11, 810 05 Bratislava, Slovak Republic; juraj.papco@stuba.sk
[2]   insar.sk s.r.o., Lesna 35, 080 01 Presov, Slovak Republic; matusbakon@insar.sk
[3]   State Geological Institute of Dionyz Stur, Department of Engineering Geology, Mlynska dolina 1, 841 03 Bratislava, Slovak Republic; pavel.liscak@geology.sk (P.L.); peter.ondrejka@geology.sk (P.O.); marian.zlocha@geology.sk (M.Z.)
*   Correspondence: czikhardt.richard@gmail.com or richard.czikhardt@stuba.sk; Tel.: +421-904-949-323

Received: 14 July 2017; Accepted: 8 September 2017; Published: 15 September 2017

Abstract: Multi-temporal synthetic aperture radar interferometry techniques (MT-InSAR) are nowadays a well-developed remote sensing tool for ground stability monitoring of areas afflicted by natural hazards. Its application capability has recently been emphasized by the Sentinel-1 satellite mission, providing extensive spatial coverage, regular temporal sampling and free data availability. We perform MT-InSAR analysis over the wider Upper Nitra region in Slovakia, utilizing all Sentinel-1 images acquired since November 2014 until March 2017. This region is notable for its extensive landslide susceptibility as well as intensive brown coal mining. We focus on two case studies, being impaired by recent activation of these geohazards, which caused serious damage to local structures. We incorporate a processing chain based on open-source tools, combining the current Sentinel Application Platform (SNAP) and Stanford Method for Persistent Scatterers (StaMPS) implementation. MT-InSAR results reveal substantial activity at both case studies, exceeding the annual displacement velocities of 30 mm/year. Moreover, our observations are validated and their accuracy is confirmed via comparison with ground truth data from borehole inclinometers and terrestrial levelling. Detected displacement time series provide valuable insight into the spatio-temporal evolution of corresponding deformation phenomena and are thus complementary to conventional terrestrial monitoring techniques. At the same time, they not only demonstrate the feasibility of MT-InSAR for the assessment of remediation works, but also constitute the possibility of operational monitoring and routine landslide inventory updates, regarding the free Sentinel-1 data.

Keywords: Multi-Temporal InSAR; Sentinel-1; slope deformation; landslide; land subsidence; undermining

1. Introduction

Slope deformations are the most significant geohazards in Slovakia which annually cause an extensive economic damage, seriously limit the rational use of land and in rare cases also threaten human lives. This is conditioned by the complex geological setting of Slovakia, especially in Flysch regions and periphery of Neovolcanic regions, where landslides threaten up to 60% of the territory. Slopes within Neogene and Paleogene depressions are also substantially affected. Moreover, the activation or re-activation of relatively large amounts of landslides has been witnessed in recent years. Since 2010, more than 700 new slope failures have been registered and their activation was driven mainly by climatic anomalies, such as extraordinary rainfalls and melting snow cover. Many of these landslides currently represent a direct threat to the lives, health and property of the residents in the affected areas [1]. This fact has initiated their inclusion in the monitoring system built within the scope of the project "Partial-monitoring system—geological factors" [2,3]. The stability condition of these sites is being observed by systematic geological, geotechnical (groundwater table level, borehole inclinometers), and geodetic (GNSS, levelling) measurements, as well as regular evaluation of climatic factors. All those conventional in-situ terrestrial monitoring methods, however, provide only point-wise information about temporal evolution of deformation phenomena, while requiring laborious and regular measurement campaigns. Another geohazard posing a threat for the residents of Slovakia is a subsidence due to undermining, related to the present or historical exploitation of brown coal, various ores or salts. Brown coal mining is still in operation in the Upper Nitra region in Central Slovakia, where a collapse due to undermining has led to the abandonment of two municipalities, or their parts—Laskar and Kos. Coal mining in the Upper Nitra region has been operated by the Hornonitrianske bane (HB) Prievidza Mining Company, which has also been monitoring the undermining subsidence using the classical geodetic survey at the surface and extensive underground monitoring by the methods of mining geodesy.

Monitoring of mass-wasting geohazards like landslides or land subsidence due to undermining has been marked by new perspective from remote sensing, especially thanks to satellite-borne synthetic aperture radar (SAR) interferometry (InSAR), which successfully attains geodetic precision at extended spatial coverage [4]. Multi-temporal SAR interferometry techniques (MT-InSAR) are nowadays a well-developed tool for ground stability monitoring. They aim to overcome the major limitations of conventional differential InSAR, employing the time series of SAR images to identify the radar targets, which do not change their backscattering characteristic over time and remain phase coherent. Those are referred to as persistent scatterers (PS) and generally correspond to man-made structures as well as stable natural reflectors (e.g., bare rocks or outcrops). For such points, temporal evolution of deformation can be retrieved, thus forming a natural, opportunistic geodetic network to monitor seismic activity, landslides, cities, or even individual structures [5–7]. This application capability has recently been emphasized by European Space Agency's Copernicus operational programme of the Sentinel-1 satellite mission, providing guaranteed and freely available SAR images of the entire Earth's landmass within a 6 day repeat cycle.

2. Study Area

The aim of this study is to test the feasibility of MT-InSAR technology within the Upper Nitra region, Central Slovakia, exploiting all to-date available Sentinel-1 data. The investigated territory spans approximately 440 km²—22 km in the longitudinal direction and 20 km in the latitudinal direction (Figure 1). This site is notable for its landslide susceptibility, as well as extensive coal mining activity. The mass movement processes are generated by the specific geological setting of the Vtacnik mountain range. Rigid volcanic rocks made of epiclastic volcanic conglomerates and breccias, pyroclastics and lava flows of Older to Middle Sarmatian age, form the summit parts and flanks of the mountain range [8]. They overlay softer complexes of Neogene sediments, prevailingly claystones, sandstones and siltstones, intercalated by brown coal seams. Slope deformations have a strong representation here, with almost all basic types of slope failures. In the summit parts of the mountain range, block ridges are present. Down the slope, detached blocks form block fields, and towards the Upper Nitra depression, landslides have evolved. Since Badenian complexes are known for rich deposits of brown coal, the unfavourable slope stability conditions have been impaired by underground mining, which at several sites has led to the acceleration of mass movements or to the subsidence of the territory above the mined-out underground spaces [9,10].

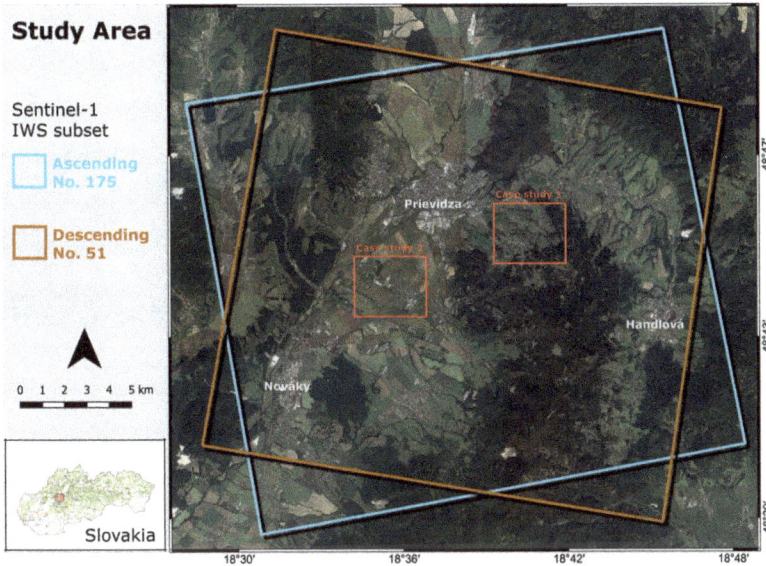

Figure 1. Area of interest (AOI) covering the Upper Nitra region, Slovakia. Basemap: Google, 2017 CNES.

We closely investigate the above-mentioned geohazards on two case studies. The first study focuses on the suburbs of Prievidza city: Hradec, Velka and Mala Lehotka (Case study 1: Figure 2). The 2012–2013 reactivation of slope deformations in Hradec and Velka Lehotka caused serious damage to local infrastructure and buildings (Figure 3), leading to a declaration of an emergency situation. Consequently, landslides survey [11,12] and remediation [13] were implemented as inevitable steps, accomplished in 2014 and followed by the establishment of a monitoring network [3]. The remedial works mostly involved underground drainage by horizontal boreholes, combined with sand piles, drainage trenches and surface drains.

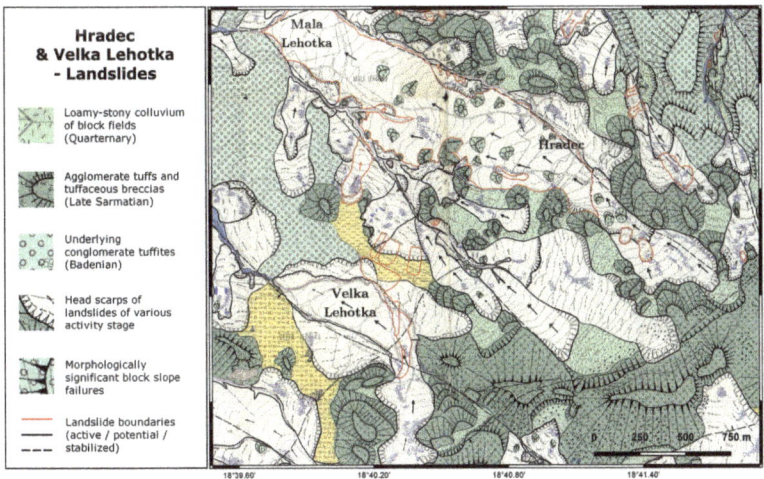

Figure 2. Case study 1: Slope deformations in Hradec, Velka a Mala Lehotka suburbs of Prievidza city, mapped by [14].

Figure 3. Photo-examples of structural damage in Hradec and Velka Lehotka (case study 1) caused by local slope deformations, (**a**) April 2012, Photo courtesy: Google Street View; (**b**) and (**c**) July 2017.

Brown coal in Upper Nitra is stored in Neogene deposits at the depth of 200–250 m with 2.5–28.4 m thickness, and other layers with 2.5–8.9 thickness are present in 120–510 m depths. Mining takes place in those depths without the fill, what is considered a technological peculiarity of this area. The large extent of long-term brown coal mining of the Cigel brown coal deposit (at a depth of around 200 m) activates surface deformations in the above area of Kos village (Case study 2: Figure 4) and spatially devastates the countryside. Since 1950, a building closure was declared in the north-western part of the Kos village, and mining began in 1988. Notable negative influences of mining are related not only to the ecosystems of the original landscape, but also to demographic changes due to the evacuation and withdrawal of former inhabitants. Population of the village has since decreased from the original 3500 to today's approximately 1200 inhabitants. The undermined areas of ≈158 ha are affected by strong subsidence movements, and the formation of several depressions (over area of ≈58 ha) is evidenced. Number of slope failures are induced over the undermined fields, which are continually subsiding during the mining activities. Therefore, the location, total area and depth of the depressions are changing dynamically over time. The predicted maximum decrease is in the range of 4–7 m with 3–5 m depth of the groundwater table level, what has resulted for example in formation of wetlands [15].

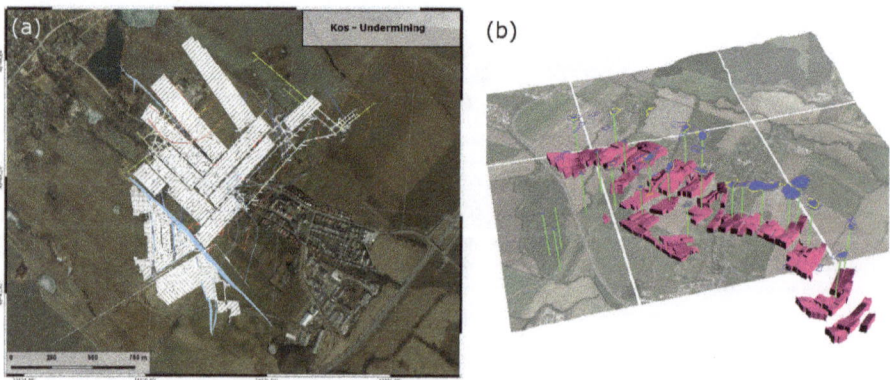

Figure 4. Case study 2: Undermined area of Kos village, Upper Nitra region. (**a**) White areas represent undermined corridors as of May 2017; (**b**) 3D model of the Cigel brown coal deposit—blue: ponds and wetlands, violet: mined out blocks (thickness exaggerated 3×) and green: perpendiculars from the lake centers.

The 3D model of the respective brown coal deposit was created on the basis of mining data of the HB Company, a.s. within the project [16]. The subsidence of the area, which was reflected in the creation of ponds, was monitored via satellite and aerial images acquired in years 1987, 2001, 2004, 2006 and 2007. In 2007, these areas were surveyed by geodetic methods, revealing that the subsidence had begun within 1–2 years after the coal seams were extracted. The extent of the worked out spaces and related sinkholes, often indicated by lakes and wetlands, is depicted in the Figure 4b.

3. Materials and Methods

3.1. Sentinel-1 Dataset

We performed MT-InSAR ground stability analysis over the wider Upper Nitra region, utilizing 150 C-band Sentinel-1 SAR images per both of its ascending (78) and descending (72) orbits, which cover the time period from November 2014 till March 2017. Figure 1 depicts subsets of Sentinel-1 IWS scenes for ascending track No. 175 and descending track No. 51 over the area of interest (AOI). Due to very small perpendicular baseline distribution secured by stable Sentinel-1 orbital tube, as well as regular temporal sampling of acquisitions, forming single-master image combinations provides stable and rigorous geodetic time series [17]. Interferometric image combinations employed in processing are represented by Figure 5. Master images were selected approximately in the barycentre of the temporal and perpendicular baseline domain, with regard to atmospheric conditions during the time of acquisition. Hence the acquisition time of master images were chosen so that no precipitation or notable overcast was reported [18].

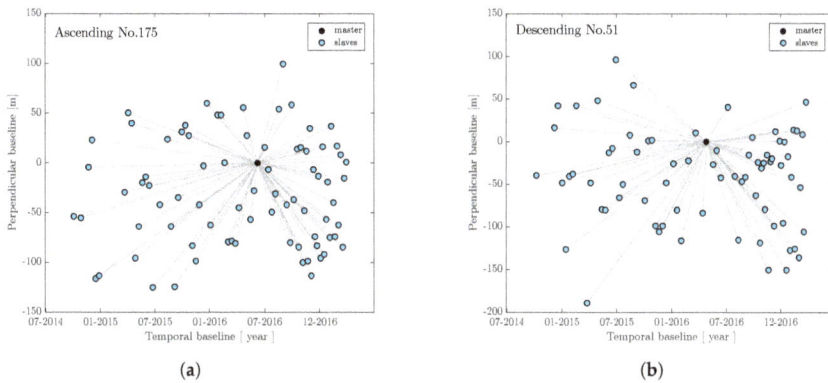

Figure 5. Interferometric combinations of Sentinel-1 SLC images from (**a**) ascending and (**b**) descending track.

3.2. MT-InSAR Processing

We incorporated the processing chain (please refer to flowchart in Figure 6) based solely on open-source tools. First processing segment was carried out under current Sentinel Application Platform (SNAP), being developed for ESA by Array Systems Computing under GPL (free, libre and open-source) license [19]. Thanks to its graph processing tool utility, we wrote series of scripts to automate the processing of whole stacks of Sentinel-1 SLC images in batch fashion.

Regarding the large size of original SLC data and associated high computational demands, the prime step involved the extraction of bursts covering the AOI. Then, the orbit state vectors of individual scenes were refined using Sentinel-1 precise orbit ephemerides, published by ESA on [20]. Sentinel-1 operates in a unique acquisition mode TOPS (Terrain Observation by Progressive Scans) [21], which captures images consisting of series of bursts with mutual overlaps. This provides wide swath width with a reasonable spatial resolution of 5 m × 20 m. Sentinel-1 TOPS SLC images, however, require a specific coregistration approach regarding the correct burst stitching (according to [22], an azimuth coregistration accuracy of 0.003 samples is required to achieve a phase error of less than

10 degrees). The main advantage of SNAP inclusion is its current state-of-the-art performance in TOPS coregistration. Note that TOPS coregistration operator in second step (consisting of back-geocoding and enhanced-spectral diversity correction) was carried out burst-wise, in order to make use of the entire burst overlap as well as retaining a sufficiently coherent area. Coregistration was assisted by SRTM 1 arc second digital elevation model (DEM) [23]. The AOI was subsequently cropped from de-bursted stacks, which was followed by differential interferogram formation, again employing SRTM 1 DEM. The output data are SLC stacks and differential interferograms coregistered onto a common master image frame, plus a geocoding mask. Geocoding based on orbital data and DEM is limited to positional accuracy within several metres, depending on the resolution of SLC data.

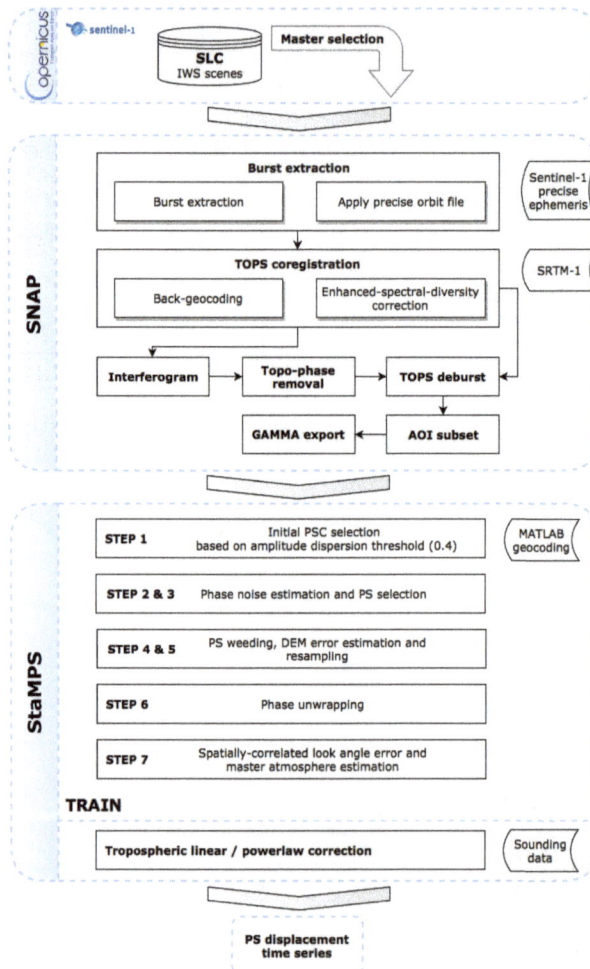

Figure 6. Multi-temporal InSAR analysis incorporated by the SNAP & StaMPS processing chain.

Multi-temporal PS processing was carried out in StaMPS (Stanford Method for Persistent Scatterers) software implementation [24]. Standard PSInSAR technique (originally presented by [5]) selects PS candidates based on their phase variation in time, presuming a temporally linear deformation model; whilst StaMPS methodology exploits the spatial correlation of their phase without prior assumptions about its temporal nature [25]. Therefore StaMPS succeeds in the identification of less bright scatterers with lower SCR (signal-to-clutter), and thus provides denser PS coverage even in sparsely urbanized and natural terrain. The processing procedure followed steps, closely described in [24,26,27]:

- Apriori PS candidates selection based on the 0.4 amplitude dispersion threshold.
- Iterative phase stability estimation of PS candidates (analyzing their phase noise term).
- PS selection based on ensemble temporal coherence in a probabilistic fashion.
- Side-lobes elimination.
- Elimination of phase contribution due to residual topography (DEM error).
- 3D phase unwrapping.
- Estimation of spatially-correlated nuisance terms (atmosphere and orbital phase contributions).

Careful selection of processing parameters was required to avoid aliasing of the deformation signal, especially the size of resampling grid for unwrapping and window sizes of temporal and Goldstein filter [28] applied prior to unwrapping. The unwrapping procedure was followed by thorough detection and correction of possible phase unwrapping errors with regard to phase time series, as illustrated by e.g., [29]. Such errors occurred where the deformation pattern was not smooth.

After reasonable mitigation of phase unwrapping errors and spatially-correlated error terms, some atmospheric signal still persisted in the estimate. To date, several strategies for the elimination of tropospheric phase delays were developed, e.g., employing weather models, GNSS measurements or spectrometer data [30,31]. However, these are often limited by the spatio-temporal resolution or a non-simultaneous acquisition of sensors. With respect to the smaller extent of our AOI, we estimated the tropospheric phase screens and their stratification for each acquisition date empirically from the relationship between the interferometric phase and the topography, as proposed by [32] and implemented in Toolbox for Reducing Atmospheric InSAR Noise (TRAIN). It is important to note that strong tropospheric turbulence, albeit not expected in our AOI, cannot be accounted for using only phase-based correction methods and would leak into the residual phase as an incorrect signal [32].

Finally, unwrapped residual phase time series were converted to displacement via sensors wavelength. Mean annual velocities of displacement were then estimated in the least squares sense by inversion of the linear movement model.

4. Results and Discussion

The basic and most commonly used representation of MT-InSAR analysis are ground deformation maps, depicting mean line-of-sight (LOS) annual velocities of individual PSs. Concerning the reliability of the estimate, its indicator is ensemble temporal coherence (as a maximum of periodogram). Hence the most common, albeit harsh practice for outlier elimination is thresholding on ensemble temporal coherence only. Though in order to preserve as much information as possible, we applied outlier elimination approach proposed by [33], which allows for the minimisation of outliers in final results while preserving spatial and statistical dependency among the observations. This approach preserved a valuable amount of 28,101 points for the ascending track and 25,957 points for the descending track (representing approximately 25 percent of whole point-cloud), that would have been otherwise discarded by standard coherence thresholding. Outlier-free mean LOS velocity deformation maps for both ascending and descending datasets are in Figure 7. Those are relative to a common reference area (depicted by a black rectangle). Reference area is situated in the old central part of Prievidza city. Its stability was discussed with geologists and verified in-situ.

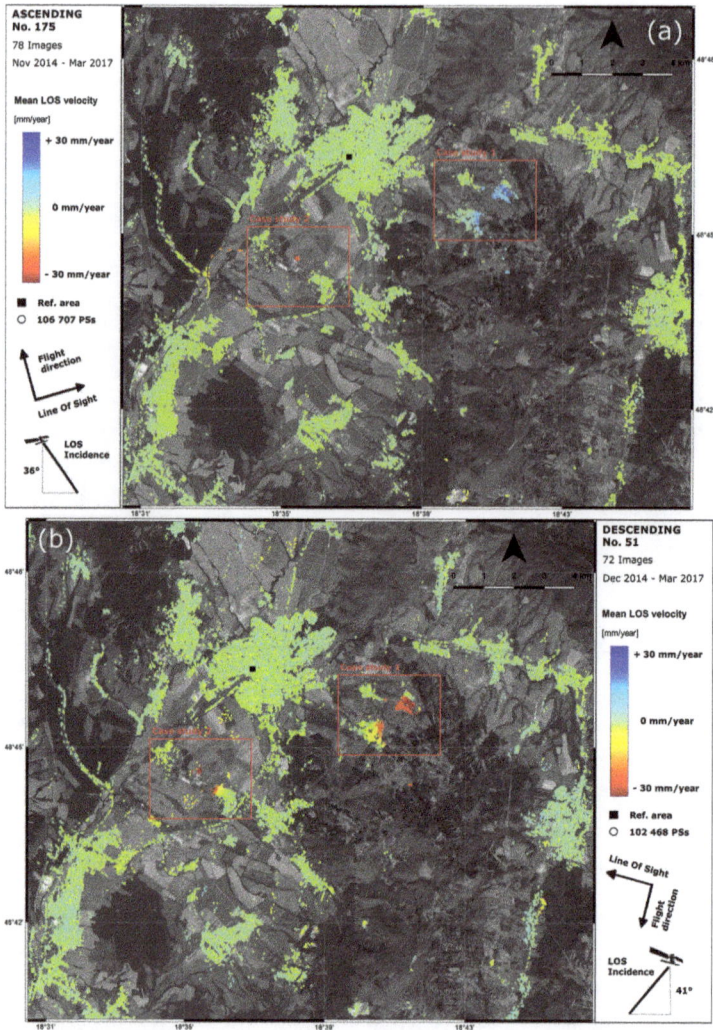

Figure 7. Mean line-of-sight (LOS) deformation velocity maps produced from (**a**) ascending and (**b**) descending Sentinel-1 stacks. Outlier elimination has already been performed. Red values indicate movement away from the satellite, whereas blue values indicate movement towards the satellite. Basemaps: Google, 2017 CNES.

Mean velocities are displayed in a common interval of ±30 mm/year. Their average values are less than 0.01 mm/year for both ascending and descending datasets, with an approximately normal distribution. Mean values of standard deviations are 0.67 mm/year for the ascending and 0.74 mm/year for the descending dataset. Although the small values of standard deviations may seem outstanding, they actually correspond to precision, not accuracy. Precision is strongly influenced by the number of images, i.e., the degree of freedom of the estimate. Higher standard deviation does not necessarily correspond to noisier scatterers, as non-linear deformation could also push this value up [4]. To assess the accuracy of the results, calibration methods employing artificial corner reflectors or compact active transponders are required [34].

We achieved solid coverage over urban and peri-urban areas, yet several clusters of points were successfully retrieved even in rural parts of the AOI (amidst arid fields or outcrops). Deformation maps provide a global outlook on the relative movements in the whole area of interest, and thus show that no large spatial wavelength deformation signal is present. At the same time, they demarcate distinct small-areas of detected spatially clustered motion. We focus on the two already introduced and most significant geohazards within the Upper Nitra region.

4.1. Case Study 1: Landslides

InSAR results reveal that Hradec and Velka Lehotka (suburbs of Prievidza city) are affected by severe slope deformation processes. While those are undergoing movement towards the satellite on ascending track (Figure 8a), respective PSs exhibit movement away from the satellite for a descending track results (Figure 8b). This feature is a consequence of prevailing horizontal direction of actual displacement vector, as illustrated by Figure 9.

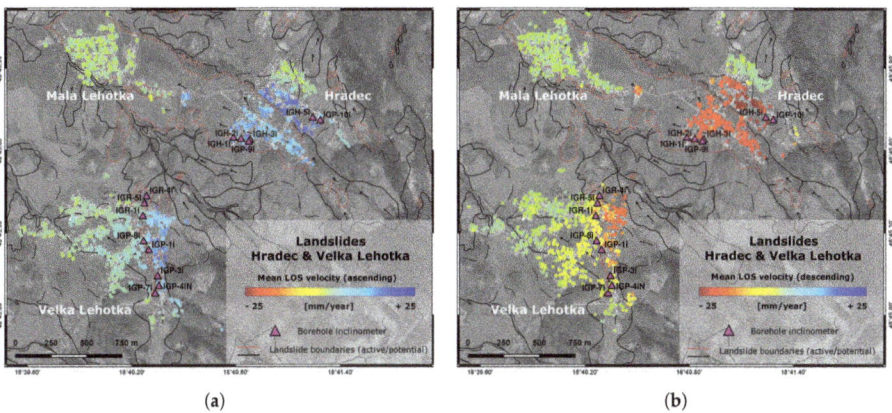

(a)   (b)

Figure 8. Hradec and Velka Lehotka landslides as observed by Sentinel-1 MT-InSAR: mean line-of-sight (LOS) deformation velocity maps from (**a**) ascending and (**b**) descending track. Purple triangles represent borehole inclinometry monitoring network. Basemap: Google, 2017 CNES.

Interpretation of surface movements is generally challenging when dealing with LOS displacement velocities. Assuming major sensitivity to vertical displacement and ignoring the presence of a potential horizontal one can introduce large errors into the final estimates if using mere projection via incidence angle [35]. Moreover, Sentinel-1 TOPS acquisition look angles are way larger than those of legacy ERS or Envisat satellites, what further expands this vulnerability. Yet, thanks to the availability of both acquisition geometries, LOS velocities can be further decomposed into vertical component and horizontal component in descending azimuth look direction (Figure 9).

Considering that ascending and descending stacks are related to different reference frames and individual PSs might actually correspond to different structures due to the strong directional dependence of the scattering mechanism, such decomposition encounters problem of PS pixel-matching. Hence we apply a simplified, three step approach:

1. Mean LOS velocities are resampled into regular 50 m × 50 m grid.
2. The grid is subsequently filled by the weighted average of all PS's velocities allocated within individual cells. Temporal coherence is assumed a weighting factor.
3. Finally, decomposition is evaluated using the system of equations in accordance to [36]. Its outcome is visualised by Figure 10.

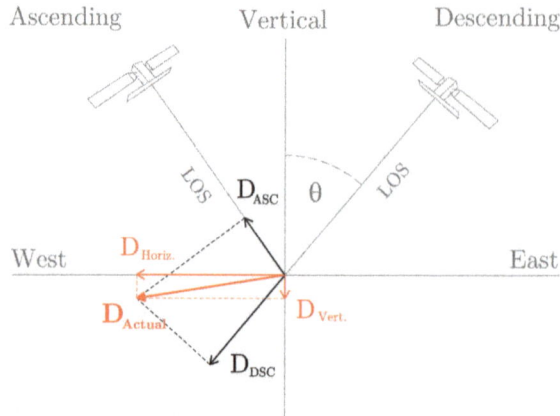

Figure 9. LOS velocities decomposition principle, specific case of Hradec landslide (Figure 8).

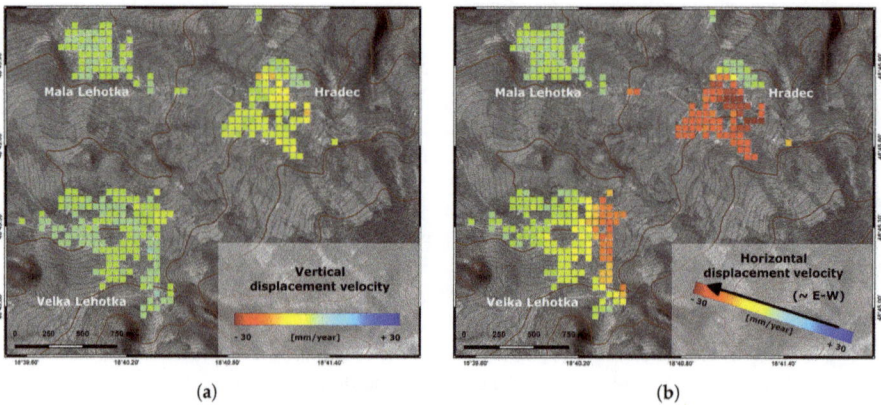

(a)   (b)

Figure 10. Decomposition of InSAR LOS velocities (Figure 8) into (**a**) vertical component and (**b**) horizontal (approximately E-W) component. Basemap: Google, 2017 CNES and ZBGIS topography, GKU Slovakia.

The detected movement activity is in a good agreement with the slope failure delineation and its direction mapped by [14]. Minor discrepancies occur only in the northern part of Velka Lehotka village (the southern one). In 2013, several buildings were damaged, which, according to the above-mentioned map, are situated beyond the boundary of the potential landslides. InSAR observations, however, show ongoing activity, especially in the eastern sector. Most unstable parts lie on the Handlova Formation (tuffs and sandstones) and correspond mainly to the landslide deluvia. Several PSs in central part of Hradec village (north-eastern one) even exceed the values of 3 cm/year. Stable northern section of the village, which rests upon a block formed by lava flows of pyroxene andesite, is clearly separated by landslide boundary. Structural failure of buildings is most common in these areas. On the contrary, in the central part of the village, where the greatest velocities are observed, the buildings are "carried" by landslide body without the impact on their construction. Similar, but smoother division is also observable in Velka Lehotka village. Results visualization in terms of displacement velocity maps is for many instances insufficient, as they do not reveal the valuable content in the displacement time series of each individual PS. In order to validate InSAR observations, PS time series were compared with independent ground measurements in borehole inclinometers. Since the measurements are of a different nature, they are displayed in individual graphs. Moreover, considering the geocoding uncertainty, collocation of points is only approximate. Therefore the closest, well coherent PSs are chosen for comparison. Figure 11 shows two examples of such comparison with precise borehole inclinometers codenamed IGH-5i and IGH-1i.

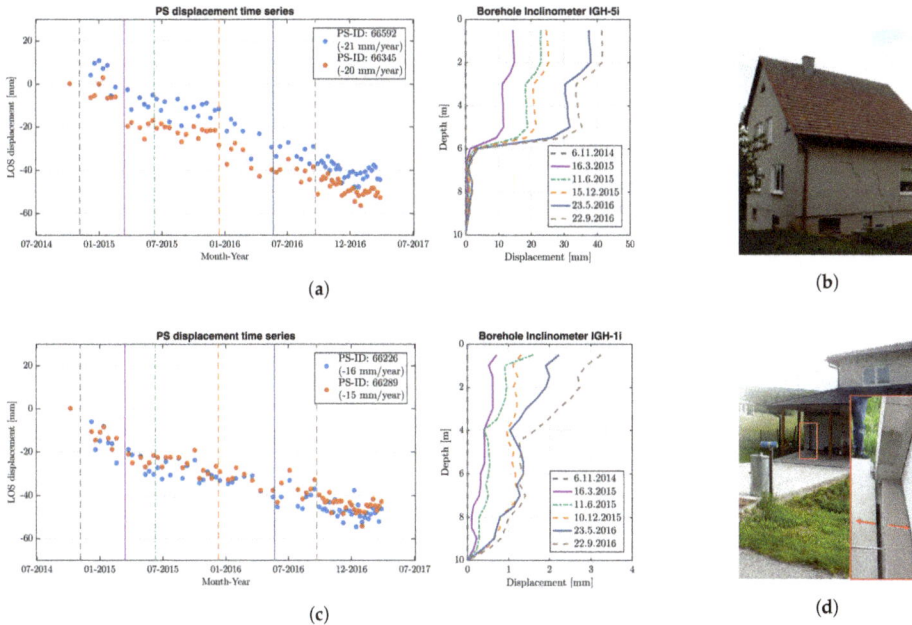

(a)

(b)

(c)

(d)

Figure 11. (**a**); (**c**) Comparison of InSAR derived LOS displacement time series (from descending track) with borehole inclinometer measurements at points IGH-5i (depth of shear zone: 4.61 m, mean displacement velocity: 18.9 mm/year, orientation: northwest-west) and IGH-1i (expected depth of shear zone: 3.44 m, mean displacement velocity: 1.2 mm/year, orientation: southwest), adjacent to (**b**); (**d**) disrupted structures.

The two example boreholes characterize distinct landslide bodies (see Figure 8). Borehole IGH-5i is located in the most active part of the slope deformation with a clearly developed shear zone (Figure 11a). Installation of a stationary inclinometer in the nearby IGP-10i borehole has confirmed that the territory is permanently in motion. Average displacement velocities range from 7 to 68 mm/year, dominantly depending upon climatic factors. Although the shear zone is less evident at borehole IGH-1i (Figure 11c), the presence of landsliding activity is recognizable at nearby buildings (Figure 11d) and fences, broken by fractures.

While the method of precise inclinometry provides information on the deformation evolution at particular depths of borehole and directly determines the shear zone, the surface associated InSAR observations enable to assess the deformation phenomena across the whole sliding areas. Therefore, those information are complementary in terms of their spatial scope. Comparison with borehole displacement values confirmed the presence, extent and magnitude of slope deformations. Periods of larger movement activity and sudden activation coincide between epochs of inclinometry measurements (vertical colour lines in PS displacement time series). Variations of slope movement activity could be presumably caused by seasonal changes of climatic factors and the related groundwater table level, being one of its major activators [3]. Besides, the magnitude of sub-surface cumulative displacement at shallow depths in IGH-5i borehole adheres to surface ones, as apparent from the time series. Displacement time series tend to gradually stabilize towards the end of an observation period. Finally, note the PS time series variance decrease, stiffened by higher sample rate since October 2016 as a benefit of Sentinel-1B integration. Altogether 30 Sentinel-1B images were employed within processing for ascending (15) and for descending (15) track.

Implementation of InSAR observations within the conventional landslide monitoring techniques is of a major significance, as it helps to better understand the surface deformation pattern throughout the sliding body, its spatial and temporal evolution, as well as to assess the success of remediation works (which have been, in this particular case, deemed insufficient). This underscores the operational capability of Sentinel-1 data for routine updates of landslide inventory maps on a weekly basis, which could even contribute to development of early warning systems.

4.2. Case Study 2: Land Subsidence Due to Undermining

Major part of Kos municipality is afflicted by significant loss of coherence due to fast deformation rates and formation of wetlands, as a consequence of land subsidence down below average groundwater level. According to [15], depth of some depressions reaches up to several meters. Theoretical maximum deformation rate observable by InSAR is half a wavelength per revisit time, i.e., 42.6 cm/year for C-band Sentinel-1 [4]. This actually also depends on the noise level and the phase unwrapping technique [37]. Therefore, identifications of PSs in northwestern part of Kos village fails.

Nonetheless, we managed to retrieve several reliable PSs in the southeastern part of village from the descending track, which exhibit coherent deformation as depicted by Figure 12. While the smaller-magnitude subsidence is noticeable even at PSs remote from mining affected areas, displacement values slowly increase towards the directly undermined territory, reaching rates higher than 3 cm/year. Since results from the ascending track are generally noisier and without coherent points over the affected area even after multi-variate outlier elimination (see Figure 7), LOS decomposition was not performed.

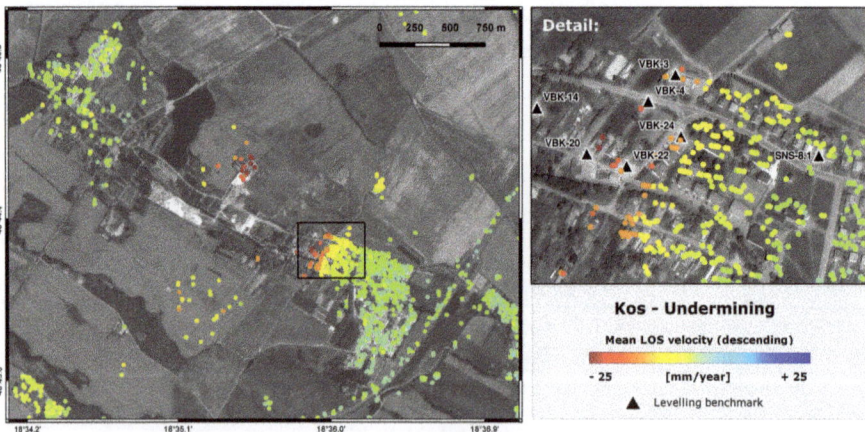

Figure 12. Kos village subsidence due to undermining as observed by MT-InSAR: mean line-of-sight (LOS) deformation velocity map from descending track with detail of area monitored by levelling network. Basemap: Google, 2017 CNES.

Detected deformation trends have been confirmed by comparison with terrestrial levelling at benchmarks coded SNS-8.1, VBK-3, VBK-4, VBK-20 and VBK-22, illustrated by Figure 13. Dividing LOS velocities by the directional cosine of the incident angle, we obtain approximate values of vertical components for comparison purposes.

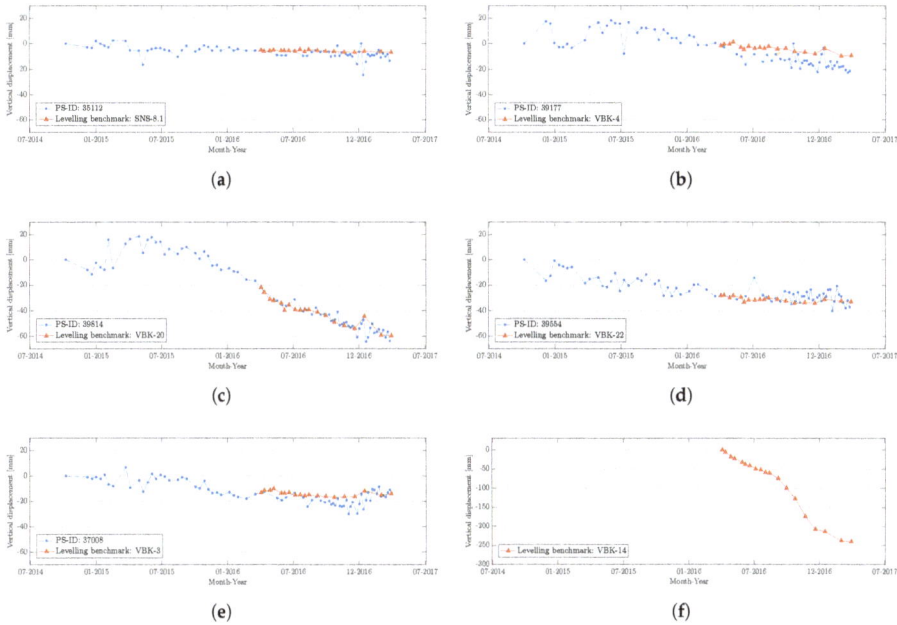

Figure 13. Comparison of PS deformation time series (from descending track, projected onto the vertical component) with levelling. Points (**a**) SNS-8.1; (**b**) VBK-4; (**c**) VBK-20; (**d**) VBK-22; (**e**) VBK-3; (**f**) VBK-14.

Independent displacements time series strongly correlate, especially at benchmark VKB-20, and thus provide the notion about the relative accuracy of InSAR observations. Sole levelling time series of benchmark VBK-14 are also presented in order to illustrate the magnitude of fast deformation rates (exceeding 3 m/year) occurring within the western part of the village, lying directly above the mining corridors. Displacement evolution of some points tends to stabilize (VBK-3, VBK-22), while for example VBK-20 exhibits a non-linear trend with activation since summer 2015, which could have been caused by the restoration of mining activities and the increase of their extent.

Proof of an extreme subsidence occurring in the old westernmost part of Kos village, devoid of PSs, is provided by differences between UAV and LiDAR-derived digital elevation models in Figure 14. The former DEM was derived from Trimble® UX5 (Geotronics Slovakia and Slovak University of Technology) measurement campaign carried out in March 2016 and processed by Agisoft® PhotoScan (Agisoft LLC, St. Petersburg, Russia). Average density of ground filtered point cloud reached 20 points/m^2. The latter DEM was generated on the basis of Trimble® Harrier 68i LiDAR system (Zilinska University and Slovak University of Technology) measurements which took place in November 2016. Through processing by LAStools (rapidlasso GmbH, Germany), we achieved average density of ground filtered point cloud of 15 points/m^2. For both cases, DEM pixel size is 50 cm, and vertical accuracy better than 10 cm was confirmed via independent ground control points. DEM comparison reveals a distinct north-western body of Kos village subsiding down 7 m over 8 months period. This gives a clear notion why the identification of coherent PSs using C-band data was impossible for this particular area. The consequences of undermining are evident not only geomorphologically, but also on structures and infrastructure (Figure 15). Several houses in northwestern and central parts of the village have already been demolished; the remaining, still inhabited houses in central part are continually exposed to risk. Sentinel-1 MT-InSAR proves to be able to secure remote and cost-effective stability monitoring of those structures, comparable to conventional monitoring techniques in terms of accuracy, yet maintaining the vast spatial coverage.

Its limitation concerning the maximum observable deformation rate could be partially overcome by different MT-InSAR techniques, e.g., Small Baseline Subset (SBAS) [38], employing data from longer wavelength sensors [39], or even by exploitation of the amplitude information of SAR images [40].

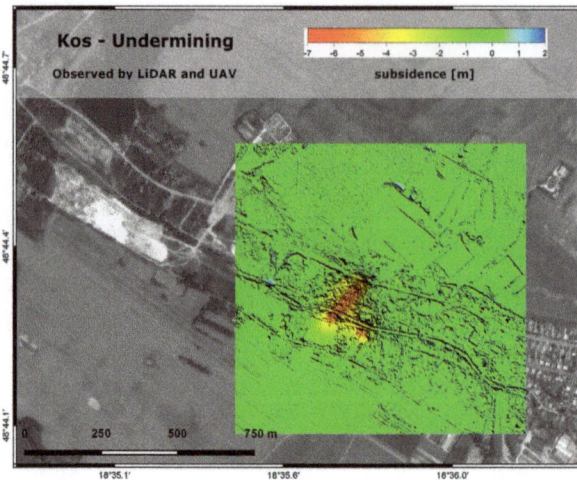

Figure 14. Kos village subsidence due to undermining as observed by digital elevation model differences between March 2016 and November 2016. Digital elevation models were generated using UAV photogrammetry and LiDAR. Basemap: Google, 2017 CNES.

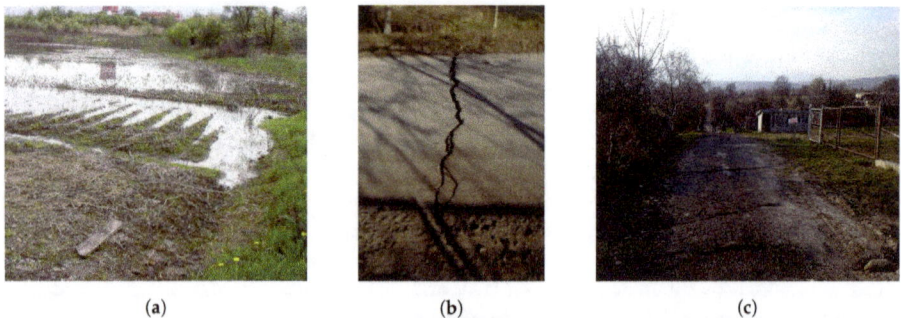

(a)   (b)   (c)

Figure 15. Consequences of Kos village subsidence: (a) Formation of wetlands (April 2004); (b) Infrastructural damage (December 2016), (c) Geomorphological changes (March 2016).

5. Conclusions

We have demonstrated the capabilities of open-source MT-InSAR tools SNAP & StaMPS in conjunction with the open-access Sentinel-1 data to monitor the most significant geohazards in the Upper Nitra region, Central Slovakia. Compared to point-wise terrestrial monitoring techniques, the advantages of MT-InSAR combine reasonable spatial coverage, assured temporal sampling and the capability to remotely assess the historic as well as ongoing deformation phenomena at low-cost merit, regarding the Sentinel data. We have investigated currently active slope deformations in Hradec and Velka Lehotka suburbs of the Prievidza city, where major part of villages exhibits coherent, dominantly westward displacements, while annual velocities of some PSs even exceed the values of 3 cm/year even after remediation works. Thanks to the application of multi-variate outlier post-processing analysis, we retained valuable PSs in affected areas, which would have been otherwise (imposing the standard coherence threshold) undersampled. Moreover, we have managed to identify several rapidly subsiding PSs in the dynamically changing environment of the undermined area of Kos municipality. Our observations have been validated via comparison with ground truth data from borehole inclinometers and terrestrial levelling. Outcomes of this analysis have enhanced our understanding of deformation patterns and their spatio-temporal progression. Assuming the detailed analysis of PS kinematic time series along with routine updates of deformation maps, those could contribute to development of early warning systems. Thus the classification of deformation trends based on conditional sequence of statistical tests (as proposed by e.g., [41] or [42]) is among the planned steps for respective case studies. Since both case study deformation phenomena can still be considered substantially active and a number of buildings are subject to continual damage, assessment of remediation works and operational routine monitoring is of utmost importance. Exploiting the free Sentinel-1 data by means of multi-temporal InSAR techniques constitutes to be capable of such task. Therefore, the main future objective involves pilot operational monitoring of at least hereby presented areas, supported by the communication with State Geological Institute of Dionyz Stur, HB mining company and local authorities. Finally, efficient installation of artificial corner reflectors or active radio transponders [43] shall further increase the density of PS at critical sites lacking any strong natural scatterers.

Acknowledgments: Sentinel-1 data were provided by ESA under free, full and open data policy adopted for the Copernicus programme. StaMPS and TRAIN software packages are freely distributed under GNU-GPL license and were developed at the University of Leeds (namely by Andrew Hooper and David Bekaert). This work has been supported by the Slovak Grant Agency VEGA under projects No. 1/0714/15 and 1/0462/16. The authors would like to thank State Geological Institute of Dionyz Stur and Hornonitrianske bane Prievidza, a.s. (namely M. Plakinger) for providing ground truth data, discussion and collaboration.

Author Contributions: Richard Czikhardt scripted the data processing chain. Richard Czikhardt, Juraj Papco and Matus Bakon performed the data processing and analyzed the results. Pavel Liscak, Peter Ondrejka and Marian Zlocha assisted with the interpretation of results, enlightened their geological background and provided geological ground truth data. Richard Czikhardt drafted the article. All authors contributed to the finalization of this paper.

Conflicts of Interest: The authors declare no conflict of interest.

References

1. Liscak, P.; Ondrejka, P.; Simekova, P.; Petro, L.; Paudits, P.; Zilka, A.; Maslarova, I.; Iglarova, L.; Madaras, J. Actual emergency landslides in Slovakia (2011–2014). *Geol. Prac. Spravy* **2014**, *124*, 71–88. (In Slovak)

2. Wagner, P.; Ondrejka, P.; Iglarova, L.; Frastia, M. Current trends in monitoring of slope movements. *Miner. Slovaca* **2014**, *42*, 229–240. (In Slovak)

3. Ondrejka, P.; Bakon, M.; Papco, J.; Liscak, P. Monitoring of sliding activity of the territory Prievidza-Hradec. *Geotechnika* **2016**, *1/2016*, 3–12. (In Slovak with English summary)

4. Crosetto, M.; Monserrat, O.; Cuevas-González, M.; Devanthéry, N.; Crippa, B. Persistent Scatterer Interferometry: A review. *ISPRS J. Photogramm. Remote Sens.* **2016**, *115*, 78–89.

5. Ferretti, A.; Prati, C.; Rocca, F. Permanent scatterers in SAR interferometry. *IEEE Trans. Geosci. Remote Sens.* **2001**, *39*, 8–20.

6. Colesanti, C.; Locatelli, R.; Novali, F. Ground deformation monitoring exploiting SAR permanent scatterers. In Proceedings of the IEEE International Geoscience and Remote Sensing Symposium, Toronto, ON, Canada, 24–28 June 2002; Volume 2, pp. 1219–1221.

7. Atzori, S.; Manunta, M.; Fornaro, G.; Ganas, A.; Salvi, S. Postseismic displacement of the 1999 Athens earthquake retrieved by the Differential Interferometry by Synthetic Aperture Radar time series. *J. Geophys. Res.: Solid Earth* **2008**, *113*, B09309.

8. Šimon, L.; Elečko, M.; Lexa, J.; Pristaš, J.; Halouzka, R.; Konečný, V.; Gross, P.; Kohút, M.; Mello, J.; Polák, M.; et al. *Geological Map of the Vtacnik Mts. and Upper Nitra Depression, 1:50,000*; Geological Survey of Slovak Republic: Bratislava, Slovakia, 1997.

9. Nemcok, A. *Landslides in Slovak Carpathians*; Veda Publishers, Slovak Academy of Sciences: Bratislava, Slovakia, 1982; p. 319. (In Slovak with English summary)

10. Malgot, J.; Nemčok, A.; Jesenák, J.; Šajgalík, J.; Baliak, F.; Mahr, T.; Čabalová, D.; Masarovičová, M.; Abelovič, J.; Gregor, V.; et al. *Engineering Geological Assessment of Geosystems, Part 2—Stability of Slopes and Subsoil of Constructions*; Final Report; SGUDS Archive: Bratislava, Slovakia, 1985. (In Slovak)

11. Ilkanič, A.; Jasovská, A.; Nigrínyová, J.; Pilko, M.; Vasiľko, T.; Smrek, M. *Engineering Geological Survey of the Affected Sites in Velka Lehotka and Hradec*; Final Report; MoE SR, SGUDS Archive; Bratislava, Slovakia, 2013; p. 50. (In Slovak)

12. Fekeč, P.; Ilkanič, A. *Realisation of Remedial Measures in Velka Lehotka, Hradec and Kraľovany. Consortium Engineering Geological Survey and Remediation of Emergency Landslides at Selected Sites of Slovakia*; Kosper, a.s.: Prague, Czech Republic, 2014; p. 33. (In Slovak)

13. Tupý, P.; Ilkanič, A.; Masiar, R.; Mišuth, K. *Remediation of Emergency Landslides in Velka Lehotka and Hradec (Part I.). Consortium "Sanácia Prievidza"*; TMG, a.s.: Prievidza, Slovakia, 2014; p. 37. (In Slovak)

14. Malgot, J.; Baliak, F.; Mahr, T. *Map of Vtacnik Slope Failures 1:10,000*; Slovak Cartography: Bratislava, Slovakia, 1983.

15. David, S. Landscape-Ecological, Environmental and Socio-Economic Consequences of Coal Mining in the Kos Cadastre. *Zivotne Prostr.* **2010**, *44*, 40–44.

16. Gregor, T.; Zlocha, M.; Vybíral, V.; Čapo, J.; Lanc, J.; Kováčik, M.; Bednarik, M.; Hrabinová, J.; Pivovarči, M. *Application of Remote Sensing at Monitoring of Environment Loads Impact Upon Geofactors in Selected Regions*; Final Report; SGUDS Archive: Bratislava, Slovakia, 2008; p. 657. (In Slovak)

17. Perissin, D.; Wang, T. Repeat-Pass SAR Interferometry With Partially Coherent Targets. *IEEE Trans. Geosci. Remote Sens.* **2012**, *50*, 271–280.

18. Weather Underground, Inc. History & Data Archive. Available online: https://www.wunderground.com/history/ (accessed on 14 September 2017).

19. Sentinel Application Platform (SNAP). Sentinel-1 Toolbox. Available online: https://qc.sentinel1.eo.esa.int/ (accessed on 14 September 2017).

20. Sentinel Payload Data Ground Segment (PDGS). Sentinel-1 Quality Control. Available online: https://github.com/senbox-org/s1tbx (accessed on 14 September 2017).

21. Zan, F.D.; Guarnieri, A.M. TOPSAR: Terrain Observation by Progressive Scans. *IEEE Trans. Geosci. Remote Sens.* **2006**, *44*, 2352–2360.

22. Prats-Iraola, P.; Scheiber, R.; Marotti, L.; Wollstadt, S.; Reigber, A. TOPS Interferometry With TerraSAR-X. *IEEE Trans. Geosci. Remote Sens.* **2012**, *50*, 3179–3188.

23. Jarvis, A.; Reuter, H. I.; Nelson, A.; Guevara, E. Hole-filled SRTM for the Globe Version 4. CGIAR-CSI SRTM 90 m Database. Available online: http://srtm.csi.cgiar.org (accessed on 14 September 2017).

24. Hooper, A.; Bekaert, D.; Spaans, K.; Arıkan, M. Recent advances in SAR interferometry time series analysis for measuring crustal deformation. *Tectonophysics* **2012**, *514–517*, 1–13.

25. Hooper, A.; Zebker, H.; Segall, P.; Kampes, B. A New Method for Measuring Deformation on Volcanoes and Other Natural Terrains Using InSAR Persistent Scatterers. *Geophys. Res. Lett.* **2004**, *31*, 5.

26. Hooper, A. Persistent Scatterer Radar Interferometry for Crustal Deformation Studies and Modeling of Volcanic Deformation. Ph.D. Thesis, Stanford University, Stanford, CA, USA, 2006.

27. Hooper, A.; Zebker, H.A. Phase Unwrapping in Three Dimensions with Application to InSAR Time Series. *J. Opt. Soc. Am. A* **2007**, *24*, 2737–2747.

28. Goldstein, R.M.; Werner, C.L. Radar interferogram filtering for geophysical applications. *Geophys. Res. Lett.* **1998**, *25*, 4035–4038.

29. Notti, D.; Calò, F.; Cigna, F.; Manunta, M.; Herrera, G.; Berti, M.; Meisina, C.; Tapete, D.; Zucca, F. A user-oriented methodology for DInSAR time series analysis and interpretation: Landslides and subsidence case studies. *Pure Appl. Geophys.* **2015**, *172*, 3081–3105.

30. Liu, S. Satellite Radar Interferometry: Estimation of Atmospheric Delay. Ph.D. Thesis, Delft University of Technology, Delft, The Netherlands, 2012.

31. Bekaert, D.P.S.; Hooper, A.; Wright, T.J. Reassessing the 2006 Guerrero slow-slip event, Mexico: Implications for large earthquakes in the Guerrero Gap. *J. Geophys. Res.: Solid Earth* **2015**, *120*, 1357–1375.

32. Bekaert, D.; Walters, R.; Wright, T.; Hooper, A.; Parker, D. Statistical comparison of InSAR tropospheric correction techniques. *Remote Sens. Environ.* **2015**, *170*, 40–47.

33. Bakon, M.; Oliveira, I.; Perissin, D.; Sousa, J.J.; Papco, J. A Data Mining Approach for Multivariate Outlier Detection in Postprocessing of Multitemporal InSAR Results. *IEEE J. Sel. Top. Appl. Earth Obs. Remote Sens.* **2017**, *10*, 2791–2798.

34. Mahapatra, P.; van der Marel, H.; van Leijen, F.; Samiei-Esfahany, S.; Klees, R.; Hanssen, R. InSAR datum connection using GNSS-augmented radar transponders. In Proceedings of the 2016 IEEE International Geoscience and Remote Sensing Symposium (IGARSS), Beijing, China, 10–15 July 2016.

35. Samieie-Esfahany, S.; Hanssen, R.; van Thienen-Visser, K.; Muntendam-Bos, A. On the effect of horizontal deformation on InSAR subsidence estimates. In Proceedings of the Fringe 2009 Workshop, Frascati, Italy, 30 November–4 December 2009; European Space Agency (Special Publication): Paris, France, 2010.

36. Ketelaar, V.B.H.G. *Satellite Radar Interferometry: Subsidence Monitoring Techniques*; Springer: Dordrecht, The Netherlands, 2009.

37. van Leijen, F. Persistent Scatterer Interferometry Based on Geodetic Estimation Theory. Ph.D. Thesis, Delft University of Technology, Delft, The Netherlands, 2014.

38. Berardino, P.; Fornaro, G.; Lanari, R.; Sansosti, E. A New Algorithm for Surface Deformation Monitoring Based on Small Baseline Differential SAR Interferograms. *IEEE Trans. Geosci. Remote Sens.* **2002**, *40*, 2375–2383.

39. Chaussard, E.; Amelung, F.; Abidin, H.; Hong, S.H. Sinking cities in Indonesia: ALOS PALSAR detects rapid subsidence due to groundwater and gas extraction. *Remote Sens. Environ.* **2013**, *128*, 150–161.

40. Huang, J.; Deng, K.; Fan, H.; Yan, S. An improved pixel-tracking method for monitoring mining subsidence. *Remote Sens. Lett.* **2016**, *7*, 731–740.

41. Berti, M.; Corsini, A.; Franceschini, S.; Iannacone, J.P. Automated classification of Persistent Scatterers Interferometry time series. *Nat. Hazards Earth Syst. Sci.* **2013**, *13*, 1945–1958.

42. Chang, L.; Hanssen, R.F. A Probabilistic Approach for InSAR Time-Series Postprocessing. *IEEE Trans. Geosci. Remote Sens.* **2016**, *54*, 421–430.

43. Mahapatra, P.; Samiei-Esfahany, S.; van der Marel, H.; Hanssen, R. On the Use of Transponders as Coherent Radar Targets for SAR Interferometry. *IEEE Trans. Geosci. Remote Sens.* **2014**, *52*, 1869–1878.

geosciences

MDPI

Article

Mapping Ground Instability in Areas of Geotechnical Infrastructure Using Satellite InSAR and Small UAV Surveying: A Case Study in Northern Ireland

Francesca Cigna [1,*], Vanessa J. Banks [1,†], Alexander W. Donald [1,2,†], Shane Donohue [3,†], Conor Graham [4,†], David Hughes [3,†], Jennifer M. McKinley [4,†] and Kieran Parker [1,2,†]

[1] British Geological Survey, Natural Environment Research Council, Nicker Hill, Keyworth NG12 5GG, UK; vbanks@bgs.ac.uk (V.J.B.); awdo@bgs.ac.uk (A.W.D.); kiepar@bgs.ac.uk (K.P.)

[2] Geological Survey of Northern Ireland, Dundonald House, Upper Newtownards Road, Belfast BT4 3SB, UK

[3] Civil Engineering, School of Natural and Built Environment, Queen's University Belfast, Stranmillis Road, Belfast BT9 5AG, UK; s.donohue@qub.ac.uk (S.D.); d.hughes@qub.ac.uk (D.H.)

[4] Centre for GIS and Geomatics, School of Natural and Built Environment, Queen's University Belfast, Elmwood Ave., Belfast BT7 1NN, UK; conor.graham@qub.ac.uk (C.G.); j.mckinley@qub.ac.uk (J.M.M.)

* Correspondence: francesca.cigna@gmail.com

† These authors are listed in alphabetical order.

Received: 7 April 2017; Accepted: 29 June 2017; Published: 6 July 2017

Abstract: Satellite Interferometric Synthetic Aperture Radar (InSAR), geological data and Small Unmanned Aerial Vehicle (SUAV) surveying was used to enhance our understanding of ground movement at five areas of interest in Northern Ireland. In total 68 ERS-1/2 images 1992–2000 were processed with the Small Baseline Subset (SBAS) InSAR technique to derive the baseline ground instability scenario of key areas of interest for five stakeholders: TransportNI, Northern Ireland Railways, Department for the Economy, Arup, and Belfast City Council. These stakeholders require monitoring of ground deformation across either their geotechnical infrastructure (i.e., embankments, cuttings, engineered fills and earth retaining structures) or assessment of subsidence risk as a result of abandoned mine workings, using the most efficient, cost-effective methods, with a view to minimising and managing risk to their businesses. The InSAR results provided an overview of the extent and magnitude of ground deformation for a 3000 km^2 region, including the key sites of the disused salt mines in Carrickfergus, the Belfast–Bangor railway line, Throne Bend and Ligoniel Park in Belfast, Straidkilly and Garron Point along the Antrim Coast Road, plus other urbanised areas in and around Belfast. Tailored SUAV campaigns with a X8 airframe and generation of very high resolution ortho-photographs and a 3D surface model via the Structure from Motion (SfM) approach at Maiden Mount salt mine collapse in Carrickfergus in 2016 and 2017 also demonstrate the benefits of very high resolution surveying technologies to detect localised deformation and indicators of ground instability.

Keywords: Synthetic Aperture Radar (SAR); Interferometric SAR (InSAR); Small Unmanned Aerial Vehicle (SUAV); Structure from Motion (SfM); ground deformation; slope stability; land subsidence; transport infrastructure; abandoned mines

1. Introduction

Ground movement is an issue of global concern and one that regularly grabs the attention of the media due to its impact on public safety, property and infrastructure networks, often necessitating expensive remedial action.

In Northern Ireland, ground movement is closely associated with slope instability, most notably on the margin and valley slopes of the Antrim plateau as well as surface subsidence in areas of historic mining [1]. While the origin of most of this movement is geologically controlled, human influences may also cause instability through urban development, historic mineral extraction and mining. The extent and form of surface motions can vary dramatically from location to location, with a number of controlling factors.

These movements are traditionally monitored by installing instrumentation around sites that have been causing persistent problems. Traditional monitoring methods are often costly and time-consuming while also limited by resources, enabling only small localised areas to be assessed over the long term. Consequently, there is a clearly an opportunity for accurate assessment of ground motion to benefit land-use planning and development across areas suspected of being susceptible to movement, particularly where this facilitates a better understanding of the instigating factors, and the development of tools that will enable early warning of catastrophic movement events.

Since the beginning of 2016, Queen's University Belfast, the British Geological Survey and the Geological Survey of Northern Ireland have been working on a research study to analyse the benefits of using satellite Interferometric Synthetic Aperture Radar (InSAR) techniques (e.g., [2]) to remotely assess ground stability and motion due to slope instability and mining subsidence in Northern Ireland and to identify, understand and quantify environmental risks to geotechnical infrastructure (i.e., embankments, cuttings, engineered fills and earth retaining structures).

The project is funded by the UK's Natural Environment Research Council under the Environmental Risks to Infrastructure Innovation Programme, and is analysing historical radar data obtained from the European Space Agency (ESA)-operated ERS-1/2 and ENVISAT satellites, in preparation for the exploitation of new Earth explorers (i.e., Copernicus' Sentinel-1). These data are processed with multi-temporal InSAR techniques which have been effectively used in other areas of interest to estimate ground motion rates from space with up to millimetre precision, for instance, to map landslides (e.g., [3–5]) and land subsidence (e.g., [6–8]).

The goal of this paper is to contribute to the discussion on the benefits that InSAR methods can provide for geotechnical infrastructure stakeholders, such as transport (e.g., roads and rails) and mining infrastructure owners. InSAR techniques have the capability to remotely monitor large areas, which would enable a step-change in techniques and data currently used by organisations to analyse risk to their infrastructure network.

To demonstrate this impact, the project team is working with five major stakeholders to examine areas of historical slope instability and subsidence, while also aiding the identification of other risk areas:

- TransportNI: a business unit within the Department for Infrastructure, playing a significant role in facilitating safe and convenient movement of people and goods throughout the province and the safety of road users, through delivery of road maintenance services and management and development of the transport network;
- Northern Ireland Railways: the publicly owned railway operator in Northern Ireland, a subsidiary of Translink;
- Department for the Economy: a devolved Northern Ireland government department in the Northern Ireland Executive, previously known as the Department for Enterprise, Trade and Investment, responsible for a number of policy areas, including economic policy development, energy, health and safety at work, mineral development and tourism;
- Belfast City Council: the local authority with responsibility for part of the city of Belfast;
- Arup: a multinational professional services firm providing engineering, design, planning, project management and consulting services for all aspects of the built environment.

These stakeholders all have a common requirement for ground deformation data either across their geotechnical assets or to assess subsidence risk (e.g., as a result of mine collapse) using the most efficient, cost effective methods, with a view to minimising and managing the geotechnical risk to their network and the general public.

Through the use of (i) InSAR to derive a regional baseline of ground stability and motion; (ii) ground truthing to provide validation; and (iii) Small Unmanned Aerial Vehicle (SUAV) surveying to derive high-resolution ortho-photography and 3D surface models, this study showcases the potential for these stakeholders to monitor ground deformation remotely from space and with aerial vehicles, in a more cost-effective, efficient, and systematic way.

The paper focuses on the analysis of InSAR in the context of the geological data for the key areas of interest identified by the stakeholders (Figures 1–3), and potentially other areas that exhibit significant ground motion rates, likely induced by consolidation settlement and dewatering.

An InSAR analysis for the period 1992–2000 is carried out to analyse historical radar data to assess the baseline ground motion scenario of the areas of interest, in preparation for the exploitation of more recent data from the ENVISAT (2002–2012) and Sentinel-1 (2014–onwards) missions.

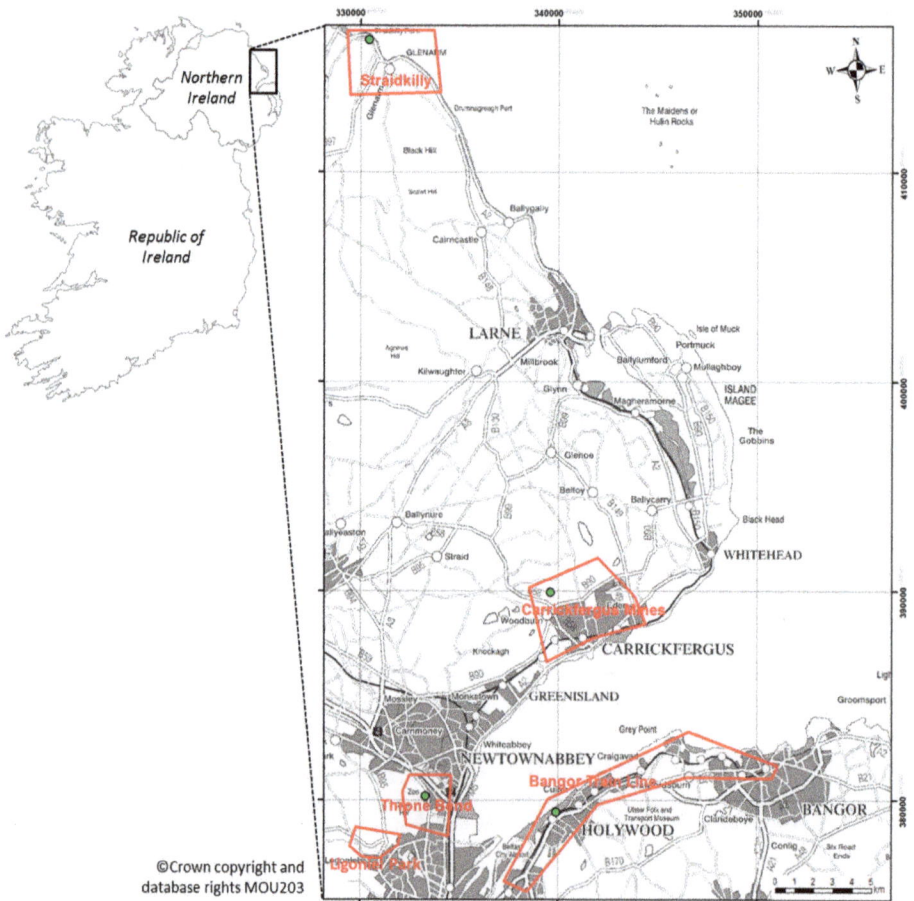

©Crown copyright and database rights MOU203

Figure 1. Location of the areas of interest onto topographic map. Reproduced from Land and Property Services data with the permission of the Controller of Her Majesty's Stationery Office, © Crown copyright and database rights MOU203.

The InSAR results are first assessed in relation to the distribution and density of identified monitoring targets across the processed area, then observed rates and spatial patterns of annual ground motion and example deformation time series are considered for selected targets, enabling any non-linearity to be highlighted.

Brief geological descriptions and tables with a summary of the local geological context and an interpretation of the observed movements are also provided for each of the focus areas.

The results are discussed with respect to embedding these technologies across the stakeholder organisations, and the potential to make a step-change in how stakeholders approach assessment and manage the resilience of their geotechnical infrastructure using network-wide data and, ultimately, reduce network wide monitoring and remediation costs.

Figure 2. Photographs of the areas of interest taken on 9 February 2016: (**a**) Maiden Mount salt mine collapse, Carrickfergus; (**b**) Belfast–Bangor railway line; (**c**) Ligoniel Park; (**d**) Garron Point; (**e**) Straidkilly Road; and (**f**) stabilisation works along the Antrim Coast Road.

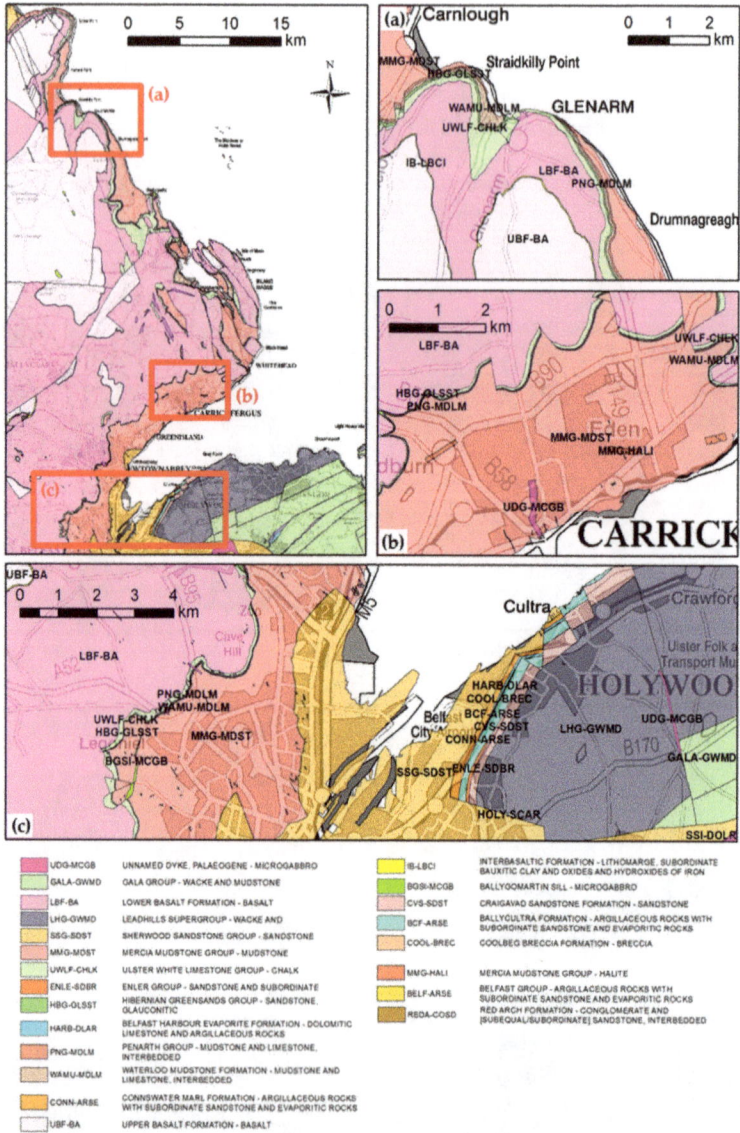

Figure 3. Bedrock geology of the study areas at 1:10,000 scale overlapped onto topography at 1:250,000 scale. (**a**) Straidkilly; (**b**) Carrickfergus mines and (**c**) Bangor Rail Line, Ligoniel Park and Throne Bend. Reproduced from Land and Property Services data with the permission of the Controller of Her Majesty's Stationery Office, © Crown copyright and database rights MOU203.

2. Study Areas

The regions of interest of this study as identified by the stakeholders focus closely on problematic sites where geotechnical infrastructure is affected by ground instability caused by land subsidence and/or landslide processes. In particular, the key areas of interest are (Figure 1):

- Carrickfergus: a residential town containing eight abandoned salt mines that display continual subsidence. Over the past two decades a number of crown holes have appeared at various locations as a result of mine collapses (Figure 2a), resulting in permanent closure of two public roads and re-routing of the main gas pipeline in an area of residential and commercial use.
- Belfast–Bangor railway line: this section of railway line is positioned within steep-sided cuttings prone to instability (Figure 2b), particularly after periods of heavy and prolonged rainfall, which occasionally cause disruption to the line.
- North Belfast: a densely populated, urban location, this area has been subject to shallow translational landslides with evidence of slope and road movement seen at Ligoniel Park (Figure 2c) and Throne Bend on the Antrim Road. At Throne Bend, a large translation slide, which reactivates during extreme rainfall events and affects a residential area, is present.
- Straidkilly and Garron Point, Antrim Coast Road (A2) (Figure 2d–f): positioned at the base of the Antrim Plateau, the A2 is a scenic route used extensively to access the many coastal towns and villages as well as a high number of tourist sites. This section of road cuts through soft Jurassic clays and debris from the slide area has frequently reached the road, increasing the risk to users and also leading to closures of this heavily trafficked strategic road, as well as the Straidkilly Road [9].

3. Materials and Methods

3.1. Satellite Data and InSAR Analysis

Sixty-eight Synthetic Aperture Radar (SAR) images acquired between 15 April 1992 and 23 December 2000 by ESA's radar satellites ERS-1 and ERS-2 were used to analyse the baseline ground motion for each area of interest. These satellite scenes cover a region of 100 km by 100 km (Figure 4a) encompassing over 3000 km^2 of land. In addition to the key areas of interest to the project, this area also includes a large portion of the rail and road networks of Northern Ireland, as well as abandoned mines north of Belfast. The entire County Antrim and portions of counties Down, Armagh Tyrone and Derry/Londonderry are covered by the satellite ground footprint.

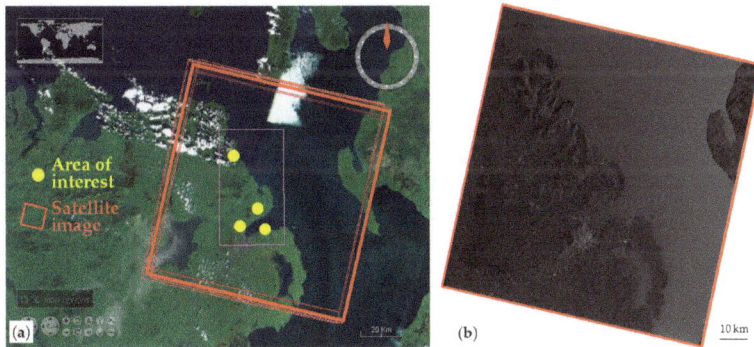

Figure 4. (**a**) Location of the areas of interest and satellite image footprints, and (**b**) average Synthetic Aperture Radar (SAR) amplitude of the ERS-1/2 data stack. ERS-1/2 data © European Space Agency (ESA) 2016.

The ERS-1/2 satellite images are characterised by medium spatial resolution (~25 m ground resolution) with a nominal repeat cycle of 35 days for each satellite (~monthly revisit). A data gap in the satellite data exists between the end of 1993 and early-1995, when ERS-1 was operated by ESA in different orbits and modes. The 68 images were acquired by SAR sensors operating in C-band (5.3 GHz frequency, 5.6 cm wavelength λ) when the satellites were flying along their descending orbits (i.e., approximately from north to south; Figure 5a,b). The Lines-Of-Sight (LOS) employed by both sensors are characterised by look angle θ of 23° with respect to the vertical direction, and observe the ground from east to west.

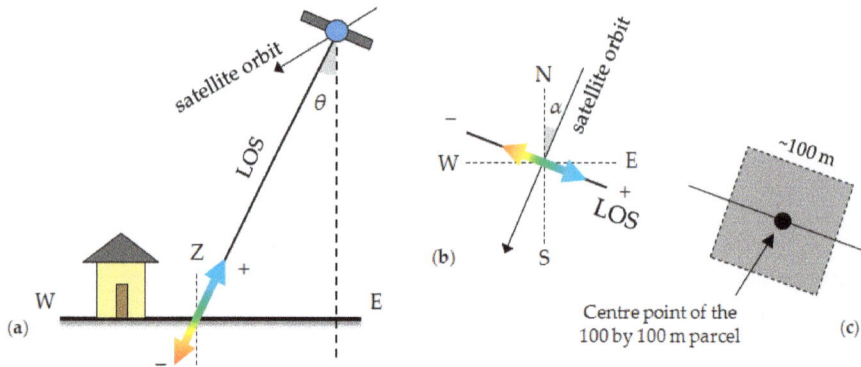

Figure 5. (a) Vertical and (b) planar view of the satellite Line-Of-Sight (LOS) in descending mode; and (c) schematic representation of each low resolution parcel for which the ground motion is estimated.

Figure 4b shows the SAR amplitude scene obtained by averaging the 68 images after precise co-registration to a single master (18 January 1997), in preparation for the multi-temporal interferometric analysis.

Multi-temporal image processing of the 68 single look complex scenes was carried out using the GAMMA SAR and Interferometry software (GAMMA Remote Sensing and Consulting AG, Gümligen, Switzerland) and using the low resolution Small Baseline Subset (SBAS) approach developed by Berardino et al. [10]. This method is based upon the generation of a set of small baseline interferograms between the satellite scenes acquired during each repeat pass, by selecting only those that are separated by less than 250 m perpendicular baseline and three years temporal baseline. This is to minimise the presence of temporal and spatial phase decorrelation components in the generated interferograms and enhance phase quality of the processed pixels (hence interferometric coherence).

Multi-looking factors of 4 in the range direction and 20 in the azimuth direction were employed to process the input single look complex scenes, in order to increase the phase signal quality and reduce radar speckle. This increased the corresponding size of the image pixels to ~85–90 m, corresponding to ~100 m on the ground. Interferograms were formed based on the small baseline pairs of scenes, and initial topographic information was subtracted from the latter using information from NASA's Shuttle Radar Topography Mission (SRTM) elevation model at 90 m resolution. A coherence threshold of 0.25 was used for the selection of the image pixels of a suitable quality for the ground motion analysis, and phase information from the stack of small baseline interferograms was processed for these pixels only. Orbital phase ramps removal and phase unwrapping of the differential interferograms followed. Interferograms with unwrapping errors were excluded from the analysis. Least squares covariance analysis of the unwrapped phases then allowed the computation of annual velocity estimates and associated standard errors, and improved height information. Full deformation time series, for the network of coherent pixels, were subsequently derived as per [10], by exploiting the singular value decomposition approach to retrieve a minimum-norm least squares solution for the phase of each pixel at each acquisition time. To remove atmospheric delay components, a cascade of a high-pass filter in time and a low-pass spatial filter was also used. The estimated phase values were finally converted to deformation via multiplication by $\lambda/4\pi$.

The reference point to which all ground deformation estimates were referred was a location within a geologically stable area in Belfast, which was also characterised by high interferometric coherence (54.59 N, −5.94 E).

The processing results were exported in the form of a point-wise database, including a set of 36,791 points, each representing a parcel of land of around 100 by 100 m on the ground (Figure 5c). For each point, the database includes: unique ID, latitude and longitude in the 1984 World Geodetic System (WGS84) datum, annual velocity estimated along the satellite LOS and its standard error, elevation expressed in metres above the Earth Gravitational Model 1996 (EGM96) ellipsoid, average coherence of the point, ranging from 0 (no coherence) to 1 (perfect coherence), plus a series of 68 values, indicating the position of the point at each date.

Accounting for the extent of the processed area (~4200 km^2), the average density of targets identified consists of approximately 10 points per square km, with sparse coverage across rural areas and very dense networks of targets over built-up areas such as Belfast and Bangor (Figure 6), where over 140 points per square km were found. This coverage reflects the presence of coherent areas from which radar backscattering is kept during the repeat passes. The resultant contrast in density in urban compared with rural areas is only to be expected.

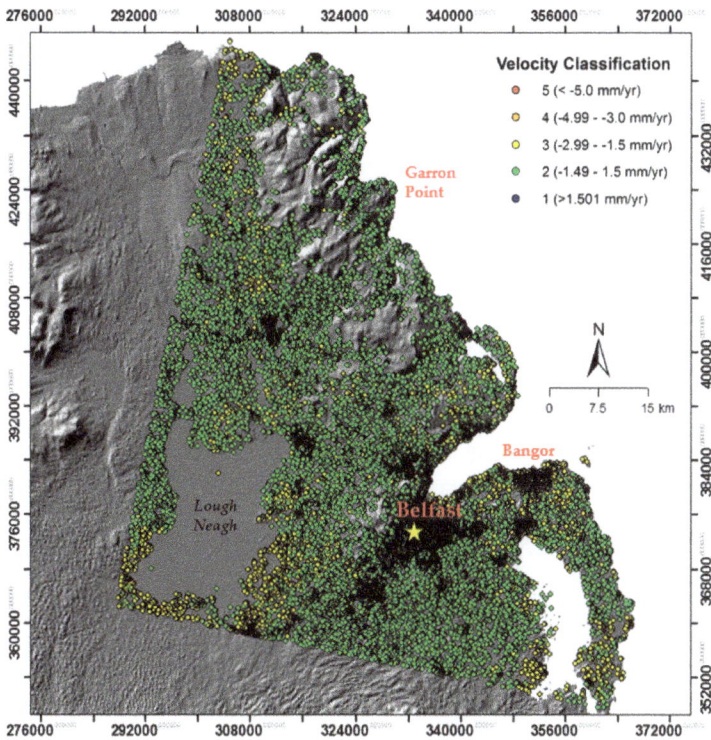

Figure 6. Ground deformation dataset obtained after Small Baseline Subset (SBAS) Satellite Interferometric Synthetic Aperture Radar (InSAR) processing of the ERS-1/2 data stack for 1992–2000, overlapped onto shaded relief of NASA's Shuttle Radar Topography Mission (SRTM) elevation model at 90 m resolution made available by CGIAR-CSI. Data are displayed according to the observed annual velocity as estimated along the satellite LOS. Negative values indicate motion away from the sensor (~subsidence), whilst positive values indicate motion towards the sensor (~uplift). The yellow star indicates the location of the reference point.

The results were converted from WGS84 to the Irish National Grid in a Geographic Information Systems (GIS) environment. A simple classification of the results based on their annual LOS velocity, *VEL*, was adopted. The symbology for the velocity classes was defined from the ERS-1/2 data range (i.e., −6.30 to 1.69 mm/year) and defined in five classes as: (1) $VEL > +1.501$ mm/year; (2) $−1.49 < VEL \leq +1.50$ mm/year; (3) $−2.99 < VEL \leq −1.50$ mm/year; (4) $−4.99 < VEL \leq −3.00$ mm/year; and (5) $VEL \leq −5.00$ mm/year.

Figure 6 provides an overview of the coverage of the processed ERS-1/2 1992-2000 SBAS results. The density of points in the area immediately to the south of Belfast Lough appears black due to the greater density of points in this area.

3.2. SUAV Surveying and VHR Surface Modeling

Unmanned Aerial Vehicles (UAV) are increasingly used in engineering geology and physical geography for the acquisition of low-altitude aerial photography and for the generation of high resolution elevation models, to study geomorphologic processes such as landslides, soil erosion and coastal processes (e.g., [11–14]). Recent applications in areas of open-pit mines have also proved the potential of UAV-derived elevation models to characterise geomorphic features (e.g., [15]).

For our study, we utilised a man-portable, namely Small UAV (SUAV) system, using an X8 rotor configuration for added flight redundancy. An autonomous flight plan campaign was designed to acquire Very High Resolution (VHR) aerial imagery of the Maiden Mount salt mine collapse at Carrickfergus, an embankment slope and surrounds at Ligoniel Park in Belfast and historic salt mine works at Salt Hill in Carrickfergus.

The first campaign took place on 10 May 2016 at Maiden Mount to collect Red-Green-Blue (RGB) and Near Infra-Red (NIR) imagery using the X8 airframe by 3DR Robotics (Berkeley, CA, USA). Two fully autonomous flights were undertaken based on a pre-programmed 'lawn-mower pattern' flight plan designed to insure image footprint over-laps of 85% and side-laps of 75% at a flight level of 60 m above ground level and capture speed of ~6 m/s.

Pre-flight ground surface distance was calculated at ~1.9 cm for RGB collection and ~2 cm for NIR image collection. Flight one collected colour imagery utilising a gimbal fixed compact digital single lens reflex camera (i.e., Sony A5000, 16 mm Pancake lens, Sony Corporation, Tokyo, Japan), the second flight acquiring supplementary 3 band NIR imagery (i.e., hot swapping a MAPIR NIR camera system, San Diego, CA, USA). RGB camera capture and shutter settings were controlled by the on-board PixHawk flight controller (ETH, Zurich, Switzerland) auto-calculated for optimum distance/overlap. NIR image capture was controlled using an 3 s intervalometer script.

To allow for accurate georeferencing of ortho-image (RGB & NIR) and surface models outputs, Ground Control Points (GCPs) in the form of temporary survey checker boards were distributed in scene and surveyed to the Irish Grid 1975 Polynomial / OSGM02 local grid and datum using Network Real Time Kinematic (NRTK) Global Navigation Satellite System (GNSS) methods with a Leica Viva GS15 receiver (Wetzlar, Germany). Six GCPs were surveyed and positioned to ~10 mm horizontal and ~15 mm vertical relative accuracies.

We collected ~140 close range aerial photographs (RGB/NIR) over the site and processed them to create a VHR ortho-photograph, a photogrammetry-derived dense point cloud and Digital Surface Model (DSM), using a Structure from Motion (SfM) computer vision approach (e.g., [16,17]) computed with AgiSoft PhotoScan Professional software (Educational Licensing; Agisoft LLC, St. Petersburg, Russia). The SfM approach is a multi-view stereo photogrammetric range imaging technique aimed to estimating three-dimensional (3D) structures from two-dimensional (2D) image sequences, usually acquired from a moving platform. The approach does not require the use of expensive equipment, specialist expertise or extremely long processing times, and can allow the generation of 3D point clouds of vertical accuracy of a few centimeters (e.g., 5 cm [17]). Its emergence has therefore revolutionised 3D topographic surveying in physical geography [16].

Our ~140 photographs were combined via SfM into a 3D structure thanks to the identification and matching of features such as corner points (e.g., edges) that were imaged from different viewpoints and could be identified in a number of photographs. 3D position of the identified features and the motion path of the camera were reconstructed, and the GCPs were incorporated into the processing steps to generate accurate georeferenced data projects in both plan and local datum (i.e., Ordnance Datum Belfast Lough).

The final SUAV-derived VHR ortho-photograph has a pixel resolution of 1 cm, and covers a 200 by 300 m area (Figure 7). A secondary NIR ortho-photo has a pixel resolution of 2 cm. Similarly, the resulting mesh based on the dense point cloud, at full resolution, includes over 4,504,000 vertices and 9,000,000 faces and has been used as the basis for the creation of the DSM, ortho-rectification of the aerial photography and 2D/3D visualisations in AgiSoft PhotoScan and a GIS. As with the RGB ortho-photograph, the output elevation model has a spatial resolution of 1 cm (Figure 8). It is worth noting that the generated DSM depicts elevation information of the top surface of the investigated area, including elements above the surface, such as trees, grass and fences, as can be seen in Figure 8.

A low resolution SUAV-derived DSM (RGB mapped) was also created and made freely available for 3D displaying and download via the SketchFab platform (New York, NY, USA). Similarly, full resolution web-views of the VHR aerial ortho-photographs have been made available via DroneLab (Drone Industries Ltd, Newcastle upon Tyne, UK) (Appendix A).

A return SUAV campaign at the same site was carried out on 16 June 2017, using a cross grid flight method and collecting ~300 close range aerial photographs that, as done for the first campaign, were processed to generate a VHR ortho-photograph and a photogrammetry-derived DSM.

Figure 7. (a) Small Unmanned Aerial Vehicle (SUAV)-derived Very High Resolution (VHR) ortho-photograph and (b) close-up of the Maiden Mount salt mine collapse, derived based on data acquired during the 10 May 2016 campaign.

Identical flights, SfM processing, and return campaigns are to be carried out at the other target sites as well, to test the SUAV method as a means to detect local ground movement and displacements between flights and data capture. Whilst the latter will be subject of future research, in this paper we use the first SUAV-derived products for Maiden Mount as a test case to demonstrate that VHR optical and topographic surveying with SUAV systems can be integrated with satellite surveying to complement regional scale analyses of ground stability when there is a lack or sparse coverage of radar targets in the investigated area, or it is necessary to identify surface indicators of instability at a much finer spatial scale—though with lower vertical accuracy than a satellite-based investigation with multi-temporal InSAR (see Section 4.1).

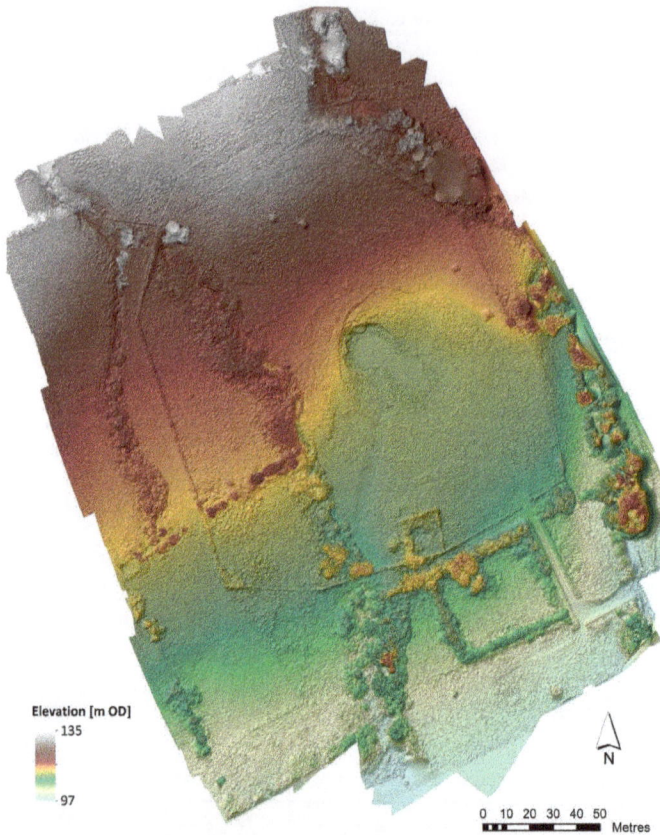

Figure 8. SUAV-derived VHR digital surface model of the Maiden Mount salt mine collapse, derived based on data acquired during the 10 May 2016 campaign.

4. Results and Discussion

By observing the areas of interest, as well as other areas that exhibit significant rates of ground deformation, we can develop a model of regional ground motion, characterising, for example, geotechnical phenomenon such as, consolidation settlement, dewatering and slope failure. In order to constrain the signatures more closely, the InSAR results can also be compared with physical monitoring datasets, such as groundwater levels, repeat ground based geomatics, or repeat scans of specific sites, as well as analysis of VHR ortho-photographs and DSM.

The following sections firstly explore the data associated with the stakeholder sites and subsequently attempt to provide a geological/geotechnical interpretation for the highest ground motion rates observed. The data are reviewed to firstly identify the density of results, and the rates and pattern of motion in the area of the key stakeholder sites (see Sections 4.1–4.5). For each site, a summary of the geological context and brief discussion is provided. Secondly, the data are assessed more broadly to identify the principal areas of ground movement revealed by this dataset and an interpretation of the causes for the observed deformation is presented. Analysis of the SUAV-derived ortho-photographs and DSM for the Maiden Mount mine collapse showcases how such VHR products can aid interpretation of localised ground instability in rural sites (Section 4.1).

Visual inspection of Figure 6 reveals variable movements in many of the known landslide areas while also highlighting motions associated within areas of historic mining activity. The results further identify a number of areas of interest that are displaying subsidence, potentially as a result of groundwater abstraction, soil compaction and shrink-swell processes.

It is worth noting that the use of a single reference point for the whole processed scene (see Section 3.1) might have caused propagation of some uncontrolled offsets across regions of the scene that are not very well connected, such as in the south-west corner of the processed scene (Figure 6). In this sector, narrow areas of land bound Lough Neagh to the west and to the south, thus increasing the risk of offsets in the estimated motion to occur. Inspection of the results against geological data, however, confirmed that the LOS deformation of -1.5 to -2 mm/year observed in this area is plausibly linked to compaction of superficial deposits (mainly peat, lacustrine deposits, till and alluvium).

4.1. Salt Mines in Carrickfergus

The town of Carrickfergus sits on the south-eastern edge of the Antrim Plateau on a low lying coastal plain rising to the west reaching a height of 275 m at Knockagh. The superficial geology is dominated by till and tidal flat deposits of clay and sand (Figure 3 and Table 1).

The bedrock geology is characterised by Triassic Mercia Mudstone Group (MMG) on the coastal plains and lava flows of the Lower Basalt Formation (LBF) at elevation towards the north-west. Between these two dominant strata are Cretaceous chalk and glauconitic sandstone sequences of the Ulster White Limestone Formation (UWLF) overlying Jurassic Waterloo Mudstone Formation (WMF) which occur at elevations above 100 m. A spring line is formed at the contact between the UWLF and impermeable WMF giving rise to extensive slope instability along the foot of the scarps. The scarps are dominated by talus aprons containing basaltic and limestone material. A number of N-S trending faults cut across the greater Carrickfergus area.

The town rests almost entirely upon MMG which, from an engineering perspective, has a high bearing strength. It is a fine grained mudstone with silty horizons. Gypsum occurs in veins and anhydrite as irregular nodules. Halite beds occur in the lower sequences of the MMG ranging in thickness from 1 to 40 m. This economic commodity was extracted by both conventional mining techniques and solution methods.

Table 1. Reference summary geology for the Carrickfergus mines area.

Age	Group/Formation	Acronym	Lithology
Quaternary	Tidal Flat deposits; Till		
Cretaceous	Ulster White Limestone Formation	UWLF	Chalk
Lower Jurassic	Waterloo Mudstone Formation	WMF	Grey mudstone with thin limestone beds
Triassic	Penarth Group	PNG	Dark grey mudstone
Triassic	Mercia Mudstone Group	MMG	Red and green mudstone and marl with thick salt bands

Rock salt mining began in the Woodburn area of Carrickfergus in the early 1850s, when a bed of rock salt at ~170 m depth was discovered at Duncrue, ~3 km north-west of the town, and a mine was developed and operated by the Belfast Mining Company. In only a few years, the mine reached a salt production rate of 20,000 tons/year. Other sites such as Burleigh Hill, French Park and Maiden Mount were later opened (in 1852, 1870 and 1877 respectively), and continued to operate after the closure of Duncrue in 1870, and until the late 1950s when the last of the old mines closed. Salt mines in this area were created using the 'pillar and stall' method, with large pillars of salt of about some metres in width and depth (e.g., 7 m by 5 m) standing at intervals to support the roof of the excavated galleries, characterised by levels of 4–5 m height. The two mines Maiden Mount and French Park became connected due to brining operations at depth in later years, hence they now technically constitute one unique mine. The overall extent of the fenced-off zone over the mine (determined from modelling the zone of influence for collapse) is around 0.14 km^2.

Currently there are eight inactive and one active salt mine within the town of Carrickfergus. The inactive mines, which were subject to solution mining, have left a legacy of instability over large areas with a large number of crown holes developing and significantly affecting public safety and infrastructure (Figure 2a). The subsurface voids formed by galleries and dissolution processes pose a hazard if upward propagation results in subsidence at the surface. In the area of Carrickfergus, unstable salt mines have been frequently subject to collapse, thus creating sinkholes on the ground and sometimes development of large ponds. An example is the collapse that occurred at the site of Maiden Mount on 19 August 2001, which generated a large surface hole of over 50 m in diameter and several metres depth (Figure 9a,c) that was subsequently filled artificially to reduce risk, but resulted in a much larger surface depression that altered the local topography.

Figure 9. Satellite images of the area of the Maiden Mount salt mine collapse at Carrickfergus from Google Earth on (**a**) 31 December 2001 © 2017 The GeoInformation Group and (**b**) 16 August 2016 © 2016 Google; (**c**) aerial view of the 2001 crown hole collapse © Crown Copyright; and (**d**) shaded relief of the VHR Digital Surface Model (DSM) with indication of the 2001 crown hole and a more recent scarp formed in the filling material.

The density of SBAS results within the area of interest of the Carrickfergus salt mine sites is limited, with only 3 targets found about 150 m east of the Maiden Mount shafts (i.e., two yellow and one green points in the centre of Figure 10a) and none around the area of the French Park shafts. They all indicate relatively stable conditions, with a maximum annual velocity of −1.7 mm/year over the monitoring period. More significant is that the time series in Figure 10b shows a steady trend of downward ground motion of at least 15 mm over the period 1992–2000.

For this study area, results from the InSAR analysis offer a somewhat limited potential due to the sparse coverage of reflectors and time series data. This poor coverage is mainly due to the rural land cover of the site and presence of vegetation, which are factors that generally cause temporal decorrelation and loss of InSAR coherence. When assessing land stability, the natural distribution and density of radar reflectors may not be sufficient to provide enough information for the problem under consideration. The low number of reflectors and limited information on ground motion in this area has therefore prevented the identification of a detailed ground stability baseline scenario for this region. Therefore, the integration with other data sources and VHR becomes necessary and useful for the analysis of localised ground instability and the assessment of hazard. In these cases, artificial reflectors (e.g., corner reflectors) to deploy across the investigated area might be used to locate monitoring targets at the desired locations.

Figure 7 shows the SUAV-derived VHR ortho-photograph and a close-up of the Maiden Mount salt mine collapse, derived based on the data acquired during the 10 May 2016 campaign. Figure 8 also shows the derived DSM, where the original crown collapse occurred in 2001 can be easily recognised around the current water pond. Elevations in the SUAV-derived DSM range between 97 and 135 m OD. It is worth noting that the SUAV campaigns focused on the Maiden Mount site only (i.e., a 0.06 km^2 survey area), whilst the overall extent of the zone of influence for collapse—including the French Park mine—is much greater than that shown in the SUAV ortho-photograph in Figure 7, as discussed above.

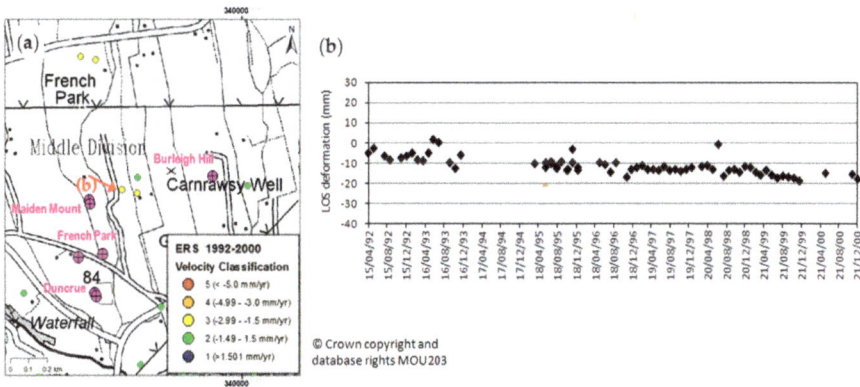

Figure 10. (a) Observed ground deformation and (b) time series for 1992–2000 in the area of the French Park and Maiden Mount salt mines in Carrickfergus. Mine shafts are indicated and labelled in pink. Reproduced from Land and Property Services data with the permission of the Controller of Her Majesty's Stationery Office, © Crown copyright and database rights MOU203.

Despite the presence of vegetation in the area of the 2001 collapse, during both the SUAV campaigns ground deformation features could be seen in the ortho-photograph and DSM (Figure 9d), as well as in a lower resolution satellite image acquired in August 2016 and available through Google Earth (Figure 9b). These data all confirm the conditions of the area which is in current restricted access state after the remediation and stabilisation works that took place after the original collapse occurred in 2001 (Figure 9a,c). Figure 11 also compares the VHR ortho-photographs obtained from the 10 May 2016 and 16 June 2017 SUAV campaigns and reveals not only significant increase in the vegetation cover and water level from 2016 to 2017 (which might be justified by the difference in the acquisition months, May and June, weather and soil moisture conditions), but also the occurrence of further ground settlement at the south-western edge of the pond, where a ~5 m wide and ~15 m long strip of deeper waters can be observed.

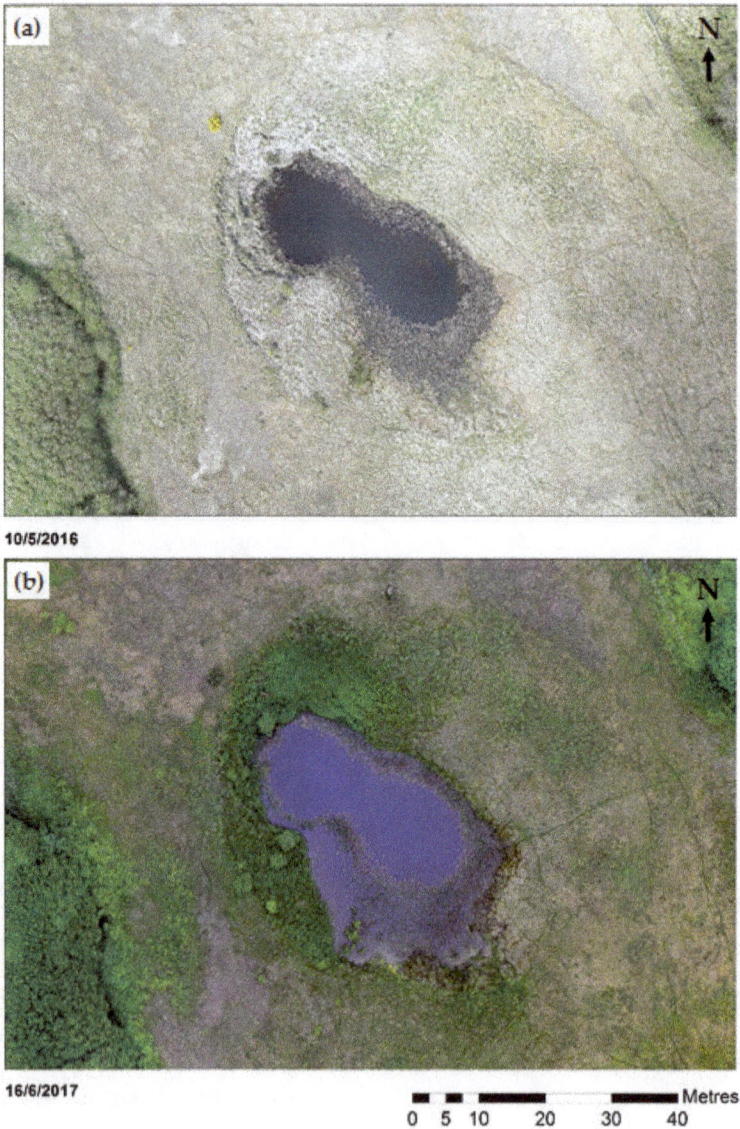

Figure 11. Comparison between VHR ortho-photographs from the (**a**) 10 May 2016 and (**b**) 16 June 2017 SUAV campaigns carried out at the Maiden Mount salt mine collapse at Carrickfergus.

Although a low density of InSAR targets was found in this area, some observations could be drawn in relation to the motion measured for the 3 targets located to the east of the site (just off the modelled zone of influence for collapse) and their time series (see Figure 10). Considering the date of the collapse (2001) and temporal span of the InSAR data, it might be possible—in theory—to relate the motion seen by the 1992–2000 InSAR analysis with the main event. In this case, however, it is difficult to identify precursors of the 2001 collapse based on analysis of the time series and the derivative of the deformation rates, as the pixels reveal an almost steady deformation trend throughout the monitoring period.

To further investigate instability in this area, other monitoring data collected from the abandoned salt mine sites at Carrickfergus in the past two decades were obtained and collated. The Department for the Economy provided extensometer, ground levelling and groundwater level data. The shafts and adits database maintained by the Geological Survey of Northern Ireland was also used to verify areas of known mine workings including recorded collapses or noted subsidence. These data were collated and time series sequences created to determine the rate of motion and water table levels at specific points within Carrickfergus. Together with the Department for the Economy sonar cavity data, they identify areas of minimal mine supports that are subsiding at a greater rate than areas of greater support while the water level is significantly influencing brine seepage events close to a public water supply intake.

Precise ground levelling station data and magnet extensometer collected from the French Park and adjacent Maiden Mount mines during the time period 1992–2000 showed differential motion across the surface with annual creep rates ranging from 1.0 mm/year to 10.00 mm/year. The greater of these movements were concentrated along the central sections of the site directly above the main underground works where sonar cavity surveys indicated the erosion of pre-existing support pillars. Groundwater level measurements for the same period showed little variation in levels indicating that. Situated 3 km to the east are the Carrickfergus/International salt mines. Precise ground levelling data collected at this site showed a trend of downward movement where significant creep rates of up to 40 mm/year were recorded. The rates of motion were steady throughout the time period 1992–2000 (as also observed from the analysis of the InSAR results) with the greatest creep rates associated with the areas close to the mine shafts.

4.2. Belfast–Bangor Railway Line

Operated by Northern Ireland Railways, the double-track Bangor-Belfast railway line (Figure 2b) is part of the key link into Belfast city centre, serving key assets of the city such as the Ulster Museum and George Best Belfast City Airport.

The line crosses a number of geological lithotypes. The area is one of low lying coastal plain with some reclaimed land between Belfast and Holywood. Superficial deposits are largely composed of diamiction sand and gravel deposits to the east with glaciofluvial, ice contact deposits of sand, gravel and boulders becoming prominent to the west (Figure 3 and Table 2). The superficial deposits rest on bedrock that is dominated, at its eastern extent, by Ordovician-Silurian greywacke and shale with a number of small blocks of Carboniferous and Triassic sandstones to the west. A number of NNW-SSE trending faults occur associated with the Carboniferous and Triassic blocks.

Table 2. Reference summary geology along the Bangor to Hollywood Railway line.

Age	Group/Formation	Lithology
Quaternary	Till; Glacial Sands and Gravels	Diamicton; sand and gravel
Triassic	Belfast Group	Marl with gypsum, dolomitic limestone and basal sandstone
Carboniferous	Holywood Group	Red conglomerate and sandstone with thin mudstones
Ordovician	Gilnahirk Group; including Helens Bay Formation	Greywacke and shale; Shale and pillow lava

Satellite InSAR monitoring of a high speed railway in China and validation with geodetic levelling has shown the potential of remotely sensed data to identify regional patterns of land motion that could affect the accessibility and safety of the lines (e.g., [18]). In this regard, the analysis of InSAR data within buffer zones around the railway track can ease the identification of critical sections of the infrastructure.

In our study, the narrow, heavily vegetated Belfast-Bangor railway line has generated a higher density of SBAS results than might have been anticipated (Figure 12a). The coverage reflects the distribution of the different land cover types across the region, with much denser networks of targets in the urbanised portions of the line, and sparser coverage where the vegetation is dense.

The rates of movement in 1992–2000 were moderate and zoned around the central area of Bangor (349,638; 381,583). In this sector, ground motion velocities reached −3.2 mm/year, but were generally in the order of −1.0 to −2.0 mm/year for most of the identified targets. Some of the time series (Figure 12b) indicated a general rise in ground level between mid-1993 and 1995, followed by stability until 1998 and then a gradual fall. This may reflect long-range seasonality in moisture in the tills and warrants comparison with historic climate data. Alternatively, there may be an underlying anthropogenic reason, for instance associated with maintenance and subsequent settlement of the track.

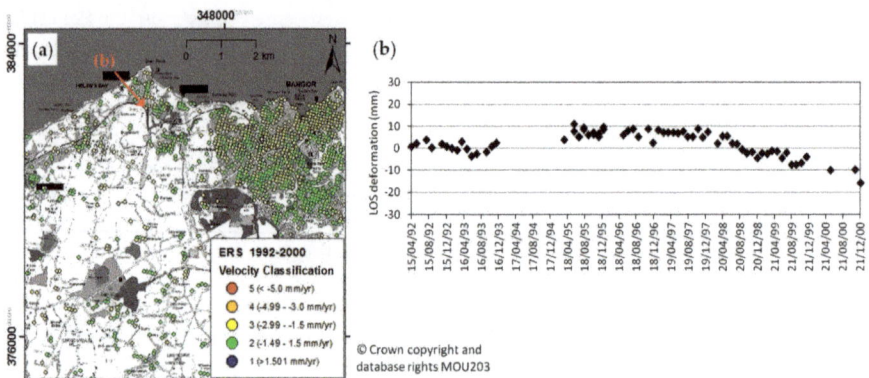

Figure 12. (a) Observed ground deformation and (b) time series for 1992–2000 in the area of the Belfast–Bangor railway line at Skelly Hill. Reproduced from Land and Property Services data with the permission of the Controller of Her Majesty's Stationery Office, © Crown copyright and database rights MOU203.

4.3. Throne Bend and Ligoniel Park

Throne Bend lies at the northern boundary of Belfast in a highly populated area and is one of the busy arterial routes into Belfast. Elevated above 80 m the area is underlain by the Triassic MMG and the Sherwood Sandstone Group close to the boundary with the overlying Jurassic mudstones, Cretaceous chalk and the LBF (Figure 3 and Table 3).

Table 3. Reference summary geology in the Throne Bend area.

Age	Group/Formation	Acronym	Lithology
Quaternary	Till		
Palaeocene	Lower Basalt Formation	LBF	Basalt
Cretaceous	Ulster White Limestone Formation	UWLF	Chalk
Lower Jurassic	Waterloo Mudstone Formation	WMF	Grey mudstone with thin limestone beds
Triassic	Penarth Group	PNG	Dark grey mudstone
Triassic	Mercia Mudstone Group	MMG	Red and green mudstone and marl with thick salt bands

Large Pleistocene landslide structures with flat areas composed of massive slump blocks displaced from the Antrim Plateau dominate the topography. Below the steep edge of the slumped blocks, the surface is dominated by landslide material composed of mudstones, sandstones, chalk and basaltic material. Incipient instability within slide debris has been recorded during excavation work on local developments [19] and roads and localised cracking of buildings and walls is evidence that some movement is still occurring within the area.

The geological setting of Ligoniel (Figure 2c) is very similar to that of Throne Bend. Lying to the north-west of the city and elevated at over 200 m, the dominant bedrock lithology is that of MMG, which forms the bedrock to much of the development in Ligoniel (Figure 3 and Table 4). The area is adjacent to a steep escarpment of UWLF and LBF with a substantial thickness of landslip debris material. An area of particular interest is in a local park, where immediately to the north of a water dam, and the impermeable MMG boundary with the Hibernian Greensands occurs within a gentle slope at the boundary between the impermeable MMG and the Hibernian Greensands. It is likely, as noted in other areas, that this boundary will be a spring line, creating conditions conducive to surface movement. Movement within the park has been monitored, at the request of Belfast City Council since the early 1990s, due to concerns regarding the stability of a water dam positioned adjacent to the area of noted movement.

Table 4. Reference summary geology for the Ligoniel Park area.

Age	Group/Formation	Acronym	Lithology
Quaternary	Till		
Palaeocene	Lower Basalt Formation	LBF	Basalt
Cretaceous	Ulster White Limestone Formation	UWLF	Chalk
Triassic	Mercia Mudstone Group	MMG	Red and green mudstone and marl with thick salt bands

There is a relatively low density of reflectors in the Throne Bend area (see Figure 6), the same as the density observed in Ligoniel Park (Figure 13a). As observed for the other areas of interest, this reflects the land cover of these regions, which are mostly dominated by the presence of vegetation and absence of surface features acting as good radar targets and showing sufficient interferometric coherence.

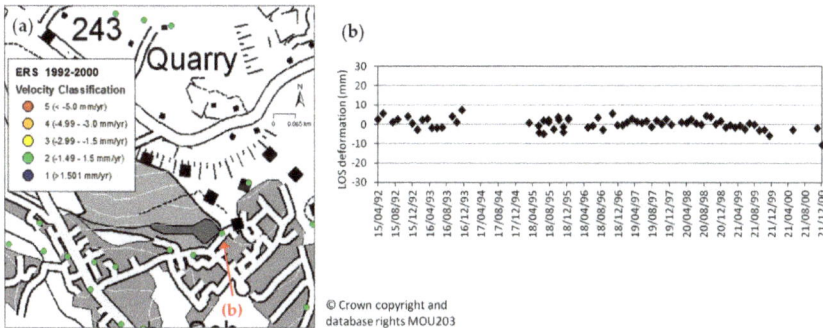

Figure 13. (a) Observed ground deformation and (b) time series for 1992–2000 in the area of Ligoniel Park. Reproduced from Land and Property Services data with the permission of the Controller of Her Majesty's Stationery Office, © Crown copyright and database rights MOU203.

The reflectors that have been identified indicate relatively stable conditions. Two points, one on the eastern perimeter of the Ligoniel Park and one to the northeast indicated −0.81 mm/year at (330,544; 377,691) and −0.45 mm/year at (330,464; 377,544) during the 1992−2000 monitoring period.

The time series in Figure 13b confirmed the small magnitudes of the motion and did not show any significant nonlinear trend, albeit there appeared to be a more pronounced settlement or motion of ground levels after April 1999. On the other hand, in situ monitoring data from Belfast City Council data for Ligoniel Park have shown the presence of seasonal effects on ground motion at the location suggesting climatic parameters have an influence on the stability of the site.

4.4. Straidkilly and Garron Point

The Garron Point-Straidkilly area sits on the eastern side of the Antrim plateau and is characterised by deep valleys incised through glacial deposits and on its eastern margin large scale rotational landslip blocks sitting at angles up to 70°. The superficial geology is predominantly glacial tills and, in upland areas, peat. Much of the slope is covered by talus composed of basalt and chalk derived from the bedrock.

The bedrock geology is dominated by sequences of Palaeogene basalt flows of the Antrim Lava Group (ALG) overlying white Cretaceous chalk (Figure 3 and Table 5). Inland, only the basalt sequences are exposed while on the coastal sections the high angle of the rotational features exposes the entire stratigraphy down to the MMG (Figure 14). The lithology and structure combine to make the area highly susceptible to landslides. Fractures within the basalt and chalk act as pathways for water infiltration to the impermeable mudstones giving rise to a number of mudslides often affecting the coastal road.

Table 5. Reference summary geology for the Garron Point area.

Age	Group/Formation	Acronym	Lithology
Palaeocene	Upper Basalt Formation	UBF	Olivine basalt lava
Palaeocene	Interbasaltic Formation	IB	Laterite, bauxite and lithomarge
Palaeocene	Lower Basalt Formation	LBF	Olivine basalt lava
Cretaceous	Ulster White Limestone Formation	UWLF	Chalk
Lower Jurassic	Waterloo Mudstone Formation	WMF	Grey mudstone with thin limestone beds
Triassic	Mercia Mudstone Group	MMG	Red and green mudstone and marl with thick salt bands

Figure 14. Slope instability at Garron Point. Photograph taken in 2012. Preventative measures that were put in place to minimise risk from rock-falls are indicated. UWLF: Ulster White Limestone Formation; LBF: Lower Basalt Formation.

The main feature identified in this area from the analysis of the InSAR data is at Caranure (328,353; 425,197), which lies to the west of Garron Point. Here a single point reached ground motion velocity of −3.1 mm/year. The surrounding points have been subject to LOS movement velocities of −1.5 to −2.9 mm/year (Figure 15a). The movement might be associated with the presence of Crearlagh Burn river, which crosses this zone. It is located in an area of mapped mass movement. The steep topography and morphology of the unstable area suggest that the majority of the movement occurs along the vertical direction with minor horizontal components. Accounting for the direction of the satellite LOS in descending mode (i.e., oriented from east to west with a look angle of 23° with respect to the vertical direction; Figure 5), the projection of the observed LOS motion onto the vertical direction would result in rates of −1.6 to 3.4 mm/year. The latter are obtained by dividing the LOS velocity values by the cosine of the look angle θ (hence, by 0.92 in this case). Alternatively, the computation of the velocity could have been performed by assuming the steepest slope direction as the most probable direction of motion, and therefore using the directional cosine of the LOS and slope to derive the correct value of the velocity for each pixel, as per [4].

The time series for Caranure appeared to provide evidence of the landslip that is known to have occurred in this area. The landslide database maintained by the Geological Survey of Northern Ireland records three landslides, one 200 m to the south and two 175 m to the east.

The movement during the period 1992 to 2000 (Figure 15b) comprises a strong trend of declining ground levels, commencing in December 1992 through to May 2000, with a number of superimposed steps of more rapid reductions in ground level. The largest of these occurred between April 1999 and May 2000 when there was a lowering of the ground surface of 16.6 mm.

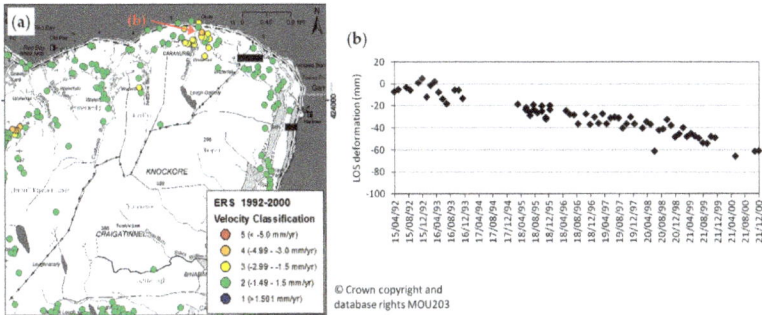

Figure 15. (a) Observed ground deformation and (b) time series for 1992–2000 in the area of Garron Point. Reproduced from Land and Property Services data with the permission of the Controller of Her Majesty's Stationery Office, © Crown copyright and database rights MOU203.

Landslide motions occurring along the east facing steep slopes of Garron Point are unlikely to be seen by the employed satellite data due to the geometry of the descending mode LOS of the ERS-1/2 satellites, and the presence of topographic distortions (mainly radar layover; e.g., [20]). In less steep but still east facing slopes, even in absence of radar layover, motions occurring along directions perpendicular to the satellite LOS are also undetectable, resulting in the presence of targets showing very low motion velocities (e.g., between ±1.5 mm/year). In such areas, the use of the ascending mode stack of ERS-1/2 data might help to complement the analysis to capture ground motion along the west to east looking direction.

Farther north, centred on (316,844; 443,189) near Doon, two additional points indicate another coastal landslide/rock fall on a steep escarpment. The points exhibit velocities of −3.0 and −3.7 mm/year. The time series plot showed the long term lowering of ground level, followed by stabilization associated with a rise in ground level. There is visual evidence of motion events at this location, but this is not yet identified in the landslide database, although landsliding was verified in a local newspaper report. The geology at this location comprises till over Carboniferous limestone.

4.5. Belfast City

The greater Belfast area can be divided into low ground of the Lagan Valley which runs south-west, the northern escarpment of the Antrim Plateau and the low hills to the south-east spreading into county Down. The low lying Lagan area is dissected by the River Lagan meandering south west in an area underlain by Triassic Sandstone Formation and overlain by superficial raised tidal flat deposits (Figure 3 and Table 6).

The northern part is dominated by the MMG and ALG rising up to the southern edge of the Antrim Plateau. To the south east there is a series of Permian mudstones and sandstones and Ordovician greywacke. Bedrock is covered by glaciofluvial sand, gravel and deposits. A number of Paleogene basalt dykes cut the strata most notably in its northern extent.

Table 6. Reference summary geology for the Belfast area.

Age	Group/Formation	Acronym	Lithology
Quaternary	Till		
Palaeocene	Lower Basalt Formation	LBF	Basalt
Cretaceous	Ulster White Limestone Formation	UWLF	Chalk
Lower Jurassic	Waterloo Mudstone Formation	WMF	Grey mudstone with thin limestone beds
Triassic	Penarth Group	PNG	Dark grey mudstone
Triassic	Mercia Mudstone Group	MMG	Red and green mudstone and marl with thick salt bands
Triassic	Sherwood Sandstone Group	SSG	Red sandstone
Permian	Connswater Marl Formation	CONN	
Permian	Enler Group	ENLE	Sandstone and subordinate breccia
Ordovician	Leadhills Supergroup	LHG	Wacke and Mudstone

During the period 1992 to 2000 Belfast ground levels appeared to have been generally stable. However, reflectors around the perimeter of the Belfast Lough (in particular on its eastern and southern sides) fall within the class defined by the −1.50 to −2.99 mm/year velocity range (Figure 16a). The time series at (334,813; 377,546) (Figure 16b) with an annual ground velocity of −3.1 mm/year indicated that in some areas this is attributable to consolidation of freshly reclaimed ground.

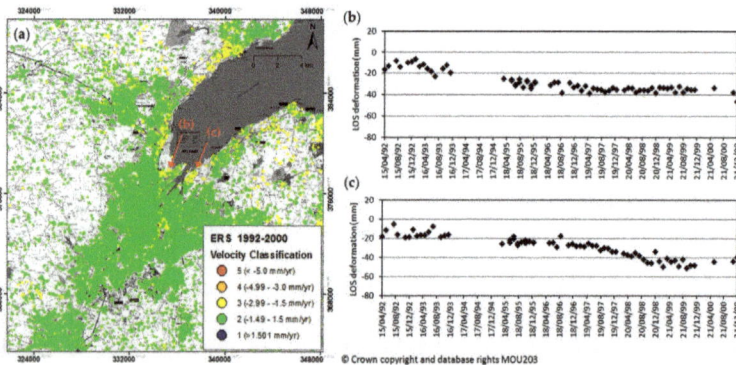

Figure 16. (a) Observed ground deformation, (b,c) time series for 1992–2000 in Belfast. Reproduced from Land and Property Services data with the permission of the Controller of Her Majesty's Stationery Office, © Crown copyright and database rights MOU203.

Similarly, other areas close to the harbour at (337,212; 377,143) and (337,187; 377,069) (Figure 16c) were subject to average velocities of −3.3 and −3.8 mm/year respectively over the period 1992–2000. It is well known in the specialist InSAR literature (e.g., [21,22]) that land subsidence movements of such type are surface descriptors of interactions between local geology and urban development.

5. Conclusions and Future Work

By analysing InSAR results for 1992–2000, together with the landslide and mine databases, geological data and historical records, this study contributes to the understanding of the geological and potential anthropogenic causes (e.g., rainfall, engineering works, mining-related drivers, lowering of the water table through water abstraction) of the observed motion in Northern Ireland. Moreover, it allows potential future hazards to the geotechnical infrastructure to be identified, by providing time series data that extend the traditional monitoring period and given an indication of how events built.

Collaborative engagement with stakeholders has been very beneficial in accessing monitoring data for ground truthing the InSAR results. The benefit for the stakeholders from such an approach is multi-facetted. TransportNI anticipated that the use of InSAR data would have provided an enhanced capability to monitor and assess landslide hazards across the whole 21,000 km of road network, and helped them to form their strategies for monitoring their geotechnical assets to feed into the existing GIS based risk assessment methods for their infrastructure assets. The site at Straidkilly was only one of many sections along the A2 coast road that is unstable and InSAR has provided an improved insight into the behaviour of geohazards that impact on the road, for instance at Garron Point, therefore providing information towards their maintenance strategies and more cost-effective and better targeted maintenance.

TransportNI also were committed to having a better understanding of the mechanisms of failure on the slow moving failures on the Throne Bend in Belfast. Unfortunately, the InSAR results provided an extremely low density of targets in this area due to its rural land cover, not allowing an enhanced understanding of the magnitude of ground movement to be achieved. In addition to similar benefits and impacts to TransportNI, Northern Ireland Railways aimed to correlate the slope instability against rainfall data on the Belfast-Bangor rail line, and to reduce the requirement to have staff on active rail tracks inspecting geohazards. This narrow, heavily vegetated line generated a higher density of SBAS results than might have been anticipated showing moderate rates of movement, zoned around the central area of Bangor.

The Department for the Economy anticipated that the study would have validated new methods of monitoring and provide baseline data of ground motion to form the basis of future strategic decisions in regards to geohazards. This would enable a greater capability to assess and communicate the risk posed by ground subsidence and support the planning process in determining zones of movement with greater accuracy. The use of InSAR at sites in Carrickfergus provided additional knowledge of subsidence rates in the eight years preceding the collapse of August 2001 (i.e., 1992–2000). Integrated at local scale with a SUAV surveying carried out in 2016, it also highlighted indicators of new surface features in the area of the 2001 collapse, thus demonstrating the need for surveying technologies at higher resolution when dealing with localised deformation such as that occurring at Maiden Mount salt mine.

Our close collaboration with the stakeholders in this work has proved the need of an effective knowledge transfer, from remote sensing experts and scientists to practitioners and infrastructure managers. Future work in the project will be focused on enhancing the stakeholders' perception of the full potential of InSAR data as well as their synergy with localised investigations of instability indicators through generation and inspection of VHR surface models.

Analysis of more recent ground deformation data derived from the processing of ENVISAT time series will be also carried out, and geostatistical analysis of the results to identify correlation, anomalies and trend deviations occurring along the transport networks and mining sites to highlight any sectors needing particular attention and maintenance for the stakeholders to focus resources.

Since satellite SAR imagery is now available almost globally and freely (for instance from the Sentinel-1 constellation), potentially any geotechnical infrastructure network and respective managers could benefit from improved network-wide monitoring and a better long-term understanding of ground behaviour from InSAR techniques.

Not only transport infrastructure but also utility networks (such as supplying water, electricity or gas) are required to monitor ground stability and this is currently particularly difficult where utilities cross remote areas. In this context, InSAR could also be used for terrain evaluation, route planning and regular monitoring for utility and transport corridors. Future efforts in the project will be devoted to extending the knowledge transfer to a wider range of stakeholders to increase the reach of these techniques.

Acknowledgments: This work is funded to the authors by the Natural Environment Research Council (NERC) under the Environmental Risks to Infrastructure Innovation Programme (ERIIP). NERC grants: NE/N013018/1 (Lead PI: D. Hughes) and NE/N013042/1 (PI: F. Cigna): *InSAR for geotechnical infrastructure: enabling stakeholders to remotely assess environmental risk and resilience* (February 2016–July 2017). ERS-1/2 and ENVISAT satellite data were provided by the European Space Agency (ESA) under grant id.32627. The authors would like to thank DroneLab (https://www.dronelab.io/) for hosting the SUAV-derived ortho-photographs at no cost. F. Cigna, V. Banks, A. Donald and K. Parker publish with the permission of the Executive Director of the British Geological Survey—NERC. Land and Property Services data are reproduced with the permission of the Controller of Her Majesty's Stationery Office, © Crown copyright and database rights MOU203.

Author Contributions: F.C., V.B., A.D., K.P., D.H., J.M, S.D. and C.G. designed the project; F.C. designed the InSAR analysis and carried out the ERS-1/2 image processing; C.G. led and carried out the SUAV campaign, and created 2D and 3D models; V.B., A.D., K.P., D.H., J.M. and S.D. analysed the InSAR data for the areas of interest; K.P. collated the mining monitoring data; F.C. wrote the paper with equal contributions from V.B., A.D., K.P., D.H., J.M. S.D. and C.G.

Conflicts of Interest: The authors declare no conflict of interest. The funding sponsors had no role in the design of the study; in the collection, analyses, or interpretation of data; in the writing of the manuscript, and in the decision to publish the results.

Appendix A

A low resolution version of our SUAV-derived digital surface model (DSM) of the Maiden Mount salt mine collapse at Carrickfergus, UK, based on the 2016 campaign is freely available for 3D displaying and download at SketchFab: https://skfb.ly/QQMG (Figure A1).

Full resolution web-views of the VHR ortho-photographs are also available for displaying at DroneLab: http://go.qub.ac.uk/SMORTHO16 and http://go.qub.ac.uk/SMORTHO17 (Figure A2).

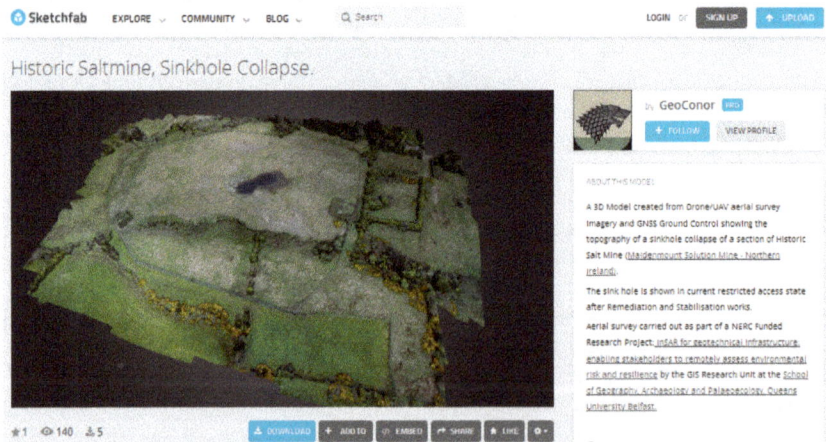

Figure A1. SUAV-derived DSM of the Maiden Mount salt mine collapse at Carrickfergus, UK, available for 3D displaying and download via SketchFab.

Figure A2. SUAV-derived VHR ortho-photographs of the Maiden Mount salt mine collapse at Carrickfergus, UK, based on the (**a**) 2016 and (**b**) 2017 campaigns and made available via DroneLab.

References

1. Mitchell, W.I. *The Geology of Northern Ireland: Our Natural Foundation*; Geological Survey of Northern Ireland: Belfast, UK, 2004; p. 318.
2. Rosen, P.A.; Hensley, S.; Joughin, I.R.; Fuk, K.L.; Madsen, S.N.; Rodriguez, E.; Goldstein, R.M. Synthetic aperture radar interferometry. *Proc. IEEE* **2000**, *88*, 333–382. [CrossRef]
3. Colesanti, C.; Wasowski, J. Investigating landslides with space-borne synthetic aperture radar (SAR) interferometry. *Eng. Geol.* **2006**, *88*, 173–199. [CrossRef]
4. Cigna, F.; Bianchini, S.; Casagli, N. How to assess landslide activity and intensity with Persistent Scatterer Interferometry (PSI): The PSI-based matrix approach. *Landslides* **2013**, *10*, 267–283. [CrossRef]

5. Herrera, G.; Gutiérrez, F.; García-Davalillo, J.C.; Guerrero, J.; Notti, D.; Galve, J.P.; Fernández-Merodo, J.A.; Cooksley, G. Multi-sensor advanced DInSAR monitoring of very slow landslides: The Tena Valley case study (Central Spanish Pyrenees). *Remote Sens. Environ.* **2013**, *128*, 31–43. [CrossRef]

6. Galloway, D.L.; Hudnut, K.W.; Ingebritsen, S.E.; Phillips, S.P.; Peltzer, G.; Rogez, F.; Rosen, P.A. Detection of aquifer system compaction and land subsidence using interferometric synthetic aperture radar, Antelope Valley, Mojave Desert, California. *Water Resour. Res.* **1998**, *34*, 2573–2585. [CrossRef]

7. Bell, J.W.; Amelung, F.; Ferretti, A.; Bianchi, M.; Novali, F. Permanent scatterer InSAR reveals seasonal and long-term aquifer-system response to groundwater pumping and artificial recharge. *Water Resour. Res.* **2008**, *44*, W02407. [CrossRef]

8. Cigna, F.; Osmanoğlu, B.; Cabral-Cano, E.; Dixon, T.H.; Ávila-Olivera, J.A.; Gardūno-Monroy, V.H.; DeMets, C.; Wdowinski, S. Monitoring land subsidence and its induced geological hazard with Synthetic Aperture Radar Interferometry: A case study in Morelia, Mexico. *Remote Sens. Environ.* **2012**, *117*, 146–161. [CrossRef]

9. Bell, A.D.F.; McKinley, J.M.; Hughes, D.A.B.; Hendry, M.; Macciotta, R. Spatial and temporal analyses using Terrestrial LiDAR for monitoring of landslides to determine key slope instability thresholds: Examples from Northern Ireland and Canada. In Proceedings of the 6th Canadian GeoHazards Conference—GeoHazards 6, Kingston, ON, Canada, 15–18 June 2014. [CrossRef]

10. Berardino, P.; Fornaro, G.; Lanari, R.; Sansosti, E. A new algorithm for surface deformation monitoring based on small baseline differential interferograms. *IEEE Trans. Geosci. Remote* **2002**, *40*, 2375–2383. [CrossRef]

11. Cook, K.L. An evaluation of the effectiveness of low-cost UAVs and structure from motion for geomorphic change detection. *Geomorphology* **2017**, *278*, 195–208. [CrossRef]

12. Lindner, G.; Schraml, K.; Mansberger, R.; Hübl, J. UAV monitoring and documentation of a large landslide. *Appl. Geomat.* **2016**, *8*, 1–11. [CrossRef]

13. Niethammer, U.; James, M.R.; Rothmund, S.; Travelletti, J.; Joswig, M. UAV-based remote sensing of the Super-Sauze landslide: Evaluation and results. *Eng. Geol.* **2012**, *128*, 2–11. [CrossRef]

14. Fernández, T.; Pérez, J.L.; Cardenal, J.; Gómez, J.M.; Colomo, C.; Delgado, J. Analysis of Landslide Evolution Affecting Olive Groves Using UAV and Photogrammetric Techniques. *Remote Sens.* **2016**, *8*, 837. [CrossRef]

15. Chen, J.; Li, K.; Chang, K.J.; Sofia, G.; Tarolli, P. Open-pit mining geomorphic feature characterisation. *Int. J. Appl. Earth Obs.* **2015**, *42*, 76–86. [CrossRef]

16. Smith, M.W.; Carrivick, J.L.; Quincey, D.J. Structure from motion photogrammetry in physical geography. *Prog. Phys. Geogr.* **2015**, *40*, 247–275. [CrossRef]

17. Ely, J.; Graham, C.; Barr, I.; Rea, B.; Spagnolo, M.; Evans, J. Using SUAV acquired photography and structure from motion techniques for studying glacier landforms: Application to the glacial flutes at Isfallsglaciären. *Earth Surf. Process. Landf.* **2016**, *42*, 877–888. [CrossRef]

18. Ge, D.; Wang, Y.; Zhang, L.; Xia, Y.; Wang, Y.; Guo, X. Using Permanent Scatterer InSAR to monitor land subsidence along high speed railway—The first experiment in China. In Proceedings of the Workshop Fringe 2009 ESA SP-677, Frascati, Italy, 30 November–4 December 2009.

19. Skempton, A.W.; Hutchinson, J.N. *Slope Problems on the North-East Side of Cave Hill*; A Report for the Ministry of Development; Imperial College London: London, UK, 1970.

20. Cigna, F.; Bateson, L.; Jordan, C.J.; Dashwood, C. Simulating SAR geometric distortions and predicting Persistent Scatterer densities for ERS-1/2 and ENVISAT C-band SAR and InSAR applications: Nationwide feasibility assessment to monitor the landmass of Great Britain with SAR imagery. *Remote Sens. Env.* **2014**, *152*, 441–466. [CrossRef]

21. Pratesi, F.; Tapete, D.; Del Ventisette, C.; Moretti, S. Mapping interactions between geology, subsurface resource exploitation and urban development in transforming cities using InSAR Persistent Scatterers: Two decades of change in Florence, Italy. *Appl. Geogr.* **2016**, *77*, 20–37. [CrossRef]

22. Cigna, F.; Jordan, H.; Bateson, L.; McCormack, H.; Roberts, C. Natural and anthropogenic geohazards in greater London observed from geological and ERS-1/2 and ENVISAT persistent scatterers land subsidence data: Results from the EC FP7-SPACE PanGeo Project. *Pure Appl. Geophys.* **2015**, *172*, 2965–2995. [CrossRef]

geosciences

MDPI

Article

Combined Use of C- and X-Band SAR Data for Subsidence Monitoring in an Urban Area

Lorenzo Solari [1,*], **Andrea Ciampalini** [1], **Federico Raspini** [1], **Silvia Bianchini** [1], **Ivana Zinno** [2], **Manuela Bonano** [2,3], **Michele Manunta** [2], **Sandro Moretti** [1] and **Nicola Casagli** [1]

[1] Department of Earth Sciences, University of Firenze, Via La Pira 4, 50121 Firenze, Italy; andrea.ciampalini@unifi.it (A.C.); federico.raspini@unifi.it (F.R.); silvia.bianchini@unifi.it (S.B.); sandro.moretti@unifi.it (S.M.); nicola.casagli@unifi.it (N.C.)

[2] Istituto per il Rilevamento Elettromagnetico dell'Ambiente, IREA-CNR, via Diocleziano 328, 80124 Napoli, Italy; zinno.i@irea.cnr.it (I.Z.); bonano.m@irea.cnr.it (M.B.); manunta.m@irea.cnr.it (M.M.)

[3] Istituto di Metodologie per l'Analisi Ambientale, IMAA-CNR, Contrada Santa Loja-Zona Industriale, Tito Scalo, 85050 Tito, Italy

* Correspondence: lorenzo.solari@unifi.it; Tel.: +39-055-275-7548

Academic Editors: Francesca Cigna and Jesús Martínez Frías
Received: 3 February 2017; Accepted: 29 March 2017; Published: 1 April 2017

Abstract: In this study, we present the detection and characterization of ground displacements in the urban area of Pisa (Central Italy) using Interferometric Synthetic Aperture Radar (InSAR) products. Thirty RADARSAT-2 and twenty-nine COSMO-SkyMed images have been analyzed with the Small BAseline Subset (SBAS) algorithm, in order to quantify the ground subsidence and its temporal evolution in the three-year time interval from 2011 to 2014. A borehole database was reclassified in stratigraphical and geotechnical homogeneous units, providing the geological background needed for the local scale analysis of the recorded displacements. Moreover, the interferometric outputs were compared with the last 30 years' urban evolution of selected parts of the city. Two deformation patterns were recorded by the InSAR data: very slow vertical movements within the defined stability threshold (± 2.5 mm/yr) and areas with subsidence rates down to -5 to -7 mm/yr, associated with high peak velocities (-15 to -20 mm/yr) registered by single buildings or small groups of buildings. Some of these structures are used to demonstrate that the high subsidence rates are related to the recent urbanization, which is the trigger for the accelerated consolidation process of highly compressible layers. Finally, this urban area was a valuable test site for demonstrating the different results of the C- and X-band data processing, in terms of the density of points and the quality of the time series of deformation.

Keywords: A-DInSAR; subsidence; urban area; C- and X-band satellites

---

1. Introduction

Subsidence is a broad term usually referring to the ground lowering induced by natural or anthropic factors. Among the anthropic causes, rapid urbanization is considered as one of the most important accelerating factors for the consolidation process, especially where unconsolidated alluvial deposits, peat-rich layers, and reclamation areas are present [1–6].

The analysis of Synthetic Aperture Radar (SAR) images with Advanced Differential Interferometry (A-DInSAR) techniques is a useful tool for monitoring different ground movements, including landslides, tectonic and volcanic activity, and subsidence [6–9]. The various analysis algorithms that have been developed in the last two decades allow reaching the millimeter accuracy, in addition to a regional-scale coverage of data at a reasonable cost [10]. These advantages have been exploited to monitor the ground subsidence of urban areas in many nations around the world, including

Italy [11–16], China [4,5,17–19], Spain [20–23], Greece [24–26], the United States of America [27–29], Vietnam [30,31], Indonesia [32,33], and Portugal [34,35].

In this study, C-band RADARSAT-2 (RSAT) and X-band COSMO-SkyMed (CSK) data have been processed with the Small BAseline Subset (SBAS) approach, to provide the deformation maps of the urban area of Pisa (Central Italy) in the time interval 2011–2014. The area of interest has been characterized by exploiting the potential of the two satellites, at a regional scale for RADARSAT-2 and at a building scale for COSMO-SkyMed. New deformation-affected areas and trend changes in the time series of deformation of selected buildings have been examined, to demonstrate the existing relation between recorded displacements, local geological setting, and urban development.

2. Geological and Hydrogeological Setting

The Arno coastal plain, which is 450 km^2 wide, is a densely-populated area with several industrial activities and diffuse cultural heritages. The city of Pisa occupies a central position in this coastal plain, which represents the southern inshore portion of the Viareggio basin, a subsiding half-graben [36,37]; this tectonically formed basin was progressively infilled by marine and alluvial sediments after the Last Glacial Maximum (LGM) [36,37]. The Quaternary sea level fluctuations produced an incised valley perpendicular to the present shoreline, with a maximum depth of 40 m [38–40]. After the eustatic sea level rise post LGM, estuarine and coastal plain sedimentation took place, filling almost half of the valley [37]. The last 20–25 m of the infilling sequence is composed of transgressive Holocene sediments, deposited after the phase of the decelerated sea level rise [41].

At its base, this succession has a level of clays and silty clays, locally known as *"pancone"*, continuously widespread in the plain to the border of the coastal beach ridges system. This level, deposited in a lagoonal environment, is characterized by the presence of layers rich in organic matter [42]. Above the *pancone*, a fluvio-deltaic sequence, 10–15 m thick, is found; it is composed of clay and silt with sporadic sand and silty sand bodies that reflect the coastal progradation of the modern Arno delta [37].

The stratigraphic structure of the Holocene succession leads to the presence of geotechnically-weak layers in the first 20 m of the underground beneath the city [43], producing a preferential predisposing factor for ground subsidence and the structural failure of buildings, as testified by the most famous cultural heritage of the city, the Leaning Tower [44].

From a hydrogeological point of view, a phreatic aquifer composed of sandy and silty-sand layers within the clayey sequence is found. This water reservoir is not usually used for water extraction because of its low transmissivity and quality, and its high vulnerability to contamination [45]. The phreatic surface level is almost the same across the northern part of the city and in the city center, while the minimum depth (2 m below sea level) is registered in the southern part of the International Airport Galileo Galilei (Figure 1). The maximum level is found in the Ospedaletto district, reaching 1.5 m above sea level [45]. The most important water resource for the city of Pisa is located at a higher depth, ranging from 50 to 100 m below the surface, and it is composed of a confined aquifer made up of Pleistocene gravels and sands, 10–20 m thick [46,47]. This aquifer, generally characterized by water of a low quality, is mainly used for gardening, irrigation, and industry, while only a few pumping wells are exploited for drinking water [46].

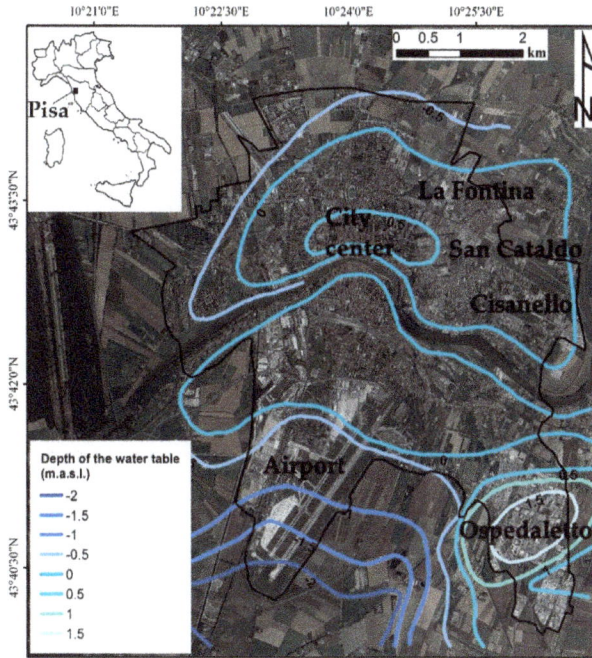

Figure 1. Geographical framework of the area of interest and the phreatic surface in the urban area of Pisa (from [45]). The black line represents the urban area. The contour map is overlaid on a 2010 digital orthophoto.

3. Available Data

3.1. Interferometric Data

The two SAR datasets exploited for this study are analyzed with the advanced DInSAR technique, referred to as the SBAS approach [48–53], to retrieve the deformation time series in the temporal range between 2011 and 2014. The data processing of the SAR images has been performed by using the full Resolution SBAS-DInSAR processing chain, developed and owned by CNR-IREA (National Research Council-Institute for Electromagnetic Sensing of the Environment, Napoli, Italy). The key feature of this algorithm's chain is the selection of the SAR images that will be used to generate the interferograms; this is done by choosing only those data pairs characterized by short temporal and perpendicular baselines [48]. This procedure allows for a minimization of the noise effects, intended as decorrelation phenomena [52], thus maximizing the number of coherent pixels in the generated interferograms. Independent acquisition datasets, separated by large baselines, are properly linked by applying the Singular Value Decomposition (SVD) method, to retrieve deformation time series and velocity maps. The SBAS algorithm is also capable of producing results at two different scales [49]: a regional-scale, for investigating diffused deformation phenomena at a medium spatial resolution (the cell resolution ranges from 100 × 100 m to 30 × 30 m) [51]; and a local scale, which exploits the full resolution interferograms (benefiting from the high-resolution capability of the SAR satellite system) for generating products ideally suited for building analysis [53,54]. The regional-scale product is used to estimate the low pass signal components, typical of the atmospheric artifacts and related to deformations over large areas [48]. Conversely, the local-scale output is exploited to investigate the high frequencies, typical of localized displacements [49,51,53].

The COSMO-SkyMed X-band ($\lambda = 3.1$ cm) dataset is comprised of 29 ascending acquisitions collected between the 26 September 2012 and the 11 May 2014. The data have been acquired in Stripmap mode with HH polarization. The CSK scenes are characterized by an incidence angle at mid-range of around 30° and by a spatial resolution of 3 m in both range and azimuth directions. The 29 CSK scenes have been firstly converted from the raw data L0 format into the Level1 SLC (Single Look Complex) format, and a quality assessment has also been applied to these images. Then, they have been co-registered with respect to the 11 July 2013 master image, chosen as reference geometry. The CSK dataset is characterized by a large spatial baseline distribution, since the orbital tube exceeds 2000 m and an average perpendicular baseline of 380 m, as well as by a quite dense temporal sequence, with an average sampling of 21 days (Figure 2a). The CSK images have been coupled, by applying the Delaunay triangulation method [55], into 88 interferometric pairs, characterized by maximum spatial and temporal baseline values of about 800m and 500 days, respectively. Following the interferogram generation step, the PhU (Phase Unwrapping) operation at the regional scale has been carried out by applying the method described in [55] and selecting the pixels that have a coherence, estimated in a box of 10 × 10 pixels, in the range and azimuth directions of greater than 0.35 in at least 30% of the interferograms [55]. The final selection of the pixels correctly unwrapped has been carried out by considering pixels with a temporal coherence value [52] greater than 0.7.

The RADARSAT-2 C-band ($\lambda = 5.6$ cm) dataset is composed of 30 ascending orbit images acquired in Standard 3 (S3) beam mode (one of eight standard beam modes of the sensor, covering a nominal stripmap swath of 108 km and acquired with a range bandwidth of 11.56 MHz) and with HH polarization, covering the time interval 27 October 2011–26 April 2014. The RSAT scenes are characterized by a mid-range incidence angle of around 30° and a spatial resolution of 12 × 5 m in the slant range and azimuth directions, respectively. The 30 RSAT images, in the Level1 SLC format, have been properly co-registered with respect to a common geometry (master image); the 14 November 2012 acquisition has been selected for this purpose. The RSAT dataset is characterized by a good spatial baseline distribution, with an orbital tube around 700 m and an average perpendicular baseline of 120 m, in addition to a temporal sequence with an average sampling of 31 days (Figure 2b). The RSAT images have been coupled into 87 interferometric pairs, characterized by maximum spatial and temporal baseline values of about 300 m and 900 days, respectively. Following the interferogram generation step, the PhU operation at the regional scale has been carried out by selecting the pixels that have a coherence, estimated in a box of 4 × 20 pixels, in the range and azimuth directions of greater than 0.35 in at least the 30% of the interferograms [55]. The final selection of the pixels correctly unwrapped has been carried out by considering pixels with a temporal coherence value [52] greater than 0.7.

The results of the data processing of both satellite datasets show a very high coverage for the urban area of Pisa, with a lack of measurement points in only some areas, near the city limits, which are largely covered by crop fields (Figure 3). The two datasets show a marked difference in terms of the density of points; in fact, the density of the measurement points within the area of interest (33 km^2) is 910 points/km^2 for RSAT and 11,027 points/km^2 for CSK. This great increase in the number of points between the C- and X-band is consistent with the results obtained from other test sites using CSK data [14,15,51,56,57]. The higher number of measurement points—i.e., coherent points—is achieved thanks to the lower revisit time and higher spatial resolution of the X-band sensors. Moreover, the shorter wavelength allows an increase in the sensitivity to LOS (Line of Sight) displacements, together with a higher capability of detecting very low displacement rates [14,15,51,56,57]. The derived displacements are projected along the line ideally joining the sensor and ground target (LOS), mainly considering vertical (up-down) and East-West movements [58,59]. Each coherent point is associated with a displacement value expressed in millimeters per year and a sign depending on the movement; positive if an uplift is registered (blue color in the deformation map), and negative if ground subsidence is affecting the point (red color in the deformation map).

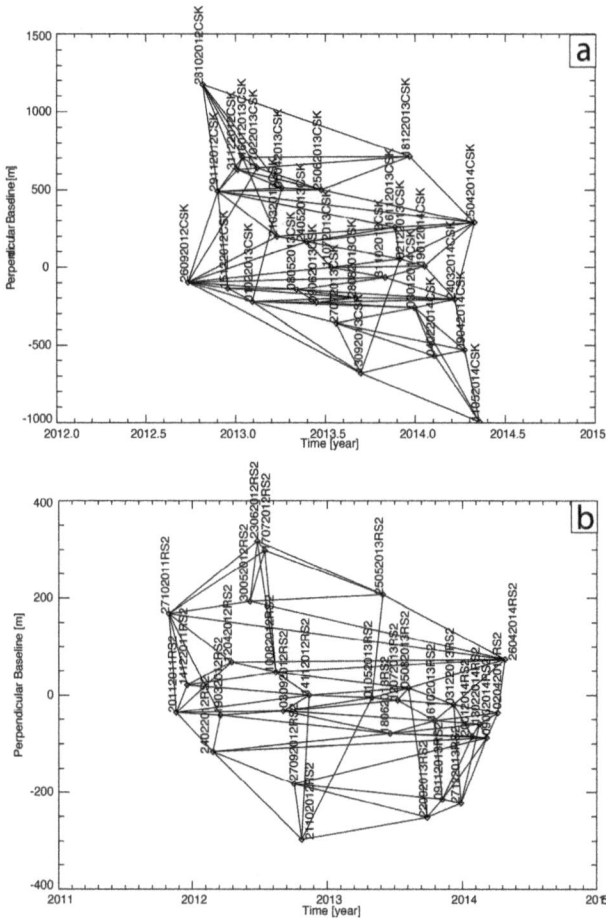

Figure 2. Synthetic Aperture Radar (SAR) data representation in the temporal/perpendicular baseline plane for the COSMO-SkyMed (**a**) and RADARSAT-2 (**b**) datasets. Each black triangle represents a SAR acquisition, with the code-name of the image, and the arcs that are the interferometric pairs selected for the Small BAseline Subset (SBAS) processing.

3.2. Stratigraphical Information and Borehole Classification System

For better understanding the recorded displacements, detailed stratigraphical information about the subsurface setting is needed. This information has been extracted from the Pisa Province borehole database, that includes both the lithological descriptions and geotechnical parameters obtained from laboratory tests. Ninety-two boreholes in the urban area of Pisa were selected according to the unique restriction of being deeper than 10–15 m, in order to intercept at least the upper limit of the *pancone* deposit. For homogenizing the subsurface information from a geotechnical point of view, a classification system was applied. This classification, proposed by Sarti et al. [43] and recently modified by Solari et al. [6,60], is composed of five facies associations. A detailed discussion on the characteristics of each class can be found in Solari et al. [6,60]. Accordingly, we will summarize the class descriptions, focusing on the geotechnical implications.

Unit 1 consists of the *pancone* very soft clay and silt clay level. The thickness of this unit varies from 3 to 12 m and can be found at a depth of 8 to 20 m, depending on the position in the plain. From a geotechnical point of view, it is considered as a highly compressible level, with a high water content that is sometimes close to the liquid limit. These characteristics strongly depend on the brackish sedimentation environment in which the unit was formed, which also favored the development of thin organic layers that further decrease the general resistance of this unit.

Unit 2 consists of dark soft clay and silty clay with a high organic matter content and occasional peat layers. It is typically found on top of Unit 1, with a thickness ranging from 2 to 6 m. This level has similar characteristics to the previous one, with a low strength and high compressibility, sometimes improved by the presence of less organic layers with a higher consistency.

Unit 3 consists of a rhythmical alternation of clayey silt, sandy silt, and silty sand, with a discontinuous distribution in the plain; where found, it lays at 2–6 m below surface, with a thickness smaller than 4 m. The geotechnical parameters, such as the water content and plasticity index, show a high variability, reflecting the chaotic structure of this sequence in which the fine sand percentage fluctuates. This unit is characterized by better mechanical properties than the previous two; however, it must be considered a potentially compressible layer because of its degree of saturation. In fact, if saturated conditions occur, even these fine-grained sands and silts can produce a high loss in volume.

Unit 4 consists of stiff clay with a very low organic matter content and high consistency due to a subaerial depositional setting of the sediments that produced resistant pedogenetic surfaces. Unit 4 is the uppermost level in the sequence and has a thickness of 5–6 m. This unit has the highest consistency and shear strength among all of the clayey layers.

Unit 5 consists of fluvial channel fine to coarse sand, with a high variability in the thickness (1–10 m) and vertical position within the stratigraphical sequence (lenticular bodies can be found at the contact with the *pancone* level). In terms of its mechanical behavior, it is considered as the unit with the lowest settlement potential.

The potential settlements in the area are produced not only by the presence of Unit 1 alone, but are related to the simultaneous presence of more compressible layers. The worst-case scenario consists of this bottom-up stratigraphy: Units 1 and 2, overlapped by Unit 4, and partially replaced by a saturated Unit 3 level. This sequence, common in the area, is considered as the most susceptible to settlements [6].

4. Results and Discussion

4.1. Regional Scale Analysis

The analysis of the two A-DInSAR datasets shows that the urban area of Pisa is not characterized by the presence of regional scale deformation patterns, such as large subsidence bowls related to water overexploitation (Figure 3). This type of large scale deformation has been observed in other sectors of the Tuscany region, such as the urban area of Pistoia, where a water overexploitation-related subsidence was recorded [61], or the geothermal area of Larderello, where a large subsidence bowl, produced by geothermal activity, has been detected [62]. The general trend of ground subsidence increases in the northern part of Pisa, where the city has developed outside of the historic center walls. In this portion of the city, a general lowering of 3.5–5 mm/yr is documented. On the other hand, the highest subsidence rates are only recorded in localized areas, corresponding to a single building or groups of buildings. The presence of these areas with high localized subsidence rates, coupled with phreatic surface reconstruction, which do not show any area of intense groundwater level lowering (Figure 1), suggests the exclusion of water overexploitation as a triggering factor for the measured displacements. Moreover, a possible role played by the second aquifer compaction was further excluded because of its high depth (>50 m below surface).

A stability threshold of ±2.5 mm/yr, which is an estimation of the accuracy of the measurements [54], has been defined on the basis of the standard deviation of the calculated velocity values of each point of the two datasets. This stability range is also coherent with various literature examples in the field of C- and X-band Interferometric Synthetic Aperture Radar (InSAR) data analysis, i.e., [12,14,16,19]. Using this threshold, two areas have been identified as stable: the Galileo Galilei International Airport (Figure 3 point A) and the historic city center, where the greatest part of the Pisa cultural heritage is present (Figure 3, point D). The airport structures and the runaways are stable in the investigated time-period, showing mean subsidence rates of −0.3 mm/yr for RSAT and −0.8 mm/yr for CSK, respectively. Within the area delimited by the ancient city walls, the LOS velocities are very low, with average values of −1.4 mm/yr and −1.2 mm/yr for RSAT and CSK, respectively. As previously mentioned, the northeastern portion of the city shows a general subsidence of 3.5–5 mm/yr in both InSAR datasets, with high localized subsidence rates. For example, in the northern part of the La Fontina district (Figure 1), ground lowering of −13 mm/yr for RSAT and −12 mm/yr for CSK datasets have been recorded by commercial buildings. In the San Cataldo district (Figures 1 and 3-point C), the peak subsidence rates measured are in agreement with those of a university residence for students, with mean values of −13 mm/yr for both sensors. In the Cisanello district (Figures 1 and 3-point C), peak velocities of −12 mm/yr for both sensors have been recorded by a group of residential buildings. Another area where localized high subsidence rates have been recorded is the commercial district recently built outside of the western border of the International airport (Figures 1 and 3-point A); values of −28 mm/yr for CSK data have been detected. The RSAT data show a very low density of points in this area; for this reason, displacement information cannot be derived. In the Ospedaletto industrial and commercial district (Figures 1 and 3-point B), the greatest part of the buildings is within the defined stability threshold (±2.5 mm/yr), but the peak velocities can also be identified, with values of −12 mm/yr for the RSAT dataset and −14 mm/yr for the CSK dataset.

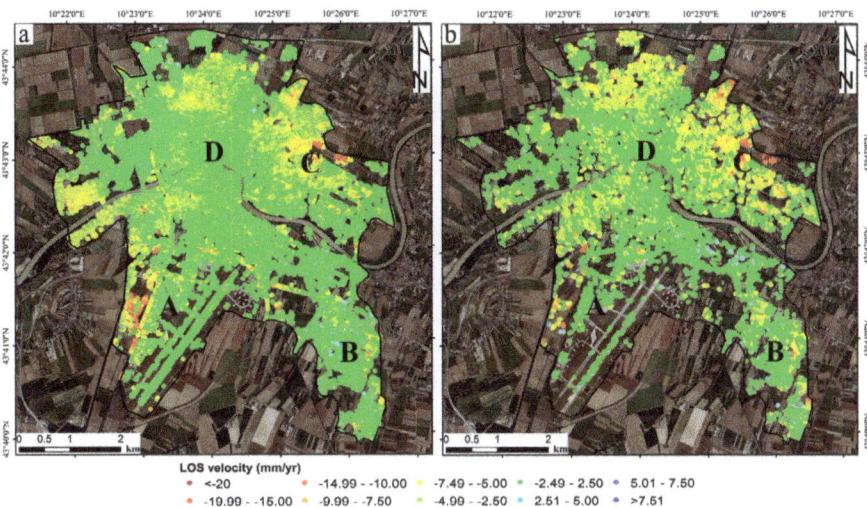

Figure 3. Registered Line of Sight (LOS) velocities between 2011 and 2014. (**a**) COSMO-SkyMed data; (**b**) RADARSAT-2 data. Deformation maps are overlaid on a 2000 digital orthophoto. The black line encloses the investigated area. (A) Galileo Galilei International Airport; (B) Ospedaletto district; (C) Cisanello–San Cataldo–La Fontina district; (D) inner city area.

4.2. Local and Building Scale Analysis

4.2.1. Movement Detection of Recently Built Buildings

The exploited datasets have allowed us to map active deformation areas, with subsidence rates higher than 10 mm/yr affecting the buildings which have been built in the last 10 years. Examples of these deformations can be found in the San Cataldo district and in the commercial area near the International Airport, where shopping malls and shipyards are present (Figure 3 points A, C).

In particular, the I Praticelli complex, a residential structure for students inaugurated in 2008, is one valuable example of a recently built building that displays high deformation rates (Figure 4a). Both SAR datasets registered similar subsidence rates; the RADARSAT-2 data ranged between −11 and −18 mm/yr with a mean value of −13.3 mm/yr for the entire complex, whereas the COSMO-SkyMed data ranged between −6 and −17 mm/yr with a mean value of −12.3 mm/yr (Figure 4b). It is interesting to note that the maximum subsidence rates for both datasets are registered in the southern part of the complex; this can be related to foundation problems or very localized variations of the geotechnical characteristics of the compressible levels that cannot be defined with the available subsurface data. This case study also shows the different performances of C- and X-band data in urban areas; in fact, the X-band data have a density of points for this single structure that is six times larger than the C-band data, passing from 75 to 500 points. Only three boreholes, localized at a mean distance of 200 m from the I Praticelli complex, are available for the geological reconstruction of the subsurface; this is reasonable considering that no significant stratigraphical variations can occur in this distance thanks to the almost homogeneous distribution of the main units. The upper limit of the most compressible succession (lagoonal clay and organic clay, Units 1 and 2) lies at a variable depth, from 4 to 5 m below the surface. This sequence, which has a total thickness of 20–25 m, is overlaid by a thin consolidated clay level (Unit 4). As mentioned by Solari et al. [6], this type of sequence is the most susceptible to settlement.

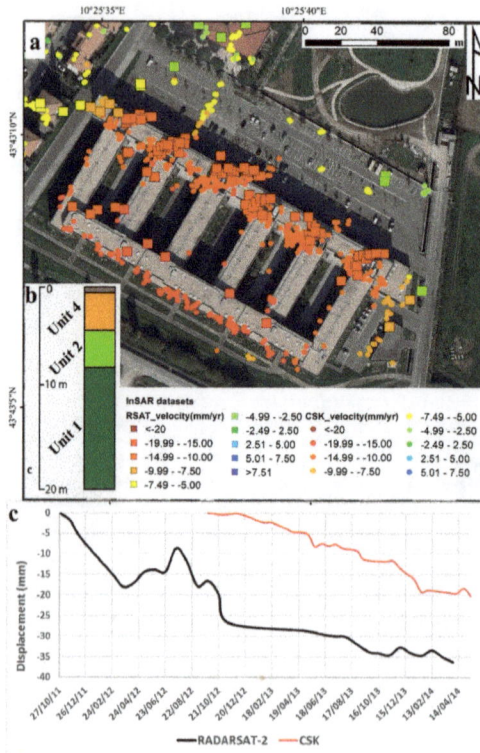

Figure 4. (**a**) LOS displacements affecting the analyzed building. The displacement map is overlaid on a 2013 digital orthophoto; (**b**) Geological reconstruction of the stratigraphical asset, the classification is the one proposed by Sarti et al. [43]; (**c**) RADARSAT-2 and COSMO-SkyMed average time series for the "I Praticelli" complex.

The registered settlements are directly related to the natural evolution of the consolidation process, as defined by Terzaghi and Peck [63]. The end of this time-dependent process is highly modified by the layers' thickness, hydraulic conductivity, and drainage conditions, as well as by the time and magnitude of the load imposition. Accordingly, the high deformation rates registered by this building are therefore related to a triggering anthropogenic factor (the recently applied load) and to a geological predisposing factor, i.e., the presence of shallow highly compressible levels (Units 1 and 2). Moreover, foundation design problems cannot be excluded.

4.2.2. Buildings with Trend Changes

The RADARSAT-2 and COSMO-SkyMed datasets have been exploited to detect possible trend changes in the temporal deformation patterns registered by buildings, in the San Cataldo and Ospedaletto districts (Figure 1), built in the eighties—early nineties. A complete time series from 1992 is available for some of these buildings, thanks to the exploitation of data from the ERS 1/2 and Envisat archives, already analyzed by Solari et al. [6].

An example of integration between historical SAR datasets and the new C- and X-band data is the analysis of the displacement rates recorded in correspondence of the CNR building complex. This three-floor construction, built at the end of the eighties, occupies a total area of 30,000 m^2. A high subsidence rate was recorded for the ERS 1/2 and Envisat C-band datasets, with a maximum magnitude of -13.4 mm/yr in the period covered by the ERS 1/2 data (1992–2000) [6]. A decrease in magnitude of the mean LOS velocity was then registered, up to the value of -9 mm/yr for the period 2003–2010, covered by the Envisat data [6]. The RADARSAT-2 data confirms this decreasing subsidence rate, recording a mean magnitude of -6 mm/yr with a maximum of -8.7 mm/yr in the central part of the complex (Figure 5a). A similar trend, with a comparable magnitude (mean velocity of about -6 mm/yr) and spatial distribution to the highest velocity values, is recorded by the COSMO-SkyMed data (Figure 5a).

The stratigraphical sequence below the analyzed building, with the presence of the Unit 1 and 2 compressible layers at a mean depth of 5m and with the only resistant Unit 4 level partially substituted by a saturated silty level (Unit 3), is again highly prone to ground settlements (Figure 5b). According to this setting, the deformation rates registered in the first part of the time series are directly related to compaction, due to the applied load of the highly compressible layers present in the sub-surface of the city (Figure 5c). The decrease of LOS velocity recorded by the CNR building represents the natural evolution of the consolidation process that slows down over time [63]. As observed by Peduto et al. [64], a decrease in the registered velocities can be related to external anthropic factors, such as restoration works to the foundation of buildings. Moreover, a deeper analysis can be performed to characterize the building settlements in further detail, through the implementation of specific indexes of structural health for each building (see for example [65,66]).

In the Ospedaletto commercial district, a different behavior of the structures has been observed in the monitored period. Several deformations, that reach velocities higher than -10 mm/yr, have been detected by ERS 1/2 and Envisat InSAR data, covering the period 1992–2010 [6]. All of the eastern part of this area has been subjected to recent urban expansion, with the construction of numerous commercial buildings since the end of the 1990s. Many of the structures built in the first part of this period, which recorded the above mentioned high deformation rates, appear now as stable or near to the defined stability threshold (± 2.5 mm/yr).

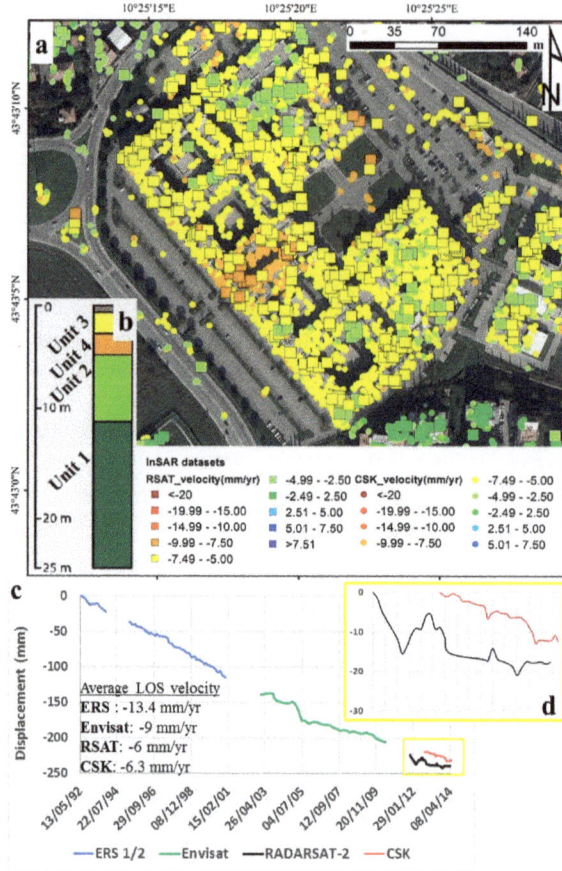

Figure 5. (**a**) LOS displacements affecting the National Research Council (CNR) complex. The displacement map is overlaid on a 2013 digital orthophoto; (**b**) Geological reconstruction of the stratigraphical asset, the classification is the one proposed by Sarti et al. [43]; (**c**) ERS 1/2, Envisat, RADARSAT-2 and COSMO-SkyMed average time series for the analyzed building. The ERS 1/2 and Envisat time series have been modified after Solari et al. [6]. The RADARSAT-2 and COSMO-SkyMed time series are shifted by a constant value for a better visualization; (**d**) RADARSAT-2 and COSMO-SkyMed average time series for the investigated period. The *y* axis units refer to the displacement in millimeters.

One example of this behavior is building A (Figure 6a); as shown by the proposed time series (Figure 6c), the average subsidence rates drastically decrease from 12.3 mm/yr for the period covered by the Envisat data (2003–2010), to 1.4 mm/yr in the RADARSAT-2 analysis and 1.0 mm/yr for the COSMO-SkyMed dataset. A stratigraphical explanation for this behavior must be considered: the most compressible sequence has its upper boundary deeper than in the previous example, passing from 5 to 10 m below surface. This ground evidence implies that the compressible sequence is less influenced by the stress increment produced by the commercial building loads, generating a lower compaction of the compressible strata. On the other hand, if we consider structures which have been built very recently, the recorded subsidence rates reach 15 mm/yr, as shown in Figure 6a by building B.

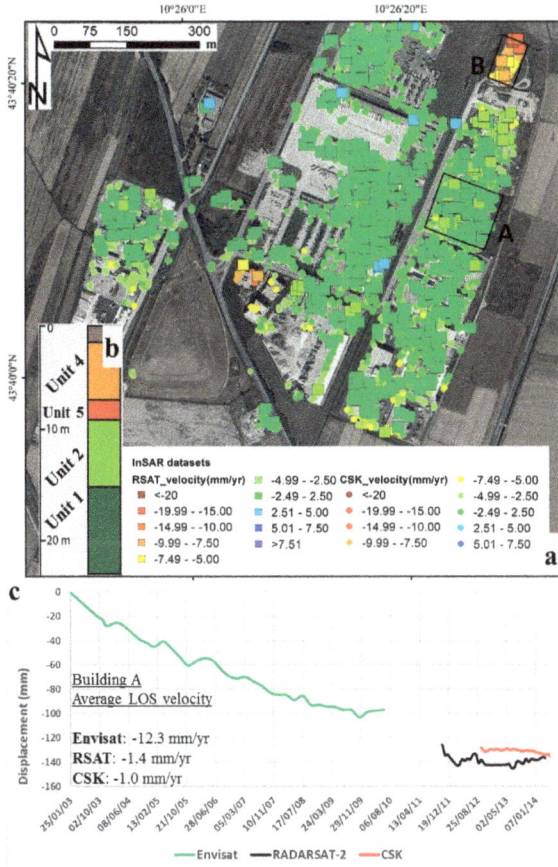

Figure 6. (**a**) LOS displacements affecting the Ospedaletto district. The displacement map is overlaid on a 2013 digital orthophoto; (**b**) Geological reconstruction of the stratigraphical asset, the classification is the one proposed by Sarti et al. [43]; (**c**) Envisat, RADARSAT-2 and COSMO-SkyMed average time series for the analyzed building. The Envisat time series have been modified after Solari et al. [6]. The RADARSAT-2 and COSMO-SkyMed time series are shifted by a constant value for a better visualization.

4.2.3. Relation between InSAR Displacements and Urbanization

For a smaller scale use of the displacement information retrieved for the single buildings, the LOS velocities have been averaged in the area occupied by 500 different structures of selected districts, in which the highest peak velocities and subsidence rates have been recorded (San Cataldo, Cisanello, La Fontina and Ospedaletto districts, see Figure 1).

In Figure 7a, the C- and X-band InSAR-derived LOS velocity values are plotted versus the age of construction of the selected buildings (Pre 1978, 1978–1988, 1988–2000, Post 2000). Four temporal classes have been defined on the basis of multi-temporal ortophotographical information. The most recent class does not present any ERS 1/2 data because the end of the acquisition period of the sensor (December 2000), that from 2001, even if still operational, does not provide reliable acquisitions because it has been operating without gyroscopes. As expected, the data show an increase in the mean LOS velocity values starting from the oldest temporal class (Figure 7b). The RSAT and CSK data show a general lower dispersion of the points for the three oldest classes (see variance values in Figure 7b); this is related not only to the general decrease in the LOS velocities of the two central classes (1978–1988 and 1988–2000), but also to the higher density of measurement points for each building. In fact, the higher redundancy of points in these two datasets allows for the minimization of the effects of anomalous velocity values within the building perimeter. The high variance (Figure 7b) of the points constituting the Post 2000 class is mainly related to the presence of outliers characterized by high subsidence rates; these points correspond to the peak velocities recorded in the analyzed districts for the studied time-periods. These evidences are coherent with the InSAR-derived deformations obtained in the city of Florence (located 80km eastern than Pisa) by Pratesi et al. [66,67], where a similar stratigraphical context and urban evolution in the last 30 years is present. As demonstrated by the I Praticelli complex example, these high subsidence rates are directly correlated to compaction, due to the recently applied load, of the highly compressible levels present in the stratigraphic setting of the city at a shallow depth. Moreover, the general decrease in the mean velocities and variance values in the two central classes demonstrate that the subsidence in the urban area is slowing down. This is testified, for example, by the CNR complex in the San Cataldo district and by many structures in the commercial area of Ospedaletto.

Considering the statistical parameters, the dispersion of the oldest class (Pre 1978) points suggests the presence of a natural subsidence, with values ranging from −2.5 to 5 mm/yr; this is more evident if only the RSAT and CSK are considered (Figure 7a). In this class, points with a displacement magnitude above 5 mm/yr are outliers related to buildings affected by the ongoing consolidation process due to geological (clayey levels with high organic matter content) or constructive (bad foundation design) characteristics.

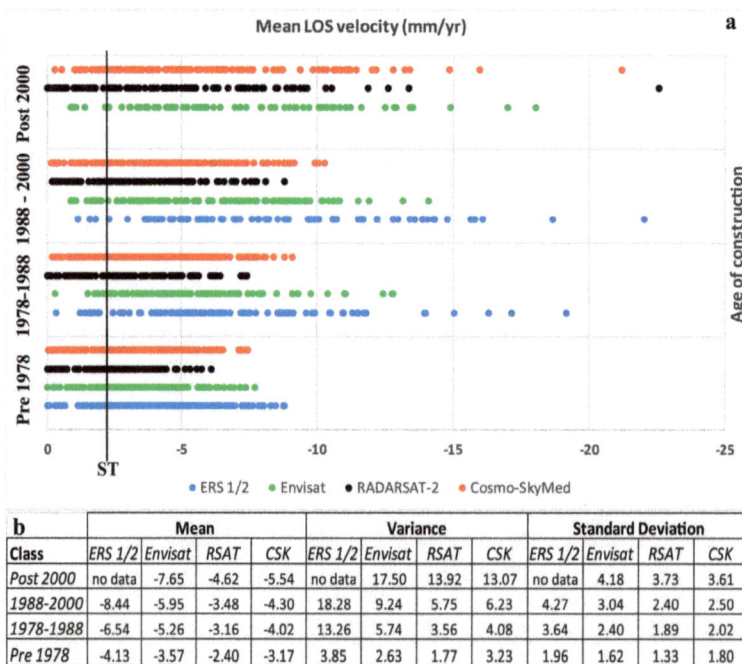

b	Mean				Variance				Standard Deviation			
Class	ERS 1/2	Envisat	RSAT	CSK	ERS 1/2	Envisat	RSAT	CSK	ERS 1/2	Envisat	RSAT	CSK
Post 2000	no data	-7.65	-4.62	-5.54	no data	17.50	13.92	13.07	no data	4.18	3.73	3.61
1988-2000	-8.44	-5.95	-3.48	-4.30	18.28	9.24	5.75	6.23	4.27	3.04	2.40	2.50
1978-1988	-6.54	-5.26	-3.16	-4.02	13.26	5.74	3.56	4.08	3.64	2.40	1.89	2.02
Pre 1978	-4.13	-3.57	-2.40	-3.17	3.85	2.63	1.77	3.23	1.96	1.62	1.33	1.80

Figure 7. (**a**) Age of construction of the selected buildings compared to the mean LOS velocity of the DInSAR data; (**b**) Mean, variance and standard deviation values for each class and each sensor. The ERS 1/2 and Envisat series are modified from Solari et al. [6]. The vertical black line represents the defined stability threshold (ST).

5. Conclusions

In this paper, A-DInSAR data from different sensors operating in C- and X-band have been exploited, to describe the temporal and spatial evolution of the land subsidence in the urban area of Pisa. The ground movements were related to a predisposing factor (presence of highly compressible layers of clays and organic clays at shallow depth), and to a triggering factor (recent urbanization of part of the city).

The C-band RADARSAT-2 and the X-band COSMO-SkyMed data, analyzed with the SBAS algorithm, show that the highest deformation registered in the investigated time-period (2011–2014) affects single buildings in the recent urban expansion areas of the city. These buildings recorded subsidence rates higher than −15 mm/yr. Considering this localized subsidence, a building-scale analysis was carried out. In the San Cataldo and Ospedaletto districts, three examples have been provided to display the different behaviors of the selected buildings. In these areas, a widespread compressible sequence was detected, mainly composed of organic rich clay and silty clay classified in two geotechnical units (Unit 1 and 2). This sequence supports the development of an accelerated consolidation process in the case of a load application; this is the case of the I Praticelli complex, a recent structure that registered subsidence rates of up to 20 mm/yr. The CNR building and the commercial structures of the Ospedaletto district show, when also exploiting older C-band data (ERS 1/2 and Envisat), the slow-down in the subsidence phenomena during the analyzed period. Finally, using the averaged LOS velocities extracted from several buildings in four selected districts, the correlation between registered subsidence rates and the age of construction of buildings was confirmed and strengthened, highlighting the role of urbanization as a triggering factor for the registered rapid subsidence.

This work has also shown the potential of DInSAR information extracted from sensors of different bands in urban areas, providing data at both a regional and building scale, with a high density of points.

Acknowledgments: The RADARSAT-2 and COSMO-SkyMed data were obtained from the COSMO-SkyMed/ RADARSAT-2 Initiative, a collaboration between the Italian Space Agency (ASI) and the Canadian Space Agency (CSA). Project #5235. RADARSAT- 2 Data and Products © MacDonald, Dettwiler and Associate Ltd. (2014)—All right reserved. RADARSAT is an official trademark of the Canadian Space Agency. COSMO-SkyMed products ASI-Agenzia Spaziale Italiana-(2014).

Author Contributions: Lorenzo Solari wrote the paper and analyzed the data. Andrea Ciampalini developed the project and helped in the geological and interferometric data interpretation. Federico Raspini and Silvia Bianchini helped in the data analysis. Michele Manunta supervised the satellite data selection, processing and analysis. Ivana Zinno and Manuela Bonano processed the RADARSAT-2 and COSMO-SkyMed SAR data and provided the SBAS results. Sandro Moretti and Nicola Casagli contributed to improve the quality of the manuscript and reviewed the paper.

Conflicts of Interest: The authors declare no conflict of interest.

References

1. Dong, S.; Samsonov, S.; Yin, H.; Ye, S.; Cao, Y. Time-series analysis of subsidence associated with rapid urbanization in Shanghai, China measured with SBAS InSAR method. *Environ. Earth Sci.* **2014**, *72*, 677–691. [CrossRef]

2. Floris, M.; Bozzano, F.; Strappaveccia, C.; Baiocchi, V.; Prestininzi, A. Qualitative and quantitative evaluation of the influence of anthropic pressure on subsidence in a sedimentary basin near Rome. *Environ. Earth Sci.* **2014**, *72*, 4223–4236. [CrossRef]

3. Polcari, M.; Albano, M.; Saroli, M.; Tolomei, C.; Lancia, M.; Moro, M.; Stramondo, S. Subsidence Detected by Multi-Pass Differential SAR Interferometry in the Cassino Plain (Central Italy): Joint Effect of Geological and Anthropogenic Factors? *Remote Sens.* **2014**, *6*, 9676–9690. [CrossRef]

4. Chen, B.; Gong, H.; Li, X.; Lei, K.; Ke, Y.; Duan, G.; Zhou, C. Spatial correlation between land subsidence and urbanization in Beijing, China. *Nat. Hazards* **2015**, *75*, 2637–2652. [CrossRef]

5. Yin, J.; Yu, D.; Yin, Z.; Wang, J.; Xu, S. Modelling the anthropogenic impacts on fluvial flood risks in a coastal mega-city: A scenario-based case study in Shanghai, China. *Landsc. Urban Plan.* **2015**, *136*, 144–155. [CrossRef]
6. Solari, L.; Ciampalini, A.; Raspini, F.; Bianchini, S.; Moretti, S. PSInSAR analysis in the Pisa Urban Area (Italy): A case study of subsidence related to stratigraphical factors and urbanization. *Remote Sens.* **2016**, *8*, 120. [CrossRef]
7. Joyce, K.E.; Belliss, S.E.; Samsonov, S.V.; McNeill, S.J.; Glassey, P.J. A review of the status of satellite remote sensing and image processing techniques for mapping natural hazards and disasters. *Prog. Phys. Geogr.* **2009**, *33*, 183–207. [CrossRef]
8. Sansosti, E.; Casu, F.; Manzo, M.; Lanari, R. Space-borne radar interferometry techniques for the generation of deformation time series: An advanced tool for Earth's surface displacement analysis. *Geophys. Res. Lett.* **2010**, *37*. [CrossRef]
9. Mohammed, O.I.; Saeidi, V.; Pradhan, B.; Yusuf, Y.A. Advanced differential interferometry synthetic aperture radar techniques for deformation monitoring: A review on sensors and recent research development. *Geocarto Int.* **2014**, *29*, 536–553. [CrossRef]
10. Crosetto, M.; Monserrat, O.; Cuevas-González, M.; Devanthéry, N.; Crippa, B. Persistent scatterer interferometry: A review. *ISPRS J Photogramm. Remote Sens.* **2016**, *115*, 78–89. [CrossRef]
11. Raspini, F.; Cigna, F.; Moretti, S. Multi-temporal mapping of land subsidence at basin scale exploiting Persistent Scatterer Interferometry: Case study of Gioia Tauro plain (Italy). *J. Maps* **2012**, *8*, 514–524. [CrossRef]
12. Tosi, L.; Teatini, P.; Strozzi, T. Natural versus anthropogenic subsidence of Venice. *Sci. Rep.* **2013**, *3*, 2710. [CrossRef] [PubMed]
13. Del Ventisette, C.; Solari, L.; Raspini, F.; Ciampalini, A.; Di Traglia, F.; Moscatelli, M.; Pagliaroli, A.; Moretti, S. Use of PSInSAR data to map highly compressible soil layers. *Geol. Acta* **2015**, *13*, 309–323.
14. Bianchini, S.; Moretti, S. Analysis of recent ground subsidence in the Sibari plain (Italy) by means of satellite SAR interferometry-based methods. *Int. J. Remote Sens.* **2015**, *36*, 4550–4569. [CrossRef]
15. Scifoni, S.; Bonano, M.; Marsella, M.; Sonnessa, A.; Tagliafierro, V.; Manunta, M.; Lanari, R.; Ojha, C.; Sciotti, M. On the joint exploitation of long-term DInSAR time series and geological information for the investigation of ground settlements in the town of Roma (Italy). *Remote Sens. Environ.* **2016**, *182*, 113–127. [CrossRef]
16. Peduto, D.; Cascini, L.; Arena, L.; Ferlisi, S.; Fornaro, G.; Reale, D. A general framework and related procedures for multiscale analyses of DInSAR data in subsiding urban areas. *ISPRS J. Photogramm. Remote Sens.* **2015**, *105*, 186–210. [CrossRef]
17. Perissin, D.; Wang, T. Time-series InSAR applications over urban areas in China. *IEEE J-STARS* **2011**, *4*, 92–100. [CrossRef]
18. Qu, F.; Zhang, Q.; Lu, Z.; Zhao, C.; Yang, C.; Zhang, J. Land subsidence and ground fissures in Xi'an, China 2005–2012 revealed by multi-band InSAR time-series analysis. *Remote Sens. Environ.* **2014**, *155*, 366–376. [CrossRef]
19. Pepe, A.; Bonano, M.; Zhao, Q.; Yang, T.; Wang, H. The Use of C-/X-Band Time-Gapped SAR Data and Geotechnical Models for the Study of Shanghai's Ocean-Reclaimed Lands through the SBAS-DInSAR Technique. *Remote Sens.* **2016**, *8*, 911. [CrossRef]
20. Tomás, R.; Márquez, Y.; Lopez-Sanchez, J.M.; Delgado, J.; Blanco, P.; Mallorquí, J.J.; Martínez, M.; Herrera, G.; Mulas, J. Mapping ground subsidence induced by aquifer overexploitation using advanced Differential SAR Interferometry: Vega Media of the Segura River (SE Spain) case study. *Remote Sens. Environ.* **2005**, *98*, 269–283. [CrossRef]
21. Herrera, G.; Tomás, R.; Monells, D.; Centolanza, G.; Mallorquí, J.J.; Vicente, F.; Navarro, V.D.; Lopez-Sanchez, J.M.; Sanabria, M.; Cano, M.; et al. Analysis of subsidence using TerraSAR-X data: Murcia case study. *Eng. Geol.* **2010**, *116*, 284–295. [CrossRef]
22. Tomás, R.; Romero, R.; Mulas, J.; Marturià, J.J.; Mallorquí, J.J.; López-Sánchez, J.M.; Herrera, G.; Gutiérrez, F.; González, P.J.; Fernández, J.; et al. Radar interferometry techniques for the study of ground subsidence phenomena: A review of practical issues through cases in Spain. *Environ. Earth Sci.* **2014**, *71*, 163–181. [CrossRef]

23. Notti, D.; Mateos, R.M.; Monserrat, O.; Devanthéry, N.; Peinado, T.; Roldán, F.J.; Fernández-Chacón, F.; Galve, J.P.; Lamas, F.; Azañón, J.M. Lithological control of land subsidence induced by groundwater withdrawal in new urban areas (Granada Basin, SE Spain). Multiband DInSAR monitoring. *Hydrol. Process.* **2016**, *30*, 2317–2331. [CrossRef]

24. Raspini, F.; Loupasakis, C.; Rozos, D.; Moretti, S. Advanced interpretation of land subsidence by validating multi-interferometric SAR data: The case study of the Anthemountas basin (Northern Greece). *Nat. Hazards Earth Syst.* **2013**, *13*, 2425–2440. [CrossRef]

25. Raspini, F.; Loupasakis, C.; Rozos, D.; Adam, N.; Moretti, S. Ground subsidence phenomena in the Delta municipality region (Northern Greece): Geotechnical modeling and validation with Persistent Scatterer Interferometry. *Int. J. Appl. Earth Obs. Geoinf.* **2014**, *28*, 78–89. [CrossRef]

26. Nikos, S.; Ioannis, P.; Constantinos, L.; Paraskevas, T.; Anastasia, K.; Charalambos, K. Land subsidence rebound detected via multi-temporal InSAR and ground truth data in Kalochori and Sindos regions, Northern Greece. *Eng. Geol.* **2016**, *209*, 175–186. [CrossRef]

27. Dixon, T.H.; Amelung, F.; Ferretti, A.; Novali, F.; Rocca, F.; Dokka, R.; Sella, G.; Kim, S.W.; Wdowinski, S.; Withman, D. Space geodesy: Subsidence and flooding in New Orleans. *Nature* **2006**, *441*, 587–588. [CrossRef] [PubMed]

28. Amelung, F.; Galloway, D.L.; Bell, J.W.; Zebker, H.A.; Laczniak, R.J. Sensing the ups and downs of Las Vegas: InSAR reveals structural control of land subsidence and aquifer-system deformation. *Geology* **1999**, *27*, 483–486. [CrossRef]

29. Samsonov, S.V.; Tiampo, K.F.; Feng, W. Fast subsidence in downtown of Seattle observed with satellite radar. *Remote Sens. Appl. Soc. Environ.* **2016**, *4*, 179–187. [CrossRef]

30. Dang, V.K.; Doubre, C.; Weber, C.; Gourmelen, N.; Masson, F. Recent land subsidence caused by the rapid urban development in the Hanoi region (Vietnam) using ALOS InSAR data. *Nat. Hazards Earth Syst. Sci.* **2014**, *14*, 657–674. [CrossRef]

31. Le, T.S.; Chang, C.P.; Nguyen, X.T.; Yhokha, A. TerraSAR-X Data for High-Precision Land Subsidence Monitoring: A Case Study in the Historical Centre of Hanoi, Vietnam. *Remote Sens.* **2016**, *8*, 338. [CrossRef]

32. Abidin, H.Z.; Gumilar, I.; Andreas, H.; Murdohardono, D.; Fukuda, Y. On causes and impacts of land subsidence in Bandung Basin, Indonesia. *Environ. Earth Sci.* **2013**, *68*, 1545–1553. [CrossRef]

33. Chaussard, E.; Amelung, F.; Abidin, H.; Hong, S.H. Sinking cities in Indonesia: ALOS PALSAR detects rapid subsidence due to groundwater and gas extraction. *Remote Sens. Environ.* **2013**, *128*, 150–161. [CrossRef]

34. Heleno, S.I.; Oliveira, L.G.; Henriques, M.J.; Falcão, A.P.; Lima, J.N.; Cooksley, G.; Ferretti, A.; Fonseca, A.M.; Lobo-Ferreira, J.P.; Fonseca, J.F. Persistent scatterers interferometry detects and measures ground subsidence in Lisbon. *Remote Sens. Environ.* **2011**, *115*, 2152–2167. [CrossRef]

35. Catalão, J.; Nico, G.; Lollino, P.; Conde, V.; Lorusso, G.; Silva, C. Integration of InSAR Analysis and Numerical Modeling for the Assessment of Ground Subsidence in the City of Lisbon, Portugal. *IEEE J-STARS* **2016**, *9*, 1663–1673. [CrossRef]

36. Pascucci, V. Neogene evolution of the Viareggio basin, Northern Tuscany (Italy). *GeoActa* **2005**, *4*, 123–138.

37. Amorosi, A.; Rossi, V.; Sarti, G.; Mattei, R. Coalescent valley fills from the late Quaternary record of Tuscany (Italy). *Quat. Int.* **2013**, *288*, 129–138. [CrossRef]

38. Aguzzi, M.; Amorosi, A.; Colalongo, M.L.; Ricci Lucchi, M.; Rossi, V.; Sarti, G.; Vaiani, S.C. Late Quaternary climatic evolution of the Arno coastal plain (Western Tuscany, Italy) from subsurface data. *Sediment. Geol.* **2007**, *202*, 211–229. [CrossRef]

39. Amorosi, A.; Sarti, G.; Rossi, V.; Fontana, V. Anatomy and sequence stratigraphy of the late Quaternary Arno valley fill (Tuscany, Italy). *Adv. Appl. Seq. Stratigr. Italy* **2008**, *1*, 55–66.

40. Amorosi, A.; Ricci Lucchi, M.; Rossi, V.; Sarti, G. Climate change signature of small-scale parasequences from Lateglacial–Holocene transgressive deposits of the Arno valley fill. *Palaeogeogr. Palaeoclimatol. Palaeoecol.* **2009**, *273*, 142–152. [CrossRef]

41. Lambeck, K.; Antonioli, F.; Anzidei, M.; Ferranti, L.; Leoni, G.; Scicchitano, G.; Silenzi, S. Sea level change along the Italian coast during the Holocene and projections for the future. *Quat. Int.* **2011**, *232*, 250–257. [CrossRef]

42. Rossi, V.; Amorosi, A.; Sarti, G.; Potenza, M. Influence of inherited topography on the Holocene sedimentary evolution of coastal systems: An example from Arno coastal plain (Tuscany, Italy). *Geomorphology* **2011**, *135*, 117–128. [CrossRef]

43. Sarti, G.; Rossi, V.; Amorosi, A. Influence of Holocene stratigraphic architecture on ground surface settlements: A case study from the City of Pisa (Tuscany, Italy). *Sediment. Geol.* **2012**, *281*, 75–87. [CrossRef]
44. Burland, J.; Jamiolkowski, M.B.; Viggiani, C. Leaning Tower of Pisa: Behaviour after stabilization operations. *Int. J. Geoeng. Case Hist.* **2009**, *1*, 156–169.
45. *Relazione Generale e Allegati Tecnici del Piano Strutturale del Comune di Pisa*; Comune di Pisa: Pisa, Italy, 1997. Available online: http://www.comune.pisa.it/doc/sit-pisa/nuovo_prg/relaz.htm (accessed on 27 October 2016).
46. Grassi, S.; Cortecci, G. Hydrogeology and geochemistry of the multilayered confined aquifer of the Pisa plain (Tuscany–central Italy). *Appl. Geochem.* **2005**, *20*, 41–54. [CrossRef]
47. Butteri, M.; Doveri, M.; Giannecchini, R.; Gattai, P. Hydrogeologic-hydrogeochemical multidisciplinary study of the confined gravelly aquifer in the coastal Pisan Plain between the Arno River and Scolmatore Canal (Tuscany). *Mem. Descr. Carta Geol. d'It.* **2010**, *XC*, 51–66.
48. Berardino, P.; Fornaro, G.; Lanari, R.; Sansosti, E. A new algorithm for surface deformation monitoring based on small baseline differential SAR interferograms. *IEEE Trans. Geosci. Remote Sens.* **2002**, *40*, 2375–2383. [CrossRef]
49. Lanari, R.; Mora, O.; Manunta, M.; Mallorquí, J.J.; Berardino, P.; Sansosti, E. A small-baseline approach for investigating deformations on full-resolution differential SAR interferograms. *IEEE Trans. Geosci. Remote Sens.* **2004**, *42*, 1377–1386. [CrossRef]
50. Casu, F.; Elefante, S.; Imperatore, P.; Zinno, I.; Manunta, M.; De Luca, C.; Lanari, R. SBAS-DInSAR Parallel Processing for Deformation Time-Series Computation. *IEEE J-STARS* **2014**, *7*, 3285–3296. [CrossRef]
51. Bonano, M.; Manunta, M.; Pepe, A.; Paglia, L.; Lanari, R. From previous C-band to new X-band SAR systems: Assessment of the DInSAR mapping improvement for deformation time-series retrieval in urban areas. *IEEE Trans. Geosci. Remote Sens.* **2013**, *51*, 1973–1984. [CrossRef]
52. Zebker, H.A.; Villasenor, J. Decorrelation in interferometric radar echoes. *IEEE Trans. Geosci. Remote Sens.* **1992**, *30*, 950–959. [CrossRef]
53. Manunta, M.; Marsella, M.; Zeni, G.; Sciotti, M.; Atzori, S.; Lanari, R. Two-scale surface deformation analysis using the SBAS-DInSAR technique: A case study of the city of Rome, Italy. *Int. J. Remote Sens.* **2008**, *29*, 1665–1684. [CrossRef]
54. Arangio, S.; Calò, F.; Di Mauro, M.; Bonano, M.; Marsella, M.; Manunta, M. An application of the SBAS-DInSAR technique for the assessment of structural damage in the city of Rome. *Struct. Infrastruct. Eng.* **2013**, *10*, 1469–1483. [CrossRef]
55. Pepe, A.; Lanari, R. On the extension of the minimum cost flow algorithm for phase unwrapping of multitemporal differential SAR interferograms. *IEEE Trans. Geosci. Remote Sens.* **2006**, *44*, 2374–2383. [CrossRef]
56. Calò, F.; Ardizzone, F.; Castaldo, R.; Lollino, P.; Tizzani, P.; Guzzetti, F.; Lanari, R.; Angeli, M.; Pontoni, F.; Manunta, M. Enhanced landslide investigations through advanced DInSAR techniques: The Ivancich case study, Assisi, Italy. *Remote Sens. Environ.* **2014**, *142*, 69–82. [CrossRef]
57. Bovenga, F.; Wasowski, J.; Nitti, D.O.; Nutricato, R.; Chiaradia, M.T. Using COSMO/SkyMed X-band and ENVISAT C-band SAR interferometry for landslides analysis. *Remote Sens. Environ.* **2012**, *119*, 272–285. [CrossRef]
58. Ciampalini, A.; Raspini, F.; Frodella, W.; Bardi, F.; Bianchini, S.; Moretti, S. The effectiveness of high-resolution LiDAR data combined with PSInSAR data in landslide study. *Landslides* **2016**, *13*, 399–410. [CrossRef]
59. Franceschetti, G.; Lanari, R. *Synthetic Aperture Radar Processing*; CRC Press: Boca Raton, FL, USA, 1999.
60. Solari, L.; Ciampalini, A.; Bianchini, S.; Moretti, S. PSInSAR analysis in urban areas: A case study in the Arno coastal plain (Italy). *Rend. Online Della Soc. Geol. Ital.* **2016**, *41*, 255–258. [CrossRef]
61. Canuti, P.; Casagli, N.; Farina, P.; Marks, F.; Ferretti, A.; Menduni, G. Land subsidence in the Arno River Basin studied through SAR interferometry. In Proceedings of the 7th International Symposium on Land Subsidence, Shanghai, China, 23–28 October 2005; Volume 1, pp. 407–416.
62. Rosi, A.; Tofani, V.; Agostini, A.; Tanteri, L.; Stefanelli, C.T.; Catani, F.; Casagli, N. Subsidence mapping at regional scale using persistent scatters interferometry (PSI): The case of Tuscany region (Italy). *Int. J. Appl. Earth Obs. Geoinf.* **2016**, *52*, 328–337. [CrossRef]
63. Terzaghi, K.; Peck, R.B. *Soil Mechanics in Engineering Practice*; John Wiley & Sons: Hoboken, NJ, USA, 1967.

64. Peduto, D.; Nicodemo, G.; Maccabiani, J.; Ferlisi, S. Multi-scale analysis of settlement-induced building damage using damage surveys and DInSAR data: A case study in The Netherlands. *Eng. Geol.* **2017**, *218*, 117–133. [CrossRef]
65. Bianchini, S.; Pratesi, F.; Nolesini, T.; Casagli, N. Building deformation assessment by means of persistent scatterer interferometry analysis on a landslide-affected area: The Volterra (Italy) case study. *Remote Sens.* **2015**, *7*, 4678–4701. [CrossRef]
66. Pratesi, F.; Tapete, D.; Terenzi, G.; Del Ventisette, C.; Moretti, S. Rating health and stability of engineering structures via classification indexes of InSAR Persistent Scatterers. *Int. J. Appl. Earth Obs. Geoinf.* **2015**, *40*, 81–90. [CrossRef]
67. Pratesi, F.; Tapete, D.; Del Ventisette, C.; Moretti, S. Mapping interactions between geology, subsurface resource exploitation and urban development in transforming cities using InSAR Persistent Scatterers: Two decades of change in Florence, Italy. *Appl. Geogr.* **2016**, *77*, 20–37. [CrossRef]

MDPI

Article

A Study of Ground Movements in Brussels (Belgium) Monitored by Persistent Scatterer Interferometry over a 25-Year Period

Pierre-Yves Declercq [1],*, Jan Walstra [1], Pierre Gérard [2], Eric Pirard [3], Daniele Perissin [4], Bruno Meyvis [1] and Xavier Devleeschouwer [1]

[1] Royal Belgium Institute of Natural Sciences, Geological Survey of Belgium, Jennerstraat 13, 1000 Brussels, Belgium; jan.walstra@naturalsciences.be (J.W.); bruno.meyvis@naturalsciences.be (B.M.); xavier.devleeschouwer@naturalsciences.be (X.D.)

[2] BATir Département, Université Libre de Bruxelles (ULB), CP194/02, Avenue F.D. Roosevelt 50, 1050 Bruxelles, Belgium; piergera@ulb.ac.be

[3] Département ARGENCO/Gemme—GEO3, Université de Liège (ULg), Allée de la Découverte, 9-Bat. B 52/, 4000 Liège, Belgium; eric.pirard@ulg.ac.be

[4] Lyles School of Civil Engineering, Purdue University, West Lafayette, IN 47907, USA; perissin@purdue.edu

* Correspondence: pierre-yves.declercq@naturalsciences.be; Tel.: +32-(0)2-788-7656

Received: 21 July 2017; Accepted: 22 October 2017; Published: 8 November 2017

Abstract: The time series of Synthetic Aperture Radar data acquired by four satellite missions (including ERS, Envisat, TerraSAR-X and Sentinel 1) were processed using Persistent Scatterer interferometric synthetic aperture radar (InSAR) techniques. The processed datasets provide a nearly continuous coverage from 1992 to 2017 over the Brussels Region (Belgium) and give evidence of ongoing, slow ground deformations. The results highlight an area of uplift located in the heart of the city, with a cumulative ground displacement of ±4 cm over a 25-year period. The rates of uplift appear to have decreased from 2 to 4 mm/year during the ERS acquisition period (1992–2006) down to 0.5–1 mm/year for the Sentinel 1 data (2014–2017). Uplift of the city centre is attributed to a reduction of groundwater extraction from the deeper (Cenozoic-Paleozoic) aquifers, related to the deindustrialization of the city centre since the 1970s. The groundwater levels attested by piezometers in these aquifers show a clear recharge trend which induced the uplift. Some areas of subsidence in the river valleys such as the Maelbeek can be related to the natural settlement of soft, young alluvial deposits, possibly increased by the load of buildings.

Keywords: Persistent Scatterer Interferometry; Radar Interferometry; InSAR; uplift; subsidence; groundwater recharge; Brussels

1. Introduction

Satellite interferometric synthetic aperture radar (InSAR) is a valuable tool for observing the deformation of the earth's surface. Such techniques use the radar phase information of SAR images acquired at different times over the same area [1,2]. Persistent Scatterer Interferometry (PSI) [2–5] is one of the Multi-Temporal InSAR (MT-InSAR) algorithms that produces deformation time series and velocity for point objects called a Persistent Scatterer (PS). PSs have, on average, identical scattering properties over time and a stronger reflection amplitude within a pixel. The procedure allows the temporal decorrelation and the geometrical decorrelation to be reduced, which are both problematic for the InSAR technique. Based on these considerations, PSI is particularly efficient in urban areas as the density of reflecting objects and the amplitude of reflection are high.

Therefore, during the international Terrafirma project [6] funded by the European Space Agency (ESA) SAR images covering a large majority of the capitals of Europe were processed using PSI.

Since then, the Geological Survey of Belgium (GSB) is involved in radar applied interferometry research. Brussels was one of the cities studied during this project. In the city centre, a previously unknown uplift has been highlighted and analysed [7] At that time, 74 scenes from the ERS satellite covering the period 1992–2003 were processed. Nowadays, thanks to new Synthetic Aperture Radar (SAR) satellites (Envisat in 2002, TerraSAR-X in 2007 and Sentinel 1 in 2014) 300 scenes covering the region of Brussels were processed using the PSI technique.

In the literature, land subsidence of large cities related to groundwater extraction highlighted by remote sensing techniques is a well-known phenomenon. Mexico city centre [8–11], Las Vegas [12], Shanghai [13,14] are just a few of many extensively studied examples. On the contrary, uplifts caused by the recharge of aquifers have received far less attention in scientific literature except for some studies dealing with the effects of fluid injections [15] or CO_2 storage [16], and uplifts related to volcanism [5,17,18]. Hence this paper monitors 25 years of the ground movements in the city centre of Brussels (Belgium) using PSI. Several piezometers giving the evolution of the water table in different aquifers were analyzed. According to the results, it has been possible to propose a model for the ground movements evolution based on the recharge of the Cenozoic and Paleozoic aquifers in the study area. The second aim highlights the duration of the recharge of the deep aquifers in Brussels after a century of groundwater withdrawal that has started during the industrialisation time interval of the city.

2. Geographical and Geological Setting

The area of Brussels, and more particularly the historical heart of the city, is crosscut by the Senne valley along a SW–NE axis (Figure 1). Some less important, parallel valleys (such as the Maelbeek and Woluwe) have incised the eastern part of the urban landscape of Brussels. The altitude difference between the top of the hills and the base of the valleys reaches more than 80 m. The large alluvial plain of the Senne River inclines gently towards the north from an altitude of 19 m in the south to 13 m in the north.

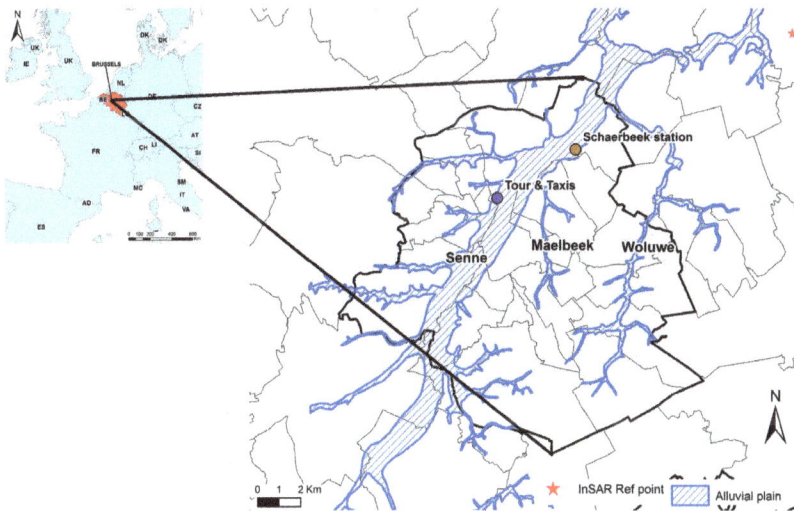

Figure 1. Localisation of the studied area around the Brussels Region (bold black line). The Quaternary alluvial plains (blue hatched surfaces) of the Senne, Maelbeek and Woluwe rivers are highlighted and the common reference point taken for all the Synthetic Aperture Radar (SAR) images is located in the NE side of the area (red star).

The region of Brussels (Figure 2) is located over a major geological structure from the Paleozoic age called the Brabant Massif, which covers large areas of Belgian territory. The Brabant Massif consists of a compressed wedge, the core of which is formed by steeply deformed Cambrian formations [19] that are flanked to the NE and SW by younger Ordovician and Silurian formations [20]. The paleotopographical surface of the Brabant Massif basement is affected along the Senne valley by a chain of NW–SE trending ridges and depressions that have a width ranging between 0.5 to 1 km and a spacing of 1 to 1.5 km [21]. The steeply dipping Lower Paleozoic deposits that occur in subcrop in the mapped area belong to the Blanmont, (quartzites), Tubize (from shales to sandstones) and Oisquercq (claystones to siltstones) formations. The Brabant Massif was a persistent positive area with only reduced sedimentation during the Late Paleozoic period. The Carboniferous was probably removed by erosion during a Jurassic uplift [22]. Late Cretaceous chalk deposits (Senonian Age) are partly preserved and were eroded by an uplift phase of the Brabant Massif during the middle Paleocene Laramide phase [23]. Sedimentary Tertiary marine series, composed of clays, silts and sands, occur in numerous eustatic cycles. The subsurface geology of the latter is detailed in the following paragraphs.

Anthropogenic fillings are present almost everywhere and have a thickness up to several meters. Holocene alluvial sediments, composed of loam, sand, clay, peat and gravel layers, are essentially located in the valleys. Their thickness ranges generally from 10 to 20 m. Pleistocene deposits (mainly loess) cover the whole area. Their thickness varies strongly and can reach up to 20–30 m. Underneath the Quaternary continental sediments, Cenozoic marine formations are present, their age spanning from the Upper Miocene down to the Upper Paleocene (Figure 2).

The Diest Formation (Upper Miocene) and the Bolderberg Formation (Lower Miocene) correspond to limited sand deposits at the top of the hills in the northern part of Brussels. The Sint-Huibrechts-Hern sandy Formation (Upper Eocene) has a reduced thickness in the north while in the southeast it can reach more than 10 m. The Maldegem Formation (Middle Eocene) contains four Members: Zomergem, Onderdale, Ursel and Asse and, finally Wemmel. The Maldegem Formation is relatively good for house foundations except when it is saturated with water. The Lede Formation (Middle Miocene) is made of sands and sandy calcareous layers. The Bruxelles Formation (Middle Eocene) is characterised by coarser sands than those of Lede. The Gent Formation (Lower Eocene) reaches a thickness of 8 m and is only observed in the north-western part of Brussels. The clay and fine sands of the Tielt Formation (Lower Eocene) are generally 20 m thick. The Kortrijk Formation (Lower Eocene) is subdivided into three Members and mainly composed of clays, silts and sands. The average total thickness of the Kortrijk Formation reaches 70 m. The Upper Paleocene Formation (Hannut) contains an upper sandy Member (Grandglise) and a lower greenish clay Member (Lincent). The thickness ranges from 15–20 m in the south, up to 28 m in the north. Cretaceous deposits are only present in the northern part of Brussels. White to grey chalk with black cherts from the Gulpen Formation is described only in drill holes. The Cretaceous disappears progressively to the southwestern part of the Brussels Region and the thickness increases towards the NE. The Tertiary and Cretaceous rocks are unconformably overlaying the Paleozoic basement.

The hydrogeological structure of the area is formed by several superimposed aquifers separated by layers of clays. An alluvial aquifer lies within the Quaternary deposits of the Senne valley. The eastern part of the Brussels Region is characterized by an important aquifer in the sands of the Lede, Bruxelles and part of the Kortrijk Formations. The glauconitic sands of the Hannut Formation (Late Paleocene) contain an aquifer separated from the artesian aquifer of the Cretaceous sediments by a clay layer a few metres thick corresponding to the Lincent Member. The Lincent Member is disappearing towards the south in the Wallonia area where the Hannut aquifer is outcropping and lying on the Cretaceous rocks when present. The Cretaceous is absent in the southern and southwestern parts of Brussels. The thickness of the Cretaceous increases from a few metres to around 20 m towards the north and to more than 40 m in the most eastern parts of Brussels [24]. The artesian aquifer of the Cambro-Silurian is sometimes separated locally from the Cretaceous aquifer by a thick clay layer originating from weathered basement rocks. Locally, it can play the role of an aquitard level, but on a regional scale it is not distinguished and these aquifers are aggregated as "Cretaceous-Paleozoic aquifer".

Table 1 gives a summary of the different aquifer and aquitard levels recognized in the Brussels Region.

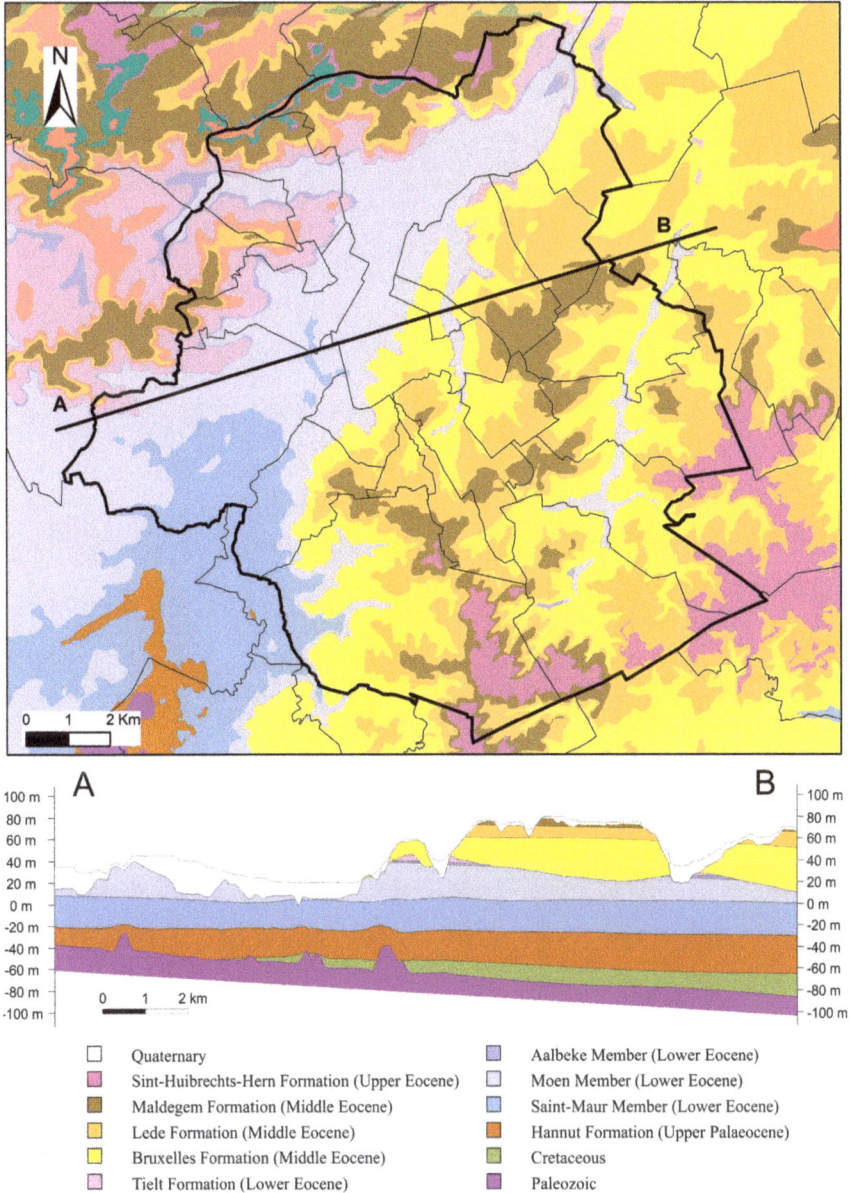

Figure 2. Geological map and cross-section profile (A–B) through the Brussels Region (limits in bold black lines) are based on the geological map 31–39 Brussel-Nijvel. The light black lines inside the Brussels Region correspond to the 19 districts of Brussels [24].

Table 1. List of the different geological formations encountered in the subcrop of the Brussels Region with the stratigraphical position and the type/nature of the different aquifers and aquitards.

Era	System	Series	Stratigraphic Formation/Member	Hydrogeological Unit	Type	Nature
CENOZOIC	QUATERNARY	HOLOCENE	-	Aquifer system of the Quaternary	AQUIFER/AQUITARD	Unconfined to confined
		HOLOCENE	-		AQUICLUDE	
		PLEISTOCENE	-		AQUIFER	
	NEOGENE	Upper MIOCENE	DIEST		AQUIFER	
		Lower MIOCENE	BOLDERBERG		AQUIFER	
		Upper EOCENE	SINT-HUILBRECHTS-HERN	Perched aquifer system with sands	AQUIFER/AQUITARD	Unconfined to locally confined
			MALDEGEM/Zomergem		AQUICLUDE	
			MALDEGEM/Onderdale		AQUIFER	
		Middle EOCENE	MALDEGEM/Ursel and Asse	Aquiclude (clays)	AQUICLUDE	
			MALDEGEM/Wemmel	Aquifer system with sands	AQUIFER	Unconfined to locally confined
	PALEOGENE		LEDE		AQUIFER	
			BRUSSEL/BRUXELLES		AQUIFER	
			GENT/Vlierzele		AQUIFER	
			GENT/Merelbeke	Aquiclude (clays)	AQUICLUDE	
		Lower EOCENE	TIELT	Aquitard with alternating sands and clays	AQUIFER/AQUITARD	Unconfined to locally confined
			KORTRIJK/Aalbeke	Aquiclude (clays) not very thick	AQUICLUDE	
			KORTRIJK/Moen	Aquifer system with sands	AQUIFER/AQUITARD	Unconfined to locally confined
			KORTRIJK/Saint-Maur	Aquiclude (clays)	AQUICLUDE	
		Upper PALEOCENE	HANNUT/Grandglise	Aquifer system with sands	AQUIFER	Confined
			HANNUT/Lincent	Aquiclude (clays)	AQUICLUDE	
MESOZOIC		Upper CRETACEOUS	NEVELE	Aquifer system combined with the top part of the Paleozoic	AQUIFER	Confined
PALEOZOIC		Lower CAMBRIAN	TUBIZE	Aquifer system of the Paleozoic with a weathered/fractured top zone	AQUIFER/AQUITARD	Confined
					AQUICLUDE	

3. Materials and Methods

Four datasets of SAR images were acquired and processed (Table 2, Figures 3 and 4), spanning a total period of 25 years. The ERS, Envisat and Sentinel-1 missions acquired data in C-band (frequency 5.7 GHz, wavelength 5.6 cm) at a ground resolution of circa 5–20 m; the TerraSAR-X data were acquired in X-band (frequency 9.6 GHz, wavelength 3.1 cm) at a ground resolution of c. 3 m (Stripmap mode). An external digital elevation model (DEM) data from the Shuttle Radar Topography Mission (SRTM 3-arc second) with 90 m horizontal resolution was used to remove the topographic component of the interferometric phase. A common reference point (Figure 1) was defined 15 km NW of Brussels (4.54° E 50.932° N).

Table 2. Characteristics of the processed datasets. PS denotes Persistent Scatterer.

Satellite	Track	Pass	Number of Scenes	Acquisition Period	Master	Processing Software	Avg PS Density (PS/km^2)
ERS 1/2	423	Descending	78	1992–2006	18 February 1998	ROI_PAC, Doris, StaMPS	181
Envisat	423	Descending	73	2003–2010	15 August 2007	ROI_PAC, Doris, StaMPS	203
TerraSAR-X	48	Descending	74	2011–2014	13 May 2013	Doris, StaMPS	2713
Sentinel 1	37	Descending	64	2014–2017	27 October 2016	Sarproz	250

The ERS1/2 scenes of descending track 423 were obtained in raw L0 format from ESA. An initial selection comprised all scenes providing full coverage of Belgium, with restrictions on the baseline difference (maximum of 1200 m relative to an arbitrarily chosen master acquisition) and Doppler shift (maximum of 1200 Hz) to ensure good coherence. During various pre-processing steps, some scenes were dropped due to failure of focusing or coregistration. The final, fully processed dataset includes 78 acquisitions between 1992 and 2006. All images were coregistered to a single master image, and selected by minimizing spatial and temporal baselines. The conversion from raw L0 to SLC (Single Look Complex) format was done using ROI_PAC software [25]. The SLC files were oversampled with a factor of two and interferograms were produced using the Doris InSAR processor [26] and recomputed ERS-1/2 ODR orbits [27]. The interferograms were then imported in StaMPS (Stanford Method for Persistent Scatterers) for Persistent Scatterer processing [5]. The default parameters were used, except a slight relaxation of the thresholds for amplitude dispersion for selecting PS candidates (0.42 instead of 0.4) and standard deviation for phase noise weeding (1.2 instead of default 1), in order to improve the number of PS points. After unwrapping, spatially-correlated errors due to atmosphere, DEM and orbit error were iteratively estimated and removed.

A dataset comprising 73 ENVISAT scenes of descending track 423 in raw L0 format was obtained from ESA. All scenes had baselines of less than 1000 m with respect to the chosen master (Table 2) and were successfully processed following the same procedures as for ERS, except for the use of DORIS Precise Orbit State Vectors and no changes made to the default parameter settings in StaMPS.

A dataset of 74 TerraSAR-X and TanDEM-X Stripmap (SM) descending scenes in SLC format was obtained from German Aerospace Centre (DLR). Creation of interferograms and PS processing were carried out using DORIS and StaMPS. Some parameter settings in StaMPS (size of spatial filters and windows) were adjusted to fit the higher spatial resolution of TerraSAR-X. This dataset is characterized by a high spatial resolution (3 m) and short baselines (maximum of 348 m).

The final dataset consisted of data acquired by the Sentinel-1A and 1B missions and accessed through the Copernicus Open Access Hub. The geometric specifications of this satellite constellation (i.e., short spatial and temporal baselines [28]) promise improved data quality (higher coherence) and enhanced performance of PS-InSAR techniques, but the processing of TOPSAR data is more challenging. The study area is covered by 64 descending scenes acquired in Interferometric Wide Swath Mode (IWSM), with a maximum baseline of 119 m. The complete PS-InSAR data processing chain was carried out in Sarproz, which can handle individual swaths of TOPSAR data [29]. Throughout the workflow the recommended/default settings were used. A significant difference with StaMPS is the use of weather data (temperature, atmospheric pressure, humidity, precipitation) for the estimation and removal of atmospheric phase.

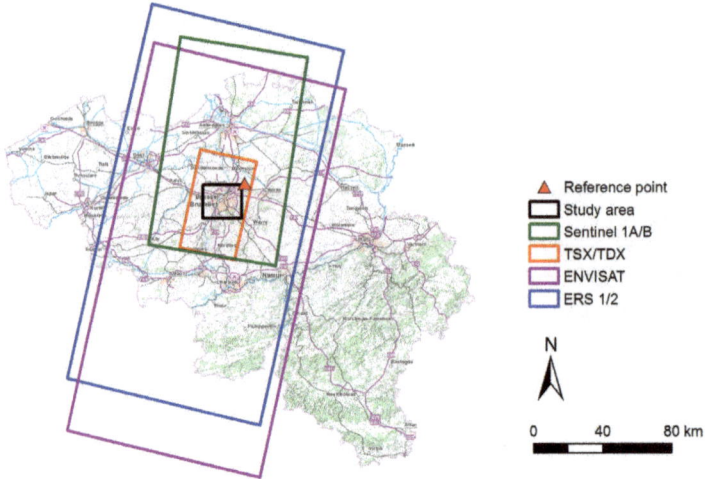

Figure 3. Map of Belgium showing the region of interest (Brussels Region), superimposed by the different footprints of the ERS, ENVISAT, TerraSAR-X and Sentinel 1 datasets used in this study and location of the reference point (red triangle).

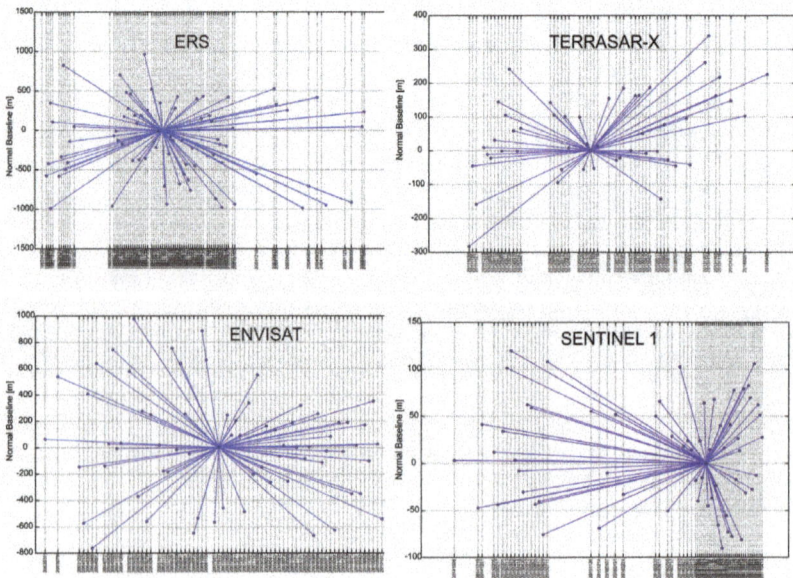

Figure 4. Normal Baselines versus Time acquisition graphs for the four SAR image datasets.

Only PS points with a coherence value larger than 0.70 were used in further analyses. If the observed displacements are predominantly in a vertical direction and considering the differences in incidence angle between the datasets, the line-of-sight (LOS) displacements were converted to vertical displacements: ddispl = dLOS/cosθ, where ddispl is the vertical displacement, dLOS is the LOS displacement and θ is the average incidence angle.

The time series of displacements were merged for common PS points by assuming a linear displacement in time gaps and minimizing the offset between datasets. PS points were considered to be common to the nearest point in other datasets within a radius of 20 m.

4. Results

Analysis of the ERS (1992–2006), Envisat (2003–2010) and TerraSAR-X (2011–2014) velocities shows that the city of Brussels is, overall, characterized by a regional uplift, with a diminishing trend through time (Figure 5A–C). The Sentinel 1 dataset (2014–2017) indicates that the uplift has faded and the area that was previously uplifting is now characterized by negative velocities (Figure 5D).

The C-band missions (ERS, Envisat, Sentinel 1) acquired data at a similar ground resolution, resulting in PS point densities of the same order of size (circa 200 PS/km^2 in the study area, see Table 2). A steady increase between the successive missions may be attributed to improved geometric specifications. Owing to its higher spatial resolution, the point density achieved by TerraSAR-X is an order of magnitude larger (circa 2700 PS/km^2 for the study area). In the urbanized centre of Brussels, point densities are highest, exceeding 800 PS/km^2 and 6000 PS/km^2 for the C-band satellites and TerraSAR-X, respectively. The Soignes Forest located to the SSE of the Brussels city centre (Figure 5A) and the Laeken Park in the northern part of the city centre do not contain PS points. Vegetation is responsible for the loss of coherence between the acquisitions and thus the capability to detect PS.

For ERS (Figure 5A), the general observed trend of uplift is most pronounced in the city centre of Brussels, along a SW–NE oriented axis (average velocity of 2.7 mm/year). The affected area roughly corresponds to the Senne river valley inside the historic heart of Brussels.

For Envisat (Figure 5B), the main uplift area has shifted along the Senne river valley towards the northern suburbs of Brussels (average velocity of 2.3 mm/year). In the SW of the city, the ground movements are reduced to the range of stable values. A notable subsidence feature is observed in Forest/Uccle districts, at the SW border of the Brussels Region: here, an area of circa 1.5 km^2 is characterized by substantial negative velocities (average velocity of −3.1 mm/year).

For TerraSAR-X (Figure 5C), the data show a more diffuse pattern in the city centre, with most of the velocities close to stable values. An uplift is still visible westwards of Brussels (average velocity of 1.8 mm/year), but the eastern part of the study area is characterized by stable velocity values. Distinct, local subsidence features are observed in some areas characterized by erratic negative velocities. Lineament "La" corresponds to the railway tracks at Schaerbeek station. Lineaments "Lb" and "Lc" correspond to the NNW–SSE oriented Maelbeek and NNE–SSW oriented Woluwe river valleys, respectively.

During the Sentinel 1 time frame (Figure 5D), the general trend of ground movements appears to have stabilized. Within a zone (which is SW–NE oriented, 200–800 m across, and corresponds to the Senne river valley), the direction of movement reversed from uplift to slight subsidence (average velocity of −1.3 mm/year). Lineament "La" is still visible as a distinct subsiding feature. However, as the range of the velocities is quite small and the observation period is short in comparison with ERS and Envisat the Sentinel 1, the results look noisy. In the future, after one or two years of new Sentinel 1 acquisition, a new processing operation will be conducted in order to verify the current observations.

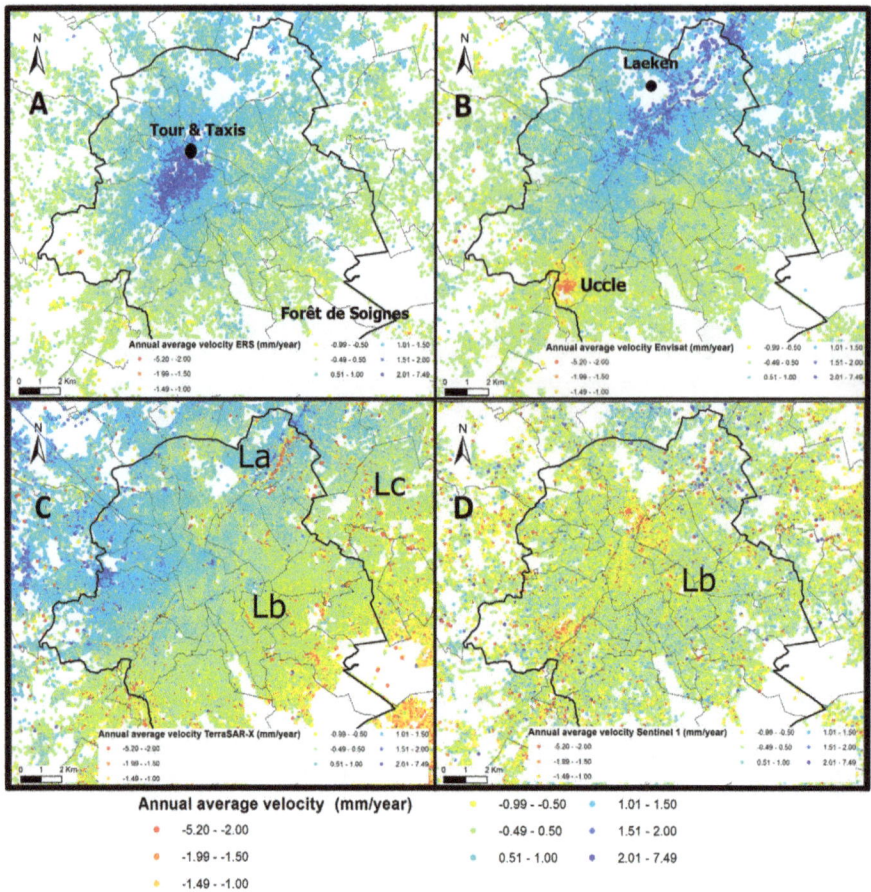

Figure 5. Colour classification based on the average annual velocities of PS points for ERS (**A**); Envisat (**B**); TerraSAR-X (**C**); and Sentinel 1 (**D**). Negative values (red/orange) indicate areas of subsidence, while positive values (blue) represent uplift.

From all datasets, histograms were plotted to analyze the frequency distribution of the average annual velocities (Figure 6A–D). Values in the range of −0.49 to 0.5 mm/year can be considered as stable during the observation period, relative to the reference point. Although the majority of PS points in Brussels appear more or less stable, the ERS and Envisat histograms show a clear asymmetry towards positive values, confirming the general trend of uplift in the area.

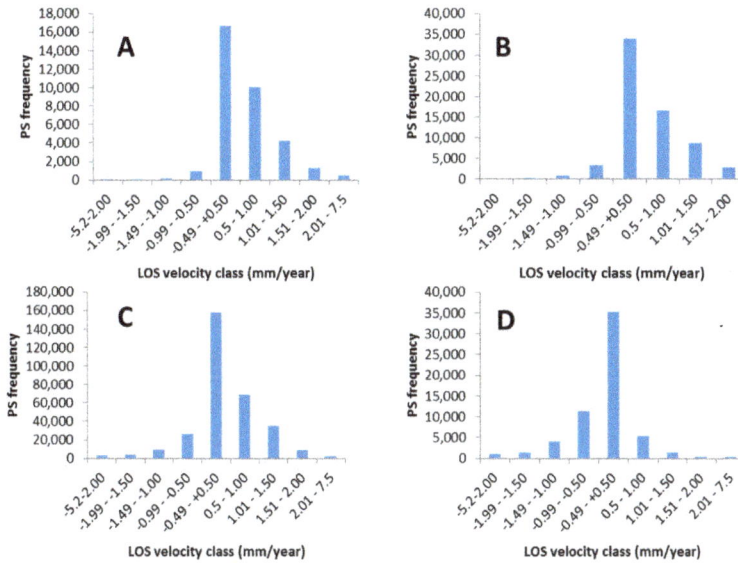

Figure 6. Histograms based on the annual average line-of sight (LOS) velocities of the PS for ERS (**A**); Envisat (**B**); TerraSAR-X (**C**); Sentinel 1 (**D**).

5. Interpretation and Discussion

PS-InSAR techniques are a powerful tool to measure displacements of moving ground objects from time series of SAR images to millimetric precision [2]. Although the lower resolution of ERS and Envisat makes it harder to exactly identify individual objects acting as persistent scatterers in the resolution cell. Therefore, for regional studies, the displacements of the objects reflecting the radar signal can be assumed to represent the ground movements of the earth surface and not only the object on top of it.

Identifying the underlying cause of regional displacement patterns can be challenging and different hypotheses should be tested. In the present study, we verified whether the observed slow, decelerating uplift in Brussels can be explained by tectonic processes or whether it is the result of human activity.

Concerning the tectonic setting, Belgium is located in an intraplate zone and the seismic activity is considered as moderate [30,31]. According to the records of historic earthquakes held by the Belgian Royal Observatory [32], the study area has not been affected by significant earthquakes during the last 25 years. Given the absence of any recent seismic activity, the hypothesis of induced tectonic uplift of Brussels can be safely rejected.

Regarding the second hypothesis, many studies have demonstrated that extraction of groundwater can have major impacts on ground stability [1–7]. To get a clear view of the evolution of groundwater levels in the study area, piezometric data were harvested from the Brussels Environment Agency database and analyzed. Figure 7 shows the location of six selected piezometers providing continuous and reliable measurements from the main aquifers located in the studied area (Figure 7). The evolution of their groundwater levels through time are reported and subdivided into the different aquifers encountered for each piezometer site (Figure 8).

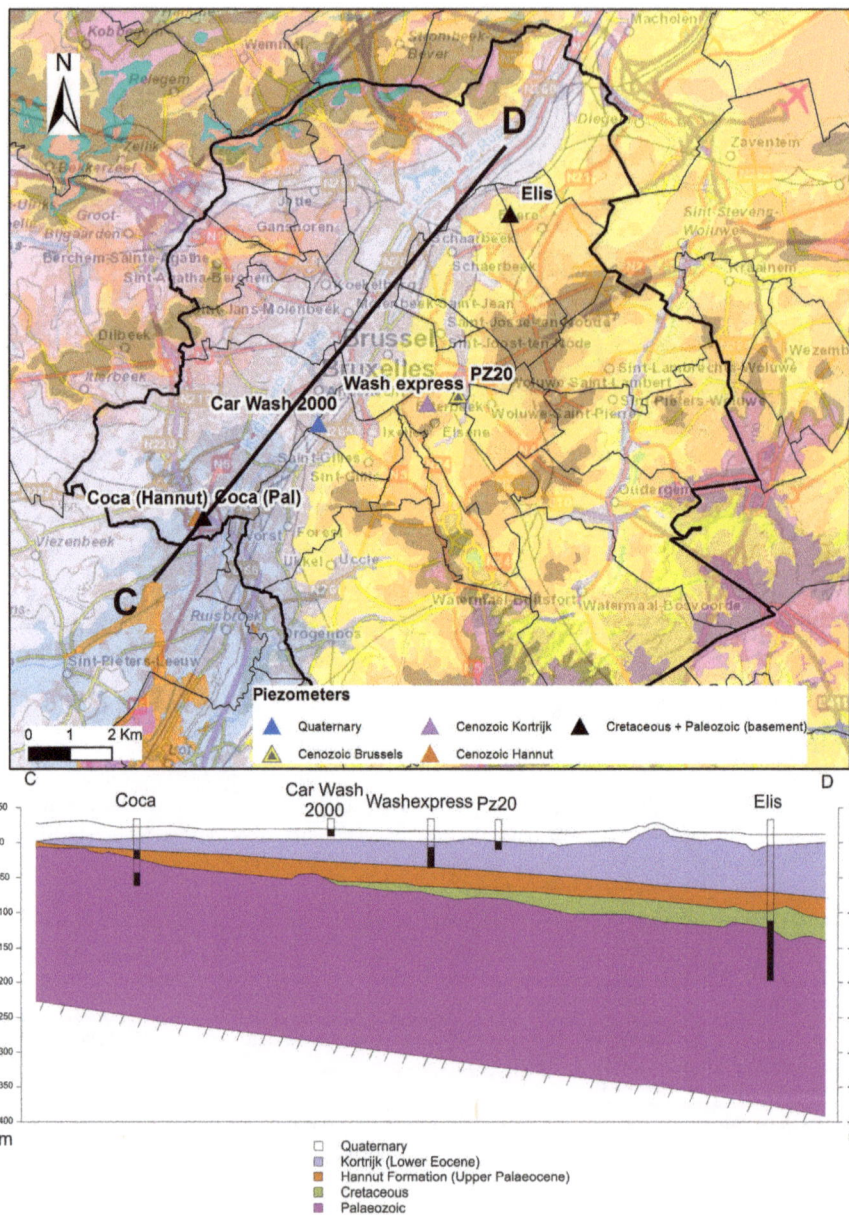

Figure 7. Location of the selected piezometers and simplified geological cross-section through the piezometers highlighting the different geological layers (from Quaternary to deep basement). The black rectangle of each piezometer corresponds to the deep extension of the strainer.

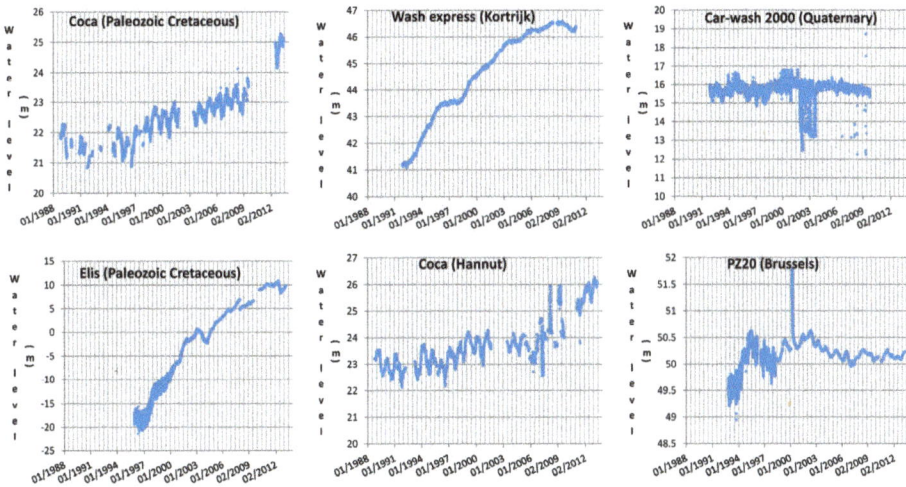

Figure 8. Evolution of the water table in the selected piezometers generally for the time interval ranging between 1988 and 2012 (depending on available data) and for the different aquifers.

Four of the selected piezometers are placed in the main aquifers below the thick aquitard of the Saint-Maur Member (Kortrijk Formation): piezometers Elis and Coca in the Paleozoic-Cretaceous aquifer, Coca in the Upper Paleocene aquifer (Hannut Formation) and Wash Express in the Lower Eocene (Kortrijk). Two piezometers are placed in the main aquifers above the Kortrijk Formation: PZ20 in the Middle Eocene aquifer (from Diest to Moen) and Car Wash 2000 in the Quaternary aquifer. A synthetic cross-section through the axis of the Senne River is presented at the Figure 7. The piezometers that are not exactly located in this axis were projected on this axis in order to simplify the understanding of the succession of aquifers/aquitards. No piezometers in perched aquifers were selected for this study as their importance in terms of areal extent and volume is limited.

The evolution of the groundwater levels through time clearly shows a recharge of the deeper aquifers in Brussels since the early/mid-1990s. In the Elis piezometer the groundwater has risen by circa 30 m between 1995 and 2014. In the two Coca piezometers, rising groundwater levels were observed since the mid-1990s. Piezometer Wash Express shows a recharge in the Lower Eocene since 1992; after 2006 the recharge rate has reduced.

In the two shallow aquifers no clear trends of rising groundwater can be observed. Piezometer PZ20 in the Brussels aquifer shows an initially rising groundwater level of circa 1 m between 1992 and 1994, which may be related to the end of a local groundwater extraction within the zone of influence around the piezometer. For the remaining period, the groundwater level is relatively stable, with short-term fluctuations reflecting seasonal effects and/or the effects of local groundwater extractions. Piezometer Car Wash 2000 is measuring in the Quaternary aquifer in the centre of the Senne river valley. Here too, the evolution of the groundwater level does not show a clear trend and appears to be mainly influenced by seasonal effects and/or local groundwater extractions.

Considering the evolution of groundwater recharge in the deep aquifers below the Kortrijk Formation aquitard, increased pore water pressures and elastic rebound of the area seem to provide a plausible mechanism for the uplift observed during the ERS and Envisat timespan (1992–2010, Figure 5A,B). The centre of the uplift is migrating from the city centre towards northern Brussels during the Envisat timespan. This evolution could be linked either to remaining groundwater pumping activities or either to an increased thickness of the Cretaceous layer or a combination of both. It is well known from drillings data that the Cretaceous is thinner in the SW and thicker towards the NE along the Senne River (Figure 7). The increased thickness implies a delay in the groundwater recharge due to a higher volume of the unsaturated aquifer. The TerraSAR-X data (2011–2014, Figure 5C) indicate a decrease in both spatial extent and rate of uplift, which corresponds with the reduced rates of groundwater recharge observed after 2006 (notably in the Elis and Wash Express piezometers). The lateral migration to the west of the uplift should be associated to groundwater recharge, which cannot be proved because this area suffers from a lack of piezometers. As for the Sentinel 1 data (2014–2017, Figure 5D), the uplift has virtually vanished inside the city centre, and has even reversed within the Senne river valley. This declining trend in uplift rates is further exemplified by averaged time series of PS points located in a 50 m buffer around the Tour & Taxis site, NW of the city centre (Figures 1 and 5A). The PS points show a steady uplift throughout the ERS and Envisat periods, which gradually declines during the TerraSAR-X timespan and reaching more stable fluctuations during Sentinel 1 timespan (Figure 9). Another possibility is that the aquifer recharge has ended and the slight negative values of the SAR data are reflecting only a natural compaction process that has been observed in alluvial plains on recent Quaternary sediments. This natural subsidence effect was never highlighted along the Senne River because it was hidden by stronger positive values related to the aquifer recharge.

A comparison between the time series of PS located in a 50 m buffer around the piezometers Elis and Wash Express (Figure 10) was realized. With respect to the scale change between the two types of measurements, it shows that the deformation trends match during the recharge period observed in the piezometers. During the TerraSAR-X period, a decrease of the velocity is visible and corresponds to the reduction of the recharge rate visible in Wash express. This decrease of the recharge rate induced the reduction of the PS velocity but it delayed over time, which is a signal that there is an offset between the recharge and its effect at the surface. The Sentinel 1 PS data show negative velocities of about −3 to −5 mm/year for both piezometers. However, this subsidence trend recorded by Sentinel-1 in the vicinity of the piezometer does not follow the schema of a recharge of the Paleozoic-Cretaceous aquifer recharge. The PS velocity trend in the city centre for the Envisat period of time and TerraSAR-X is characterized by a reduction of the rates towards stable ground movements. Negative PS velocities of Sentinel-1 could be explained by a short-term subsidence event included inside a longer uplifting trend visible since the first ERS SAR image. Unfortunately, the piezometric measurements after 2012 are not available and consequently it is not possible to confront the decreasing velocity trend of PS with the evolution of the deep aquifers.

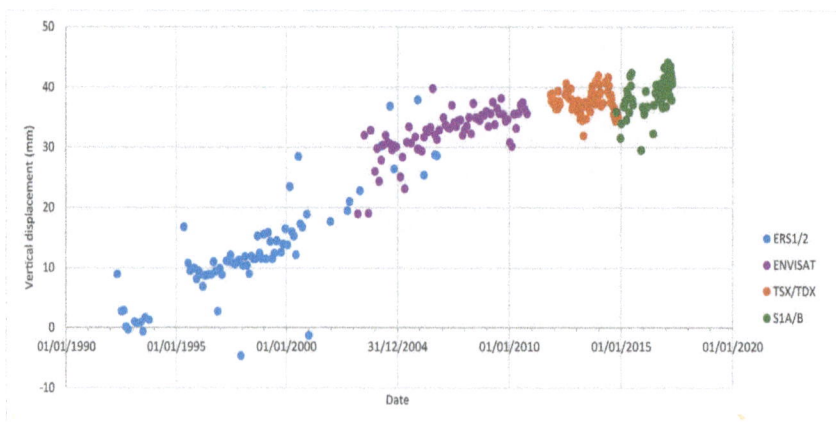

Figure 9. Time series of a PS (mm) located on the Tour & Taxis building.

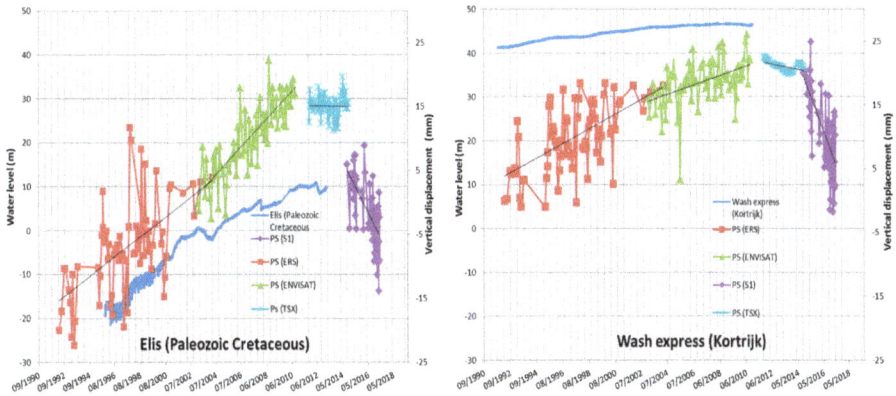

Figure 10. Comparison of the evolution of the water table (blue) and time series of averaged velocity of PS located in a buffer of 50 m around the selected piezometers. The different colours refer to the different radar satellites (red for ERS, green for ENVISAT, light blue for TerraSAR-X and purple for Sentinel 1).

It should be noted that the rising groundwater levels in the Paleozoic and Mesozoic aquifers since the early/mid-1990s are part of a general trend that started in the 1970s. From historical records [33] it is known that, due to the abundant presence of water-demanding industries in the city centre of Brussels, groundwater levels had dropped drastically between the 1880s and 1960s (circa 50 m, see Figure 11). It can be reasonably argued that overexploitation of the aquifers would have led to wide-spread ground subsidence in the period of heavy industrialisation. Since the 1970s, heavy industries moved out of the city centre [34] and groundwater levels began to recover (circa 30 m by the end of the 1980s, Figure 11).

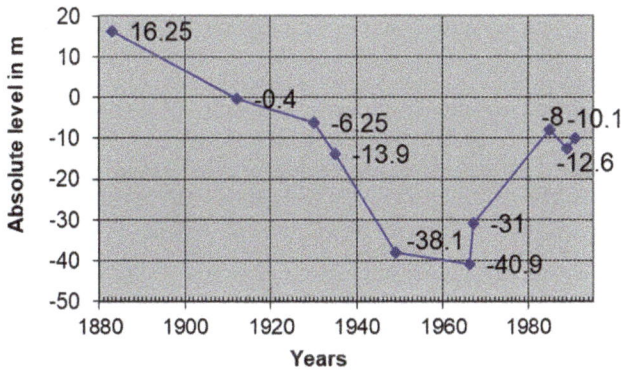

Figure 11. Historical groundwater levels in the Cretaceous aquifer.

Several local subsidence features can be identified on the maps of ground displacements (Figure 3A–D) and may correspond to local groundwater extraction activities or ground settlements. The subsiding phenomenon in Forest/Uccle is observed only in the Envisat data and is probably related to intense groundwater pumping, which lowers the Quaternary aquifer level for civil engineering works locally (confidential report, Brussels Environment Agency). Subsidence along lineament "La" in the TerraSAR-X and Sentinel 1 datasets may be related to engineering works or settlement of the railway embankments at Schaerbeek station, although conclusive records are not available to the authors. Subsidence in the Maelbeek and Woluwe river valleys (Figure 1) (TerraSAR-X data: linear features "Lb" and "Lc"), as well as in the Senne river valley (Sentinel 1 data: apparent reversal of the direction of movement from uplift to subsidence), may be attributed to the compaction of alluvial deposits [35–37]. The natural settlement of the river deposits is intensified by the load of buildings. The ground settlements probably existed during previous observation periods [38], but may have been obscured by the stronger signal of uplift in the area. As for local phenomena observed in the TerraSAR-X data, their absence in the other datasets may be related to differences in ground resolution.

6. Conclusions

The analysis of 25 years of PSI measurements in the Brussels region showed that the city centre was affected by an uplift of more than 4 cm along the Senne valley. Following our model, the recharge of the deeper aquifers has been the driving force since the 1970s. The end of large-scale groundwater exploitation since the end of the industrialization of the city is responsible for the groundwater rise. This is attested by several modern (1988–2014) piezometric measurements in these aquifers as well as historical levels. From 1880 to 1970 the city centre was probably subsiding due to the drop of the pore water pressures in the Cenozoic and the Paleozoic aquifers resulting in a compaction of the rocks. The observed uplift rate in the city centre is at its maximum during the ERS and Envisat periods, and tends to reduce during TerraSAR-X and Sentinel 1. The main results of this study demonstrate also that an inelastic rebound produced by the recharge of the deep aquifers has lasted almost 45–50 years. Groundwater exploitations have increased during the industrial era (at least from 1880 to the 1960s) implying a progressive overexploitation of the shallow aquifers and the progressive use of the artesian aquifers such as the deep ones related to the Cretaceous and Paleozoic rocks. The SAR data presented here suggest that the area of Brussels is again stable at a large scale and is no longer affected by uplifting conditions. These results need to be confirmed by new Sentinel 1 SAR images over a long interval (i.e., several years).

On the contrary, localized subsidence ground movements seem to affect the alluvial plain of the Senne River suggesting natural compaction of recent Quaternary sediments. This process was never before observed along the Senne River on ERS, Envisat and TerraSAR-X images due to the positive PS velocities associated to the aquifer recharge. Subsidence patterns in the Maelbeek, Woluwe and Senne river valleys were also highlighted. The subsidence can be related to the natural settlement of soft, young alluvial deposits, possibly intensified by the load of buildings.

Acknowledgments: TerraSAR-X data were provided by Deutsches Zentrum für Luft-und Raumfahrt (DLR) under the proposal GEO3185. The ERS and Envisat are provided by ESA. Sentinel 1 data were provided by the Copernicus program. Recomputed ERS 1/2 orbital data were obtained from DEOS, TUDelft. The research is funded by the Geological Survey of Belgium and partly by the Belspo GEPATAR project.

Author Contributions: Pierre-Yves Declercq globally wrote this paper being part of his research activities at the Geological Survey of Belgium, a department of the Royal Belgian Institute of Natural Sciences. Jan Walstra provided the InSAR results of ERS and TerraSAR-X and wrote the materials and methods part. Xavier Devleeschouwer (acting also as the PhD supervisor—official co-promotor of the PhD thesis of Pierre-Yves Declercq) and Bruno Meyvis helped with the geological and hydrogeological data interpretation. Pierre Gérard and Eric Pirard (acting also as the PhD supervisor—official promotor of the PhD thesis of Pierre-Yves Declercq) contributed to improve the quality of the paper. Daniele Perissin provided the SAR software support for processing the Sentinel 1 data.

Conflicts of Interest: The authors declare no conflict of interest.

References

1.   Gabriel, A.K.; Goldstein, R.M.; Zebker, H.A. Mapping small elevation changes over large areas: Differential radar interferometry. *J. Geophys. Res.* **1989**, *94*, 9183. [CrossRef]

2. Crosetto, M.; Monserrat, O.; Cuevas-González, M.; Devanthéry, N.; Crippa, B. Persistent Scatterer Interferometry: A review. *ISPRS J. Photogramm. Remote Sens.* **2016**, *115*, 78–89. [CrossRef]
3. Ferretti, A.; Prati, C.; Rocca, F. Nonlinear subsidence rate estimation using permanent scatterers in differential SAR interferometry. *IEEE Trans. Geosci. Remote Sens.* **2000**, *38*, 2202–2212. [CrossRef]
4. Ferretti, A.; Prati, C.; Rocca, F. Permanent scatterers in SAR interferometry. *IEEE Trans. Geosci. Remote Sens.* **2001**, *39*, 8–20. [CrossRef]
5. Hooper, A.; Zebker, H.; Segall, P.; Kampes, B. A new method for measuring deformation on volcanoes and other natural terrains using InSAR persistent scatterers. *Geophys. Res. Lett.* **2004**, *31*. [CrossRef]
6. Terrafirma Project. Available online: http://www.terrafirma.eu.com/ (accessed on 24 October 2017).
7. Devleeschouwer, X.; Pouriel, F.; Declercq, P. Vertical displacements (uplift) revealed by the PSInSAR technique in the Centre of Brussels, Belgium. In Proceedings of the IAEG2006 Engineering Geology for Tomorrow's Cities, Nottingham, UK, 6–10 September 2006.
8. Cigna, F.; Osmanoğlu, B.; Cabral-Cano, E.; Dixon, T.H.; Ávila-Olivera, J.A.; Garduño-Monroy, V.H.; DeMets, C.; Wdowinski, S. Monitoring land subsidence and its induced geological hazard with Synthetic Aperture Radar Interferometry: A case study in Morelia, Mexico. *Remote Sens. Environ.* **2012**, *117*, 146–161. [CrossRef]
9. Sowter, A.; Amat, M.B.C.; Cigna, F.; Marsh, S.; Athab, A.; Alshammari, L. Mexico City land subsidence in 2014–2015 with Sentinel-1 IW TOPS: Results using the Intermittent SBAS (ISBAS) technique. *Int. J. Appl. Earth Obs. Geoinf.* **2016**, *52*, 230–242. [CrossRef]
10. Chaussard, E.; Wdowinski, S.; Cabral-Cano, E.; Amelung, F. Land subsidence in central Mexico detected by ALOS InSAR time-series. *Remote Sens. Environ.* **2014**, *140*, 94–106. [CrossRef]
11. Castellazzi, P.; Arroyo-Domínguez, N.; Martel, R.; Calderhead, A.I.; Normand, J.C.L.; Gárfias, J.; Rivera, A. Land subsidence in major cities of Central Mexico: Interpreting InSAR-derived land subsidence mapping with hydrogeological data. *Int. J. Appl. Earth Obs. Geoinf.* **2016**, *47*, 102–111. [CrossRef]
12. Bell, J.W.; Amelung, F.; Ferretti, A.; Bianchi, M.; Novali, F. Permanent scatterer InSAR reveals seasonal and long-term aquifer-system response to groundwater pumping and artificial recharge. *Water Resour. Res.* **2008**, *44*. [CrossRef]
13. Hu, R.L.; Yue, Z.Q.; Wang, L.C.; Wang, S.J. Review on current status and challenging issues of land subsidence in China. *Eng. Geol.* **2004**, *76*, 65–77. [CrossRef]
14. Dong, S.; Samsonov, S.; Yin, H.; Ye, S.; Cao, Y. Time-series analysis of subsidence associated with rapid urbanization in Shanghai, China measured with SBAS InSAR method. *Environ. Earth Sci.* **2014**, *72*, 677–691. [CrossRef]
15. Teatini, P.; Gambolati, G.; Ferronato, M.; Settari, A.T.; Walters, D. Land uplift due to subsurface fluid injection. *J. Geodyn.* **2011**, *51*, 1–16. [CrossRef]
16. Newell, P.; Yoon, H.; Martinez, M.J.; Bishop, J.E.; Bryant, S.L. Investigation of the influence of geomechanical and hydrogeological properties on surface uplift at In Salah. *J. Pet. Sci. Eng.* **2017**, *155*, 34–45. [CrossRef]
17. Wnuk, K.; Wauthier, C. Surface deformation induced by magmatic processes at Pacaya Volcano, Guatemala revealed by InSAR. *J. Volcanol. Geotherm. Res.* **2017**. [CrossRef]
18. Pinel, V.; Raucoules, D. The Contribution of SAR Data to Volcanology and Subsidence Studies. In *Land Surface Remote Sensing*; Elsevier: London, UK, 2016; pp. 221–262, ISBN 978-1-78548-105-5.
19. Sintubin, M.; Everaerts, M. A compressional wedge model for the Lower Palaeozoic Anglo-Brabant Belt (Belgium) based on potential field data. *Geol. Soc. Lond. Spec. Publ.* **2002**, *201*, 327–343. [CrossRef]
20. Piessens, K.; De Vos, W.; Herbosch, A.; Debacker, T.; Verniers, J. Lithostratigraphy and geological structure of the Cambrian rocks at Halle-Lembeek (Zenne Valley, Belgium). *Prof. Pap. Geol. Surv. Belg.* **2004**, *300*, 1–142.
21. Mathijs, J.; Debacker, T.; Piessens, K.; Sintubin, M. Anomalous topography of the lower Palaeozoic basement in the Brussels region, Belgium. *Geol. Belg.* **2005**, *8*, 69–77.
22. Vercoutere, C.; Van Den Haute, P. Post-Palaeozoic cooling and uplift of the Brabant Massif as revealed by apatite fission track analysis. *Geol. Mag.* **1993**, *130*, 639. [CrossRef]
23. Deckers, J.; Matthijs, J. Middle Paleocene uplift of the Brabant Massif from central Belgium up to the southeast coast of England. *Geol. Mag.* **2017**, *154*, 1117–1126. [CrossRef]
24. Buffel, P.; Matthijs, J. *Kaartblad 31–39 Brussel-Nijvel*; Departement Omgeving, Vlaams Planbureau voor Omgeving: Brussels, Belgium, 2009; ISBN 1370-3803.

25. Rosen, P.A.; Hensley, S.; Peltzer, G.; Simons, M. Updated repeat orbit interferometry package released. *Eos Trans. Am. Geophys. Union* **2004**, *85*, 47. [CrossRef]

26. Kampes, B.; Usai, S. Doris: The delft object-oriented radar interferometric software. In Proceedings of the 2nd International Symposium on Operationalization of Remote Sensing, Enschede, The Netherlands, 16–20 August 1999; Volume 16, p. 20.

27. Rudenko, S.; Otten, M.; Visser, P.; Scharroo, R.; Schöne, T.; Esselborn, S. New improved orbit solutions for the ERS-1 and ERS-2 satellites. *Adv. Space Res.* **2012**, *49*, 1229–1244. [CrossRef]

28. Agence Spatiale Européenne. *Sentinel-1: ESA's Radar Observatory Mission for GMES Operational Services*; ESA Communications Production: Oakville, ON, Canada, 2012; ISBN 92-9221-418-7.

29. Perissin, D.; Wang, Z.; Wang, T. The SARPROZ InSAR tool for urban subsidence/manmade structure stability monitoring in China. In Proceedings of the 34th International Symposium on Remote Sensing of Environment, Sydney, Australia, 10–15 April 2011.

30. Camelbeeck, T.; Vanneste, K.; Alexandre, P.; Verbeeck, K.; Petermans, T.; Rosset, P.; Everaerts, M.; Warnant, R.; Van Camp, M. Relevance of active faulting and seismicity studies to assessments of long-term earthquake activity and maximum magnitude in intraplate northwest Europe, between the Lower Rhine Embayment and the North Sea. In *Special Paper 425: Continental Intraplate Earthquakes: Science, Hazard, and Policy Issues*; Geological Society of America: Boulder, CO, USA, 2007; Volume 425, pp. 193–224, ISBN 978-0-8137-2425-6.

31. Camelbeeck, T.; Alexandre, P.; Vanneste, K.; Meghraoui, M. Long-term seismicity in regions of present day low seismic activity: The example of Western Europe. *Soil Dyn. Earthq. Eng.* **2000**, *20*, 405–414. [CrossRef]

32. Camelbeeck, T.; Alexandre, P.; Sabbe, A.; Knuts, E.; Moreno, D.G.; Lecocq, T. The impact of the earthquake activity in Western Europe from the historical and architectural heritage records. In *Intraplate Earthquakes*; Talwani, P., Ed.; Cambridge University Press: Cambridge, UK, 2014; ISBN 978-1-107-04038-0.

33. Nuttinck, J.-Y. *Evolution de la Nappe Artésienne du Crétacé Sous L'agglomération Bruxelloise*; Mémoire Ing. Civil des Mines, Université Libre de Bruxelles: Brussels, Belgium, 1991.

34. De Beule, M. *Bruxelles: Une Ville Industrielle Méconnue: Impact Urbanistique de L'industrialisation*; Dossiers de La Fonderie; La Fonderie, Centre D'histoire Sociale et Industrielle de la Region Bruxelloise: St-Jan-Molenbeek, Belgium, 1994; ISBN 2-930048-00-X.

35. Stramondo, S.; Bozzano, F.; Marra, F.; Wegmuller, U.; Cinti, F.R.; Moro, M.; Saroli, M. Subsidence induced by urbanisation in the city of Rome detected by advanced InSAR technique and geotechnical investigations. *Remote Sens. Environ.* **2008**, *112*, 3160–3172. [CrossRef]

36. Chen, F.; Lin, H.; Zhang, Y.; Lu, Z. Ground subsidence geo-hazards induced by rapid urbanization: Implications from InSAR observation and geological analysis. *Nat. Hazards Earth Syst. Sci.* **2012**, *12*, 935–942. [CrossRef]

37. Herrera, G.; Tomás, R.; Monells, D.; Centolanza, G.; Mallorquí, J.J.; Vicente, F.; Navarro, V.D.; Lopez-Sanchez, J.M.; Sanabria, M.; Cano, M.; et al. Analysis of subsidence using TerraSAR-X data: Murcia case study. *Eng. Geol.* **2010**, *116*, 284–295. [CrossRef]

38. Devleeschouwer, X.; Declercq, P.-Y.; Pouriel, F. Radar interferometry reveals subsidence in Quaternary alluvial plains related to groundwater pumping and peat layers (Belgium). In Proceedings of the 5th European Congress on Regional Geoscientific Cartographic and Information Systems, Barcelona, Spain, 13–16 June 2006; Volume 315–317.

geosciences

MDPI

Article

Subsidence Trends of Volturno River Coastal Plain (Northern Campania, Southern Italy) Inferred by SAR Interferometry Data

Fabio Matano [1,*], Marco Sacchi [1], Marco Vigliotti [2] and Daniela Ruberti [2]

[1] Istituto per l'Ambiente Marino Costiero (IAMC), Consiglio Nazionale delle Ricerche (CNR), 80133 Naples, Italy; marco.sacchi@cnr.it
[2] Department of Civil Engineering, Design, Building and Environment (DICDEA), University of Campania "L. Vanvitelli", 81103 Aversa (Caserta), Italy; Marco.VIGLIOTTI@unicampania.it (M.V.); Daniela.RUBERTI@unicampania.it (D.R.)
* Correspondence: fabio.matano@cnr.it; Tel.: +39-081-542-3834

Received: 3 September 2017; Accepted: 27 December 2017; Published: 2 January 2018

Abstract: The Volturno Plain is one of the largest alluvial plains of peninsular Italy, which is one of the most susceptible plains to coastal hazards. This area is characterized by both natural and human-induced subsidence. This present study is based on the post-processing, analysis and mapping of the available Persistent Scatterer interferometry datasets. The latter were derived from the combination of both ascending and descending orbits of three different radar satellite systems during an observation period of almost two decades (June 1992–September 2010). The main output of this study is a map of vertical deformation, which provides new insights into the areal variability of the ground deformation processes (subsidence/uplift) of Volturno plain over the last few decades. The vertical displacement values obtained by the post-processing of the interferometric data show that the Volturno river plain is characterized by significant subsidence in the central axial sectors and in the river mouth area. Moderate uplift is detected in the eastern part of the plain, whereas other sectors of the study area are characterized by moderate subsidence and/or stability. On the basis of the analyzed subsoil stratigraphy, we inferred that the subsidence recorded in the Volturno plain is mainly a consequence of a natural process related to the compaction of the fluvial and palustrine deposits that form the alluvial plain. The anthropic influences (e.g., water exploitation, urbanization) are substantially considered to be an additional factor that may enhance subsidence only locally. The uplift mapped in the eastern sector of the plain is related to the tectonic activity. The study of the subsidence in the Volturno plain is a valuable tool for river flood analyses and the assessment of the coastal inundation hazards and related risk mitigation.

Keywords: SAR; alluvial plain; coastal areas; subsidence; Volturno; Italy

1. Introduction

Most Quaternary alluvial coastal plains of the Mediterranean are affected by significant subsidence due to simultaneous ground sinking caused by natural and human causes [1–4] and acceleration of the global sea level rise [5]. The main potential drivers of subsidence include tectonics [6], compaction of alluvial/coastal plain deposits [7] and fluid extraction [8,9]. The major consequences of subsidence include inundation of low lands, coastal erosion and aquifer salinization, increased vulnerability to flooding and storm surges as well as structural damage to infrastructures.

In the last few decades, several techniques based on the satellite-based Synthetic Aperture Radar (SAR) Interferometry (InSAR) have been used to assess ground-surface deformation related to several dynamic processes, such as subsidence, volcanism, seismicity and tectonics [10–16]. Two main different approaches have been developed to extract relevant information from the phase values of SAR images: Differential SAR Interferometry (DInSAR) and multi-interferograms SAR Interferometry (Persistent Scatterer Interferometry, PSI) techniques [17]. The DInSAR technique depends on the processing of two SAR images of the same target area acquired at different times [18] to detect phase shifts related to surface deformations (i.e., after an earthquake) that occur between the two reference acquisitions. This approach also requires an external digital elevation model of the analyzed morphological surfaces, from which the topographic phase contribution is estimated and removed from the interferogram. The PSI approach is based on the use of a long series of co-registered, multi-temporal SAR imagery, for which a larger number of images allows for more precise and robust results [17,19]. The PSI techniques allow for the production of maps of displacement velocity, which are measured along the straight line between target and radar sensor (Line Of Sight) and follow the temporal evolution of the displacement at each SAR acquisition epoch.

An overall, retrospective view on the spatial and temporal distribution of ground deformation can be derived for the area of the Campania Region by space-borne multi-temporal DInSAR techniques, based on the analysis of the C-band sensors onboard ERS-1/2, ENVISAT and RADARSAT satellites. A number of Italian national and regional remote sensing projects have recently reported several PSI-processed datasets related to Campania territory [20–26]. These various datasets have been acquired as the result of implementation of different processing techniques, such as Permanent Scatterers (PS-InSAR) [27], Persistent Scatterers Pairs (PSP) [19,28,29] and Small BAseline Subset (SBAS) [30–32]. Integrated results from these techniques provide very accurate displacement measurements along the SAR Line of Sight (LOS), which yield accuracy in the order of millimeters [27–33].

The use of SAR interferometry measurements to accurately analyze ground subsidence and/or uplift processes of different origins has already been tested and validated in several sectors of Campania Region [16,33–38]. In this research, we use the PS-InSAR and PSP datasets, which rely on the analysis of pixels that remain coherent over a sequence of interferograms (PSI techniques). These datasets are available for the entire study area over a relatively long time period (1992–2010).

The aim of the present study is the quantitative analysis of the ground deformation trends across a span of about two decades (1992–2010), which characterize the alluvial plain of the Volturno River located in the northern sector of Campania (Italy) (Figure 1).

This research was based on the post-processing, mapping and interpretation of the available ground deformation datasets [13,20–24], which was obtained from interferometric processing of ERS-1/2 (1992–2001), RADARSAT (2003–2007) and ENVISAT (2003–2010) radar satellite scenes, the latter provided by European Space Agency (ESA), using PS-InSAR and PSP techniques.

The Volturno plain has been largely investigated in recent years as a representative case of main land-use and hydro-geomorphological changes characterizing the urbanized coastal plains of the Mediterranean Sea during the last 50 years. Environmental changes documented for the Volturno coastal plain include the dramatic alteration of alluvial channels, floodplains and the deltaic environment as well as coastline retreat [37,39–42]. Previous studies have suggested that the considerable subsidence affecting the Volturno plain area may be interpreted as the result of a negative balance between the groundwater recharge of the Volturno alluvial plain and the water drainage by the channel system created for industrial and intense agricultural activities [16,43]. Based on the GIS overlay, Riccio [44] has pointed out that there is no spatial correlation in the Volturno plain between subsidence rates and the agricultural land use and/or livestock farming and cattle breeding areas. Aucelli et al. [37] elaborated the risk inundation maps for the coastal area of the plain by comparing sea level rise forecasts and local subsidence trends.

Our study presents a new quantitative scenario for the ground deformation in the whole Volturno plain following the method developed by Vilardo et al. [16]. Previous studies [16,37] have referred only to ERS-1/2 dataset for the time interval of 1992–2000. The main purpose of this research was to obtain a map of cumulated ground deformation for the study area over the period of 1992–2010, relying on three different SAR datasets provided by various satellites. The outcomes of our analysis are also discussed on the basis of the geological setting of Volturno Plain and subsurface stratigraphic data.

2. Geological, Geomorphological and Land Use Framework

The Campania Plain is part of a large extensional sedimentary basin, which mostly formed during the Quaternary between the western flank of the southern Apennines and the eastern Tyrrhenian Sea margin [45–50]. The plain is surrounded by calcareous—dolomitic mountain ridges (Figure 1), which is delimited by NW–SE trending regional normal faults, whereas its carbonatic and volcanic substratum is affected by extensional tectonics with NE–SW, NW–SE and E–W trending faults [51–55]. Some of these lineaments correspond to normal faults that have been active since the Late Pleistocene and display a vertical slip rate in the order of 0.2–2.5 mm/year [51].

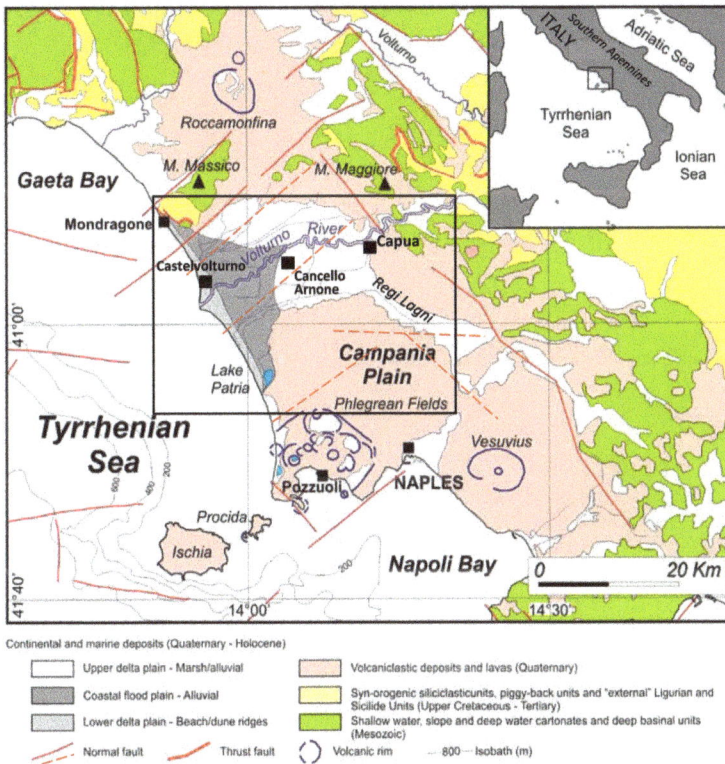

Figure 1. Geological map of the Campania Plain, with the black frame indicating the study area.

The area was largely submerged by the sea water since mid-late Pleistocene, when extensional tectonics locally developed across the continental margin and accompanied the onset of an intense volcanic activity [56]. At approximately 39 ky BP, a highly explosive eruption covered the entire Campania Plain with an ignimbrite deposit, up to 55 m thick, which is known as Campanian Ignimbrite or Grey Tuff (CGT) [57–60]. This unit represents the substrate for the uppermost Pleistocene–Holocene and recent sedimentation. A subsequent eruption, dated at approximately 15 ky BP [61], led to deposition of another ignimbrite deposit, the Neapolitan Yellow Tuff (NYT), which is primarily exposed in the urban area of Naples and at the Campi Flegrei (or Phlegrean Fields area). Tuff units

(CGT and NYT) and related ash deposits include most of the Quaternary volcaniclastic products cropping out in the Campania Plain (Figure 1).

The latest Pleistocene-early Holocene (approximately 16.0–6.0 ky BP) sea-level rise caused a rapid flooding of the lower paleo-Volturno incised valley, which was accompanied by the formation of a pronounced embayment at the mouth of paleo Volturno estuary and a landward widening of the continental shelf. Since approximately 6.5 ky BP, a coastal progradational phase was established, allowing for the onset of the present-day alluvial plain. The coastal area evolved as a wave-dominated delta system, with flanking strand plains forming beach-dune ridges partially enclosing lagoon-marsh areas [62,63] (Figures 1 and 2). Beach and lagoon environments persisted up until 2.0 ky BP. Since then, the crevasse splay and overbank deposits filled most of the swamp areas. Evidence for a persisting progradational trend of the shoreline has been documented since the 3rd century BC [39,64]. During the last century, a significant shoreline retreat affected the Volturno River mouth, particularly at the delta cusp, which was related to the decrease in river sediment load [39,41,65].

The studied sector of the Volturno plain is characterized by a flat morphology (Figure 3), with a slope <5° and an elevation of 0–40 m above sea level (a.s.l.) from the coastline to the north of Capua. The plain is bordered seaward by a 40-km long sandy beach, behind which there is a wide lowland area with an elevation between 0–2 m a.s.l. (Figure 3). This area is regarded as the remnant of an ancient barrier lagoon system [62,66–69], which is characterized by clayey and silty alluvial deposits (Figure 2) locally interbedded with peat layers [42].

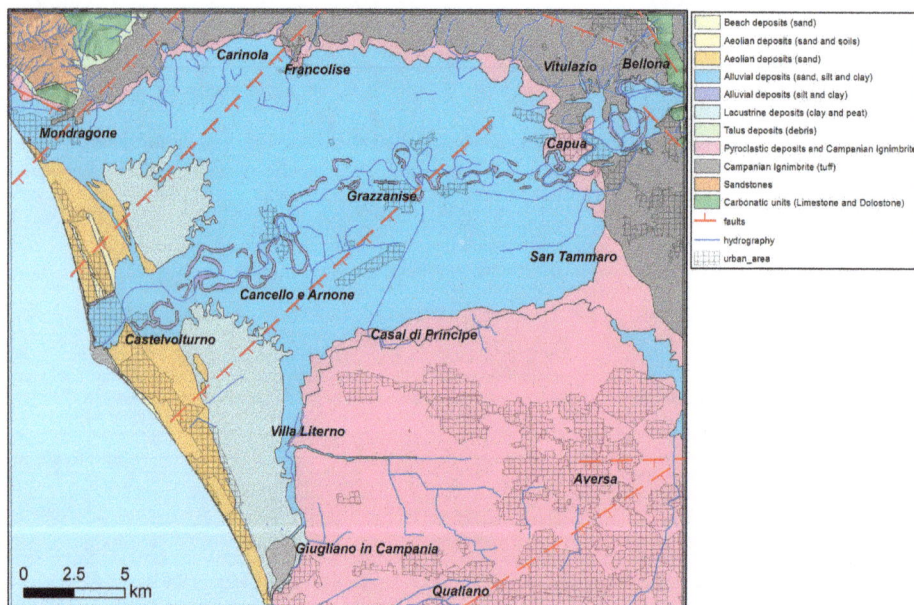

Figure 2. Geological map of the Volturno Plain (based on Geological Map of Italy and stratigraphic data [62,63,70]).

Most of the coastal plain marshy areas were reclaimed starting in the XVI century, which allowed for the development of agriculture and farming as well as severe urbanization along the coastal zone at the expense of the beach-dune system and along the river course since the mid-1900s [42]. Human activity often resulted in dramatic changes of the landscape, loss of coastal wetland and accelerated coastal erosion [39–41,65]. The effects of the above anthropic impacts are currently adding complexity to a scenario characterized by long-term land subsidence and predicted sea-level rise [69].

The study area is incised by a network of natural and artificial streams belonging to two major hydrographic catchments, which are namely the Volturno River and Regi Lagni channel. The relatively flat morphology of the coastal plain is characterized by the occurrence of a topographic high (up to 8–10 m a.s.l.) formed by Holocene dune ridge, which develops in NW–SE direction at approximately 4 km inland with respect to the present-day current shoreline. The ridge is locally interrupted by the linear erosion processes associated with the course of the Volturno River, Regi Lagni channel and Patria Lake emissary [37].

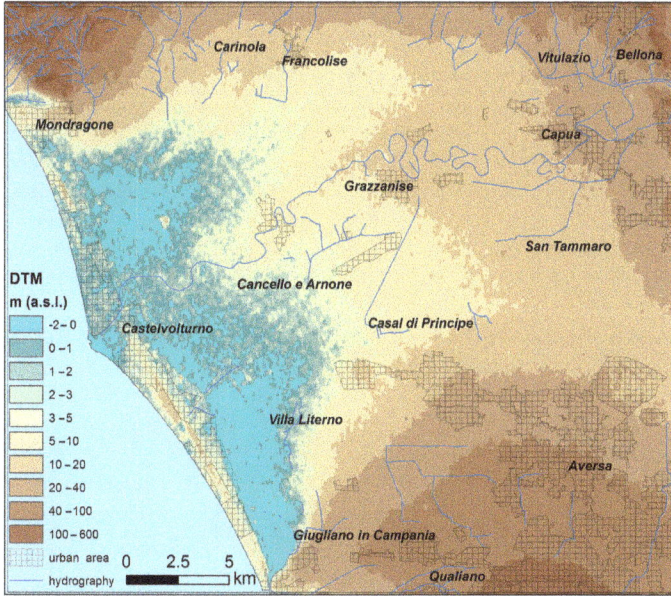

Figure 3. Altimetry classes of the Volturno Plain from Digital Terrain Model with 70-m cell size.

The Volturno plain is characterized by a locally high concentration of infrastructure (roads, railways, etc.) and inhabited areas both in coastal sectors (Castelvolturno and Mondragone towns) and in inland areas (Capua, Cancello Arnone), where several farming, agricultural and cheese production activities are located.

Land use shows a strong agricultural activity in the Volturno plain, which indicates that fields are almost completely farmed (81.1%) all year round. The urbanized environment represents a relevant percentage of the area (7.4%), and extends mainly along the coast. Grain cereals and grass cultivations prevail, mostly in association to the numerous zootechnical activities of the area. Most of these are produced by buffalo farming companies. Both cattle breeding and agricultural activities require high amounts of water.

3. Materials and Methods

Several interferometrically-processed (PSI approach) datasets, derived by some Italian national [23–25] and Campania regional [13,20–22] remote-sensing projects, are available for research activity and territorial planning and monitoring.

Recent trends in ground deformations of the Italian territory have been established and have been made available by processed sets of archive SAR images of ERS-1/2, RADARSAT 1 and ENVISAT satellites. The interferometric processing was performed both at the national scale, within the frame of the Not-Ordinary Plan of Environmental Remote Sensing [23,24] by means of the Persistent

Scatterer Pair (PSP) technique [19,28,29], and at a regional scale, within the activities of the TELLUS Project [13,20–22] by means of the Permanent Scatterers (PS-InSAR) standard technique [27].

The PS-InSAR processing allows for detection of the displacements through the time of man-made or natural reflectors (Permanent Scatterers, PS) along the radar LOS. This is obtained by separating time-dependent surface motions, atmospheric delays and elevation-error components of the measurement [27,71–73]. The main results for each PS consist of a time series of the variation in the sensor-target distance along the LOS (ground deformation) over the entire acquisition period, which includes the annual average velocity of the displacement calculated by linear fitting. This technique requires a master scene and a stable reference point, assumed motionless, to which the zero in the time series and the relative measurements of deformation are respectively referred [27].

The PSP technique [19,28,29] overcomes some limitations of other PS standard methods. Due to the pair-of-point approach, the PSP outputs are not affected by artifacts that are slowly variable in space, similar to those depending on atmosphere or orbits. This approach guarantees very dense and accurate displacement measurements both for anthropic structures and natural terrains [19].

The ERS-1/2, RADARSAT and ENVISAT satellites provided a comprehensive archive of imagery, which is currently used as a retrospective tool for historical interferometric monitoring of ground deformations. With reference to the Volturno plain study area, which extends for about 770 km^2, 354 images of ERS-1/2, RADARSAT 1 and ENVISAT satellite were used for PSI interferometric processing in the source projects (Table 1).

The average incidence angles of LOS are in the range of 22–34°, with a relevant difference between ERS-1/2, ENVISAT (22–25°) and RADARSAT (32–34°) angles (Table 1). The ERS-1/2 reference time range is from June 1992 to December 2001, whereas ENVISAT and RADARSAT acquisition periods range from November 2002 to July 2010 with overlapping datasets between March 2003 and September 2007 (Table 1).

For the study area, we have collected and post-processed six PS datasets (Table 2): (a) ENVISAT ascending orbit; (b) ENVISAT descending orbit; (c) RADARSAT ascending orbit; (d) RADARSAT descending orbit; (e) ERS-1/2 ascending orbit and (f) ERS-1/2 descending orbit. The PS datasets include about 506.000 PSs, which were identified within the study area with a coherence higher than 0.65. The selected datasets are characterized by negative mean values of displacement velocities in both ascending and descending orbits (Table 2). Specific PSI techniques used for each dataset are also listed (Table 2).

Table 1. Orbital parameters of interferometrically processed satellite images [20–25].

Satellite-Orbit	Track/Frame	Line of Sight Incidence Angle	Used Scenes	Time Range
ENVISAT-Descending	36/2781	22°	41	5 June 2003–3 June 2010
ENVISAT-Ascending	358, 129/819	25°	60	13 November 2002–30 July 2010
RADARSAT-Ascending	104/S3	34°	52	4 March 2003–15 September 2007
RADARSAT-Descending	11, 111/S3	32–33°	51	5 March 2003–23 August 2007
ERS-1/2-Ascending	129/801	22°	69	14 June 1992–13 December 2000
ERS-1/2-Descending	36, 265/2781, 2783	23°	81	8 June 1992–23 December 2001

PS datasets have been georeferenced to the projection WGS-84 UTM Zone 33N and spatially processed using GIS software. Each point of the four PS shapefile datasets is identified by coordinates (North, East) and a set of attributes, including: (a) identifier code; (b) average velocity of entire acquisition time period, expressed in mm year^{-1}; (c) standard deviation of the average velocity; (d) coherence; and (e) a subset of measurements (expressed in mm) of the displacement along the LOS of each PS. The PS coherence is a normalized index of the local signal-to-noise ratio of the interferometric phase and reflects the accuracy of PS measurements [14,74–76]. PSI processing allows us to obtain very accurate measurements within 1.0 mm/year for the PS average velocity along the

LOS by assuming a threshold value of 0.65 for coherence [33,71,75,77,78] thanks to the large number of SAR images used for the processing (Table 1).

Due to the side-looking view of SAR satellite sensors, PS displacements are measured along the LOS, which is in the range of 22–34° in the study area (Table 1). The availability of both ascending and descending datasets allows for the evaluation of the vertical components of the deformation for those areas common to both acquisition geometries, based on simple trigonometric calculations [16,33]. We could extract only two spatial components of ground deformation (vertical and E–W horizontal components), as the N–S horizontal component cannot be valued by the SAR satellite acquisition system [16].

The mean displacement rates have been used to derive the annual average ground deformation LOS velocity maps for ascending and descending orbits of ERS-1/2, RADARSAT and ENVISAT PS data by the Inverse Distance Interpolation Weighted (IDW) method. The IDW approach is commonly used to interpolate scattered points, thus allowing preservation of local variability of the data and reliable results [37,79,80]. An interpolation method, consisting of a quadratic weighting power of 2 within a 500-m radius neighborhood [16,33], was used to obtain 50-m regularly spaced grids. The vertical components of the mean annual velocity of the ground deformation have been computed using the ascending and descending LOS velocities (vLOS$_{asc}$ and vLOS$_{desc}$) raster maps derived from ERS-1/2, RADARSAT and ENVISAT data.

Table 2. Summary of Permanent Scattere (PS) datasets used in this study.

Satellite-Orbit	Persistent Scatterer Interferometry Technique	Count	PS Velocity-Mean (mm/year)	PS Stand. Dev.-Mean	PS Coherence-Mean	PS Density (Num./km^2)
ENVISAT-Ascending	PSP	105.319	−0.88	0.31	0.73	136.8
ENVISAT-Descending	PSP	120.557	−0.40	0.39	0.77	156.6
RADARSAT-Ascending	PS-InSAR	57.563	−1.32	1.31	0.84	74.8
RADARSAT-Descending	PS-InSAR	55.324	−0.96	0.72	0.85	71.8
ERS-1/2-Ascending	PSP	73.694	−1.08	0.73	0.73	95.7
ERS-1/2-Descending	PSP	93.480	−0.80	0.50	0.68	121.4

The vertical component of the velocity was calculated by combining the vLOS$_{asc}$ and vLOS$_{desc}$ raster maps on pixels common to both maps [13,16,32,33,81,82], which assumed that the ascending and descending LOS belong to the East–Z plane and the look-angle is the same for both ascending and descending geometries. Based on these assumptions, the following equation [16,33] was used:

$$vVert = \frac{(vLOSdesc + vLOSasc)/2}{\cos(q)} \quad (1)$$

where "v" is the displacement velocity vector of an investigated PS; vVert is the projection along the Cartesian vertical axes; vLOSdesc and vLOSasc are the projections of velocity along different LOSs; and q is the look-angle. A value of q given by the average of the two LOS incidence angles of the ascending and descending orbits was used for the calculation of vVert related to each satellite dataset, which was 22.5° for ERS-1/2, 33.25° for RADARSAT and 23.5° for ENVISAT. The obtained 50-m spaced grid maps show the distribution of the vertical components of ground deformation velocity in the Volturno Plain for the datasets in the time periods of 1992–2000 (ERS-1/2), 2003–2007 (RADARSAT) and 2003–2010 (ENVISAT).

In order to obtain a quantitative assessment of the subsidence process, the amount of vertical ground deformation was calculated based on the average yearly vertical velocity for each satellite dataset and the number of years of the time interval to which they are referred (i.e., 9 years for ERS-1/2, 2.5 years for RADARSAT and 5.5 years for ENVISAT, as the overlapping years of RADARSAT and ENVISAT datasets count 0.5 year each). The result is a 50-m spaced grid map, showing the cumulated distribution of vertical movements in the Volturno Plain (expressed in millimeters) for the period of 1992–2010.

4. Results

4.1. LOS Velocity Fields

The LOS velocity data distribution of used PS datasets is mainly characterized by negative values in both ascending and descending orbits for all satellite datasets (Table 3). The mean and median LOS velocity values are negative (respectively ranging from −1.32 to −0.40 mm/year and from −0.83 to −0.01 mm/year) for all datasets, showing an asymmetrical distribution toward negative values. This is due to general subsidence affecting the plain (skewness ranging from −3.62 to −1.49). The Quartile (Q1, Q3) values and the interquartile range (IQR) are similar for all datasets (Table 3). The ranges of "normal" values (less and more than two standard deviations away from the mean) that account for approximately 95% of all data show very similar values for all datasets (Table 3). This implies that the distribution of most (95%) of the data in each dataset is homogenous, which describes a constant and persistent process of ground deformation at the regional scale. Only the extreme maximum and minimum values of velocity (representing approximately 5% of all data) show significant differences, which are usually due to very local causes (e.g., instabilities of anthropic structures or buildings, water extraction from the subsurface or infiltration).

Table 3. Data distribution statistics of Line of Sight (LOS) velocity data in the used datasets.

Satellite-Orbit	Mean	Std. Dev.	Skewness	Q1	Median	Q3	IQR	−2 s.d.	+2 s.d.	Min	Max
ENVISAT-Ascending	−0.88	2.14	−3.32	−1.20	−0.20	+0.30	1.50	−5.16	+3.40	−31.30	+7.80
ENVISAT-Descending	−0.40	1.88	−3.16	−0.80	−0.01	+0.60	1.40	−4.16	+3.36	−28.10	+5.70
RADARSAT-Ascending	−1.32	2.60	−1.49	−2.25	−0.83	+0.10	2.35	−6.52	+3.88	−34.25	+26.09
RADARSAT-Descending	−0.96	2.45	−2.52	−1.49	−0.57	+0.21	1.70	−5.86	+3.94	−35.02	+16.51
ERS-1/2-Ascending	−1.08	2.41	−3.62	−1.26	−0.46	+0.08	1.34	−5.90	+3.74	−38.87	+8.26
ERS-1/2-Descending	−0.80	2.23	−3.47	−1.05	−0.26	+0.31	1.36	−5.26	+3.66	−37.22	+6.71

The spatial distribution of PS yearly average LOS velocity values is shown in different maps for both ascending and descending orbits of ERS-1/2, RADARSAT and ENVISAT datasets (Figure 4). All maps show a similar geographical distribution of PSs. Most of the PS are concentrated along linear infrastructures (roads, railway, etc.) and over urban areas, where the concentration of buildings and other manufactured structures is very high (Figure 4). Even if there is no data in local areas, the overall distribution of PS is sufficiently homogenous for a regional analysis of ground deformation. Considering the PS density (Table 2), we observed that ERS-1/2 and ENVISAT datasets are characterized by higher values of 96–157 PS/km^2, whereas RADARSAT datasets are characterized by lower values of PS density (72–75 PS/km^2). This difference depends on the adopted procedure. For example, using the PSP technique for ERS-1/2 and ENVISAT data processing results in a denser distribution of points for the measurement of displacement [19].

The spatial distribution of the mapped velocity classes displays a common general trend in the six datasets, even though some differences are present in the different maps (Figure 4). The lowest LOS velocity values (lower than −10 mm/year) are recorded in the central sector of the study area (around Grazzanise) and along the Volturno River (Figure 4). The Volturno river mouth shows moderate negative LOS velocity values (−2 to −10 mm/year), whereas positive values (+2 to +10 mm/year) are recorded in the eastern sector (eastward Capua and San Tammaro cities).

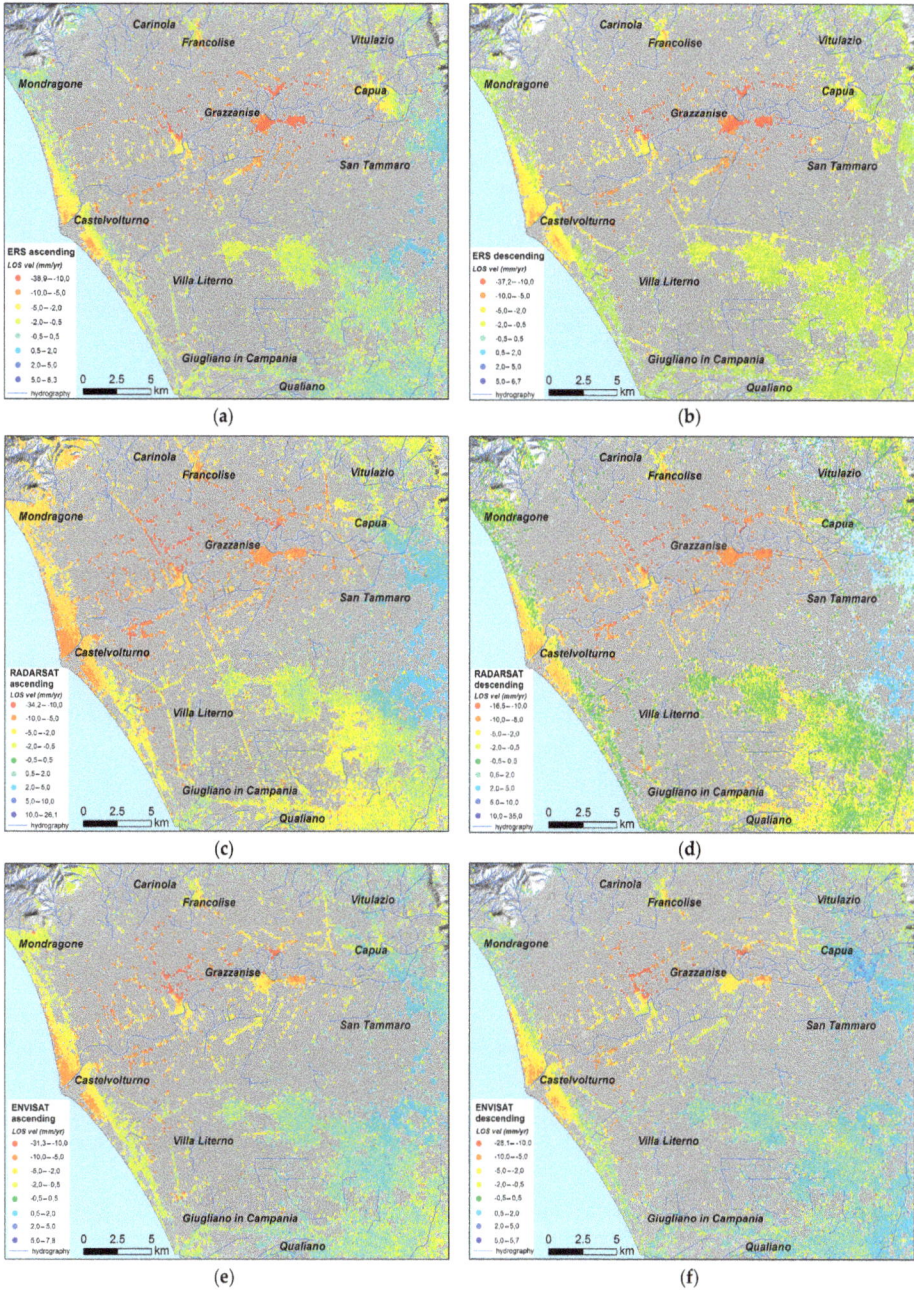

Figure 4. Map view of range-change rate measurements (LOS velocity) of PS with coherence ≥0.65 of the six used datasets: (**a**) ERS-1/2 ascending orbit; (**b**) ERS-1/2 descending orbit; (**c**) RADARSAT ascending orbit; (**d**) RADARSAT descending orbit; (**e**) ENVISAT ascending orbit; and (**f**) ENVISAT descending orbit.

In more detail, RADARSAT and ENVISAT datasets display higher positive values in the eastern sector of the study area and less negative values in the central sector of the study area when compared to ERS-1/2 datasets, showing some changes in the intensity of subsidence/uplift velocity during the two decades of observation. Moreover, some datasets show differences in the displacement rates obtained from ascending and descending geometries. RADARSAT datasets show some differences along the coastal area near Castelvolturno, which display higher subsidence in ascending datasets. RADARSAT and ERS datasets are characterized by lower positive values in descending datasets in the eastern uplifting sector. Along the southern sector of the study area, minor differences were found between ascending and descending RADARSAT and ERS datasets. These differences are interpreted as the effect of non-vertical movements affecting those sectors, such as local displacement from East to West along the horizontal component of ground deformation for the coastal area and the eastern sectors.

It is important to note that the difference between the LOS angles for the two orbital systems (32–34° compared to 22–25°; Table 1) can only justify a difference of a few mm/year in the values of LOS velocity related to the same ground deformation phenomenon. Consequently, only a minor part of the discrepancy between RADARSAT and ERS/ENVISAT datasets are ascribed to the geometrical differences in the satellite acquisition geometry.

4.2. Vertical Components of Ground Deformation Velocity

The maps of the vertical component of the displacement velocity, derived by the processing of ascending and descending LOS velocities maps, demonstrate that the study area has been characterized by similar vertical velocity patterns over the 1992–2000 (Figure 5a), 2003–2007 (Figure 5b) and 2003–2010 (Figure 5c) time intervals.

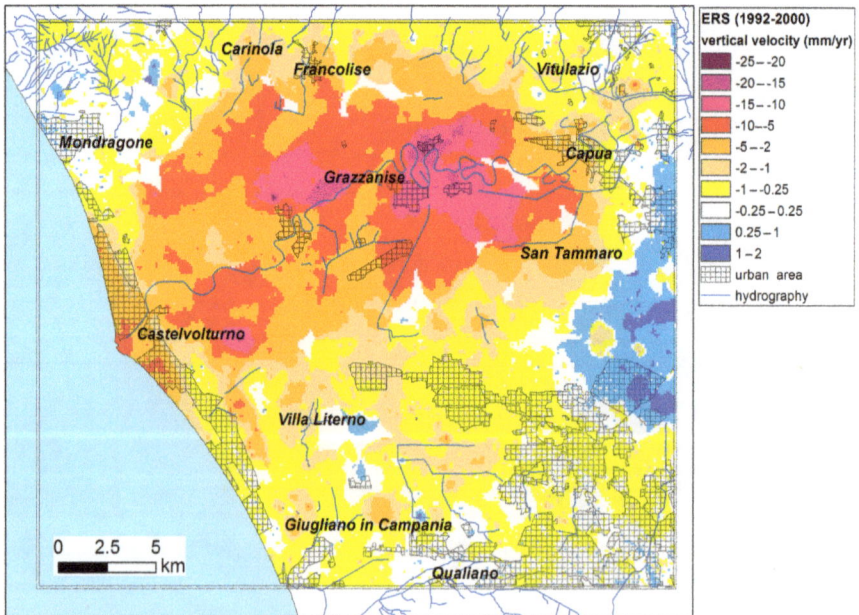

(a)

Figure 5. *Cont.*

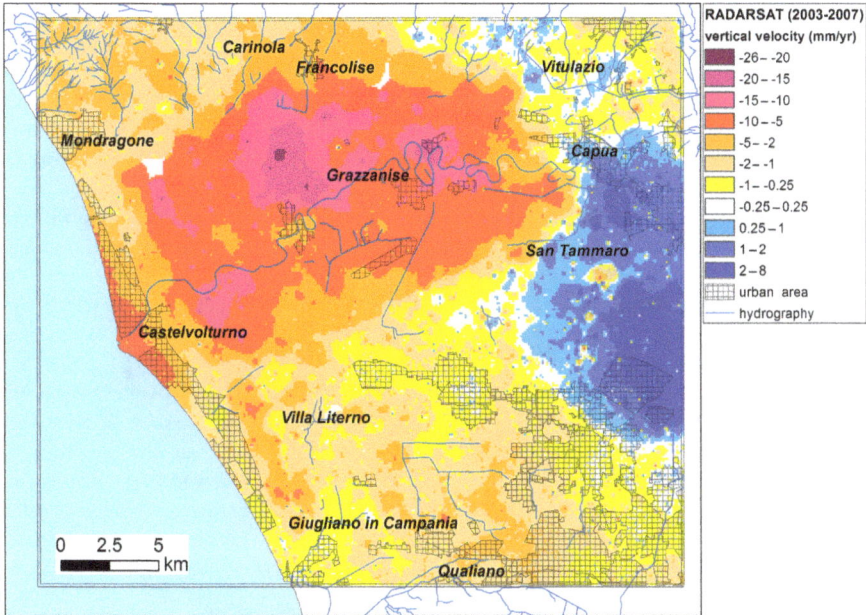

(b)

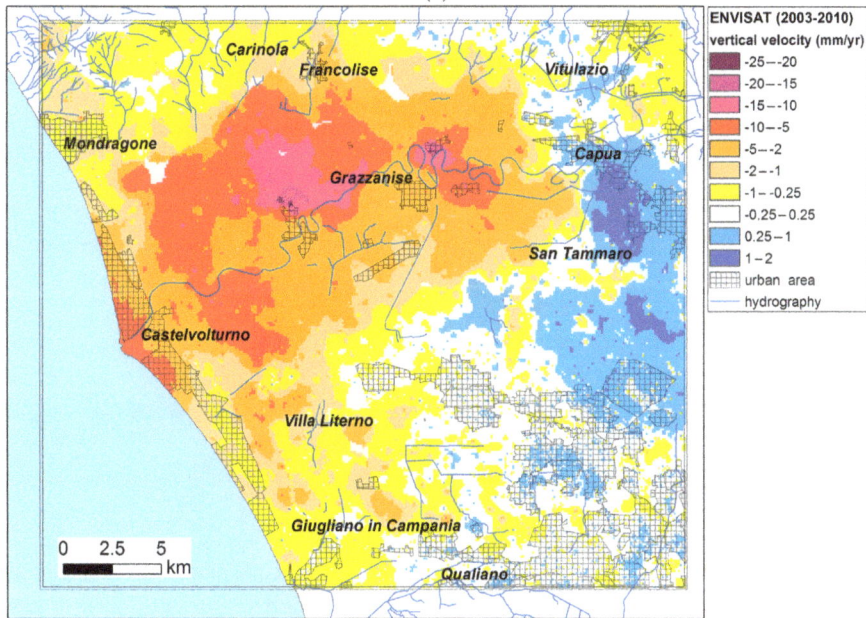

(c)

Figure 5. Vertical components of ground deformation velocity detected by interferometric processing of (**a**) ERS-1/2; (**b**) RADARSAT; and (**c**) ENVISAT datasets.

All the vertical velocity maps (Figure 5) indicate significant ground deformation in central-northern part of the Campania plain, which is mostly concentrated along the course of the

Volturno River and around Grazzanise since the early 1990s. The subsidence rates are in the order of −33 to −15 mm/year. Wider areas that are also affected by significant subsidence processes with rates of −15 to −5 mm/year occur along the Volturno river channels in several areas: the central sector of the plain, along the lateral flood plains, on the lowland area at the back of the dune system (Villa Literno) and at the estuary area of Castelvolturno. An uplift sector with positive (uplift) rates of +0.25 to +8 mm/year is detected to the East of San Tammaro, which is located in the eastern sector of the study area.

Even though the main ground deformation trends are consistent throughout the processed datasets, the differences between subsiding and uplifting sectors appear more marked and extended on the 2003–2010 maps (Figure 5b,c), which show uplifting sectors displaying rates up to +8 mm/year. In contrast, subsiding areas of the intermediate class (−5 to −10 mm/year) appear wider on the 2003–2007 map (Figure 5b).

The differences in vertical velocity rates between RADARSAT and ENVISAT datasets in the uplifting sector (Figure 6) are mainly limited to 1–3 mm/year, with RADARSAT velocity rates being higher than the ENVISAT datasets. These differences are likely to depend on the different values of LOS incidence angle of the two orbital systems and also reflect a net change in the intensity of subsidence/uplift velocity between the two periods. This change is observed in the displacement time series trends in Figure 7a. The same considerations can be made for areas affected by subsidence, which are characterized by different negative values in RADARSAT and ENVISAT datasets (Figure 7b).

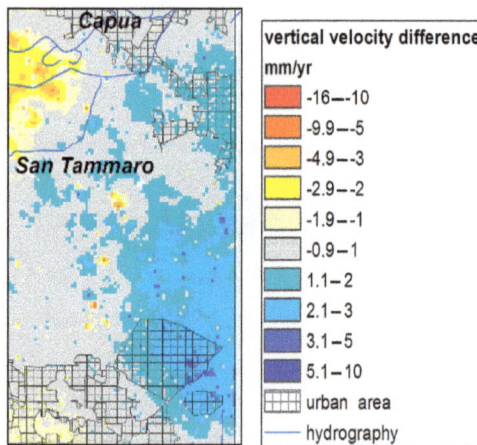

Figure 6. Differences in vertical velocity ranges between 2003–2007 RADARSAT and 2003–2010 ENVISAT datasets in the uplifting eastern sector of the study area.

4.3. Subsidence Assessment

A quantitative assessment (expressed in mm) of subsidence and uplift processes affecting the Volturno plain has been obtained by calculating the cumulative amount of the vertical deformation of ground surface, which was derived by the average annual vertical velocity of the three processed satellite datasets. The map of Figure 8 shows that approximately 89% of the study area (685 km^2) is characterized by negative values (from 0 to −417.4 mm). Only the remaining part, representing approximately 11% of the area (85 km^2), is characterized by positive values (from 0 to +40 mm) in the analyzed time interval (1992–2010).

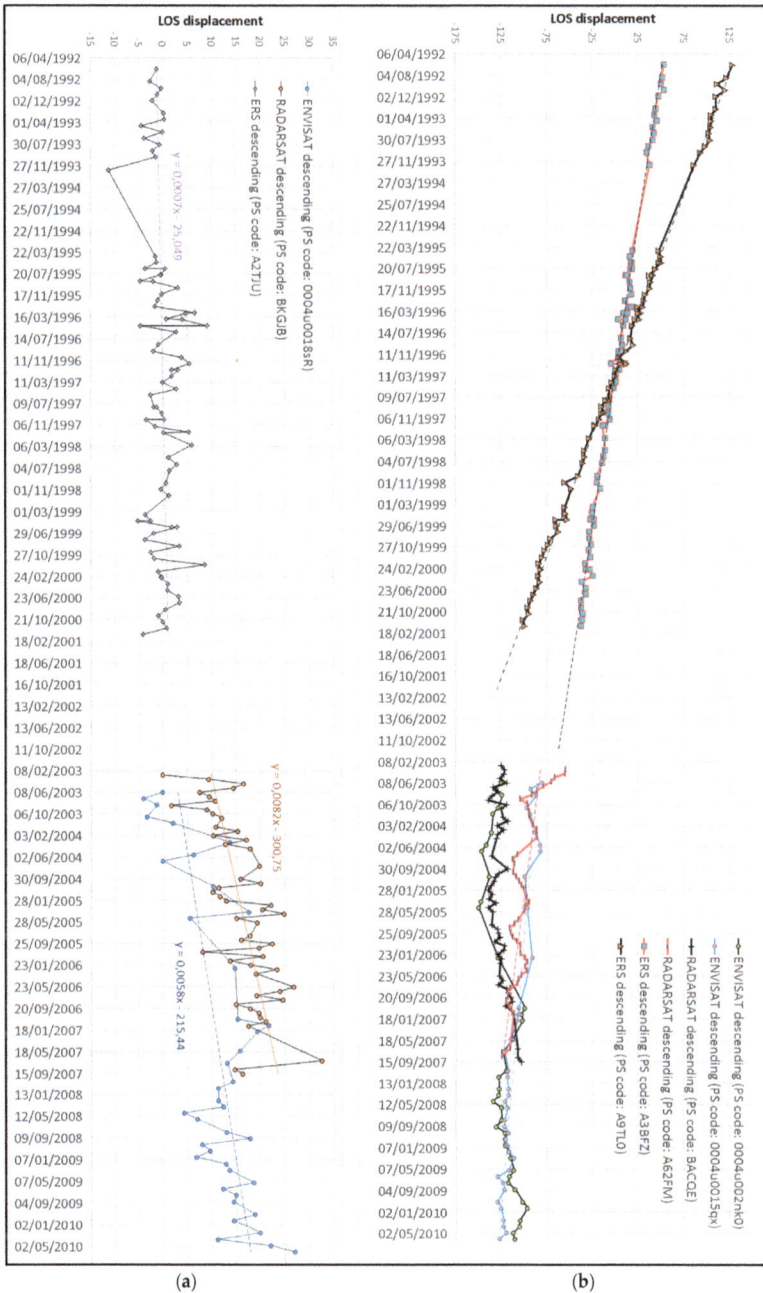

Figure 7. Time series of exemplative PS located in the study area: (**a**) comparison between time series of PSs located near San Tammaro in the uplifting sector (PS codes: A2TJU, BKGJB, 0004u0018sR); and (**b**) comparison between the time series of PSs located near Grazzanise in subsiding sectors (PS codes: A3BFZ, A62FM, 0004u0015qx) and PS in Phlegrean Fields area (PS codes: A9TL0, BACQE, 0004u002nk0). The PSs location is shown in Figure 9, except for the PS of Phlegrean Fields area, which is located within the harbor of the Pozzuoli city.

If we regard the coastal plain sectors characterized by vertical ground movements with values between +10 and −10 mm as being stable (green, gray and light blue colors in the map of Figure 8), almost 200 km^2 (26% of the study area) shows no significant evidence of subsidence or uplift during the period of 1992–2010. Conversely, the central part of the Volturno Plain shows a large area of 215 km^2, corresponding to approximately 30% of the investigated area, which displays subsidence values ranging from −50 to −418 mm. This sector is characterized by a complex deformation pattern (Figure 8) with the highest negative values (lower than −200 mm) along the course of the Volturno River around the city of Grazzanise. High subsidence values (−50 to −200 mm) are found in the more external sectors of the alluvial plain and the back-dune depressions, which extend for approximately 8 km^2, in the vicinity of Villa Literno. The coastal sector close to the Volturno river mouth (approximately 10 km^2) is characterized by moderate subsidence values (−50 to −150 mm), whereas the dune ridge system shows very slight subsidence or stability. Low subsidence (−10 to −50 mm) was found in the remaining sectors of the plain, which are located to the North and to the South of the Volturno River. The eastern sector (East of San Tammaro and Capua cities), extending for approximately 40 km^2 and corresponding to approximately 5% of the investigated area, displays moderate uplift with vertical ground movements between +10 mm and +40 mm (Figure 8).

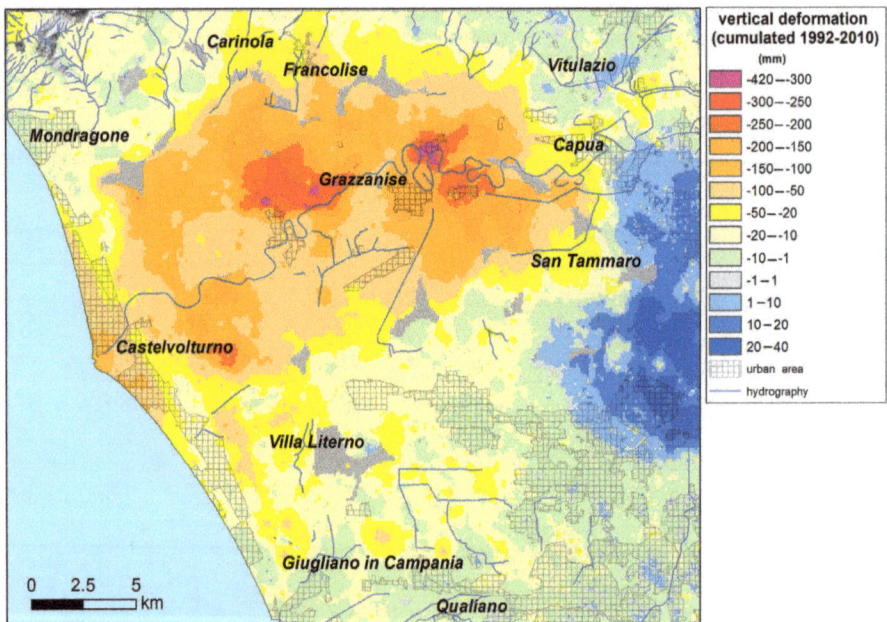

Figure 8. Cumulative amount of vertical ground displacement estimated during 1992–2010.

5. Discussion

Vertical deformation values derived by the post-processing of selected PSI datasets indicate that the Volturno river plain is characterized by a complex trend of vertical ground deformation. A 19-year period of observation (1992–2010) revealed that a moderate subsidence (between −50 mm and −10 mm) has characterized most of the investigated area. Severe subsidence has been detected in the central sectors of the plain with ground deformation lower than −200 mm along the course of the Volturno River and values lower than −150 mm close to the river mouth. Conversely, the eastern part of the study area, East of San Tammaro and Capua, is characterized by the occurrence of a slightly uplifting sector (Figure 8). The maps created in this study, which are based on results from different datasets (Figure 5), indicate that the study area is generally characterized by consistent vertical velocity patterns during the two periods of observation (1992–2000 and 2003–2010). LOS displacement time series (Figure 7) of some exemplative PS located (Figure 9) both in the subsiding sector near Grazzanise and in the uplifting sector near San Tammaro display mainly linear trends, which are characterized by continuous subsidence or uplift with similar rates through time. This behavior is substantially different from that observed in the adjacent structural domain of the Campi Flegrei (Figure 1). In fact, LOS displacement time series (Figure 7b) derived from the selected PS located near Pozzuoli clearly indicate that the Campi Flegrei district is characterized by remarkably non-linear behavior of ground deformation, due to repeated alternation between phases of uplift and subsidence. This is ostensibly related to active volcano–tectonic processes occurring in the volcanic caldera [12].

The interpretation of mechanisms controlling ground deformation patterns in the Volturno Plain is quite complex, as the entire coastal area has suffered from severe land use and profound landscape modifications during the last three centuries [42]. On the coastal zone, especially along the shoreline [37,42], a wild growth of both touristic and residential housing over the last decades has had a dramatic environmental impact. In turn, the increasing construction load and water pumping activity have contributed to enhance the longer-term subsidence rates due to tectonics [47,48], as suggested by a comparison of the 1992–2000 ERS-1/2 (Figure 4a) and 2003–2010 RADARSAT-ENVISAT datasets (Figure 4b,c).

Vilardo et al. [16] have proposed that the subsidence characterizing the Volturno plain is mostly due to the negative balance between the water recharge rates of the Volturno basin and the drainage operated by the artificial channeling, which is related to industrial and intense agricultural activities. The lack of an unequivocal correlation of subsidence trends with agricultural land use and zootechnical farm location [44] indicates that these human activities do not exert a significant impact on regional subsidence rates.

In contrast, the overlay of the main trends of subsidence with a geological map of the study area (Figure 9) suggests that the vertical ground deformation is partly controlled by the lithology and stratigraphic architecture of geological units forming the subsurface of the alluvial plain.

The occurrence of several tens of meters thick of Holocene alluvial deposits in the axial zone of the Volturno plain between the cities of Capua, Grazzanise, Cancello Arnone and Castelvolturno, and subordinately at the Volturno River mouth suggests that the high subsidence values (from −30 mm to −418 mm) recorded in this area are affected by the nature of the subsurface geological units (Figure 9). Particularly, the highest subsidence values (lower than −170 mm) detected in the vicinity of Grazzanise and Castelvolturno are correlated with the occurrence of thick peat layers in the subsurface (Figure 10). These layers are characterized by successions of organic clays, peat layers and alluvial brackish deposits (Coastal plain, Swamp and Lagoon deposits in Figure 10), which are easily compressible by a lithostatic load [37,62,63]. They were formed as a result of the accumulation of flood plain sediments by the Volturno river during the transition between the late phases (Early-Middle Holocene) of the post-glacial marine transgression and the Late Holocene progradation of the Volturno delta system [62,63]. In these areas, subsidence has been ostensibly driven by enhanced sediment compaction following the rapid progradation of the coastal environments during the Holocene [37,63].

The stratigraphic section of Figure 10 shows this geological phenomenon, which is also common to other major deltas of the Mediterranean region and Asia. Recent studies [7,83–87] have suggested that the natural compaction of alluvial/coastal plain deposits may cause subsidence of several millimeters/year, especially in organic-rich deposits, although there is still a lack of reliable information on the magnitude of the expected compaction rates.

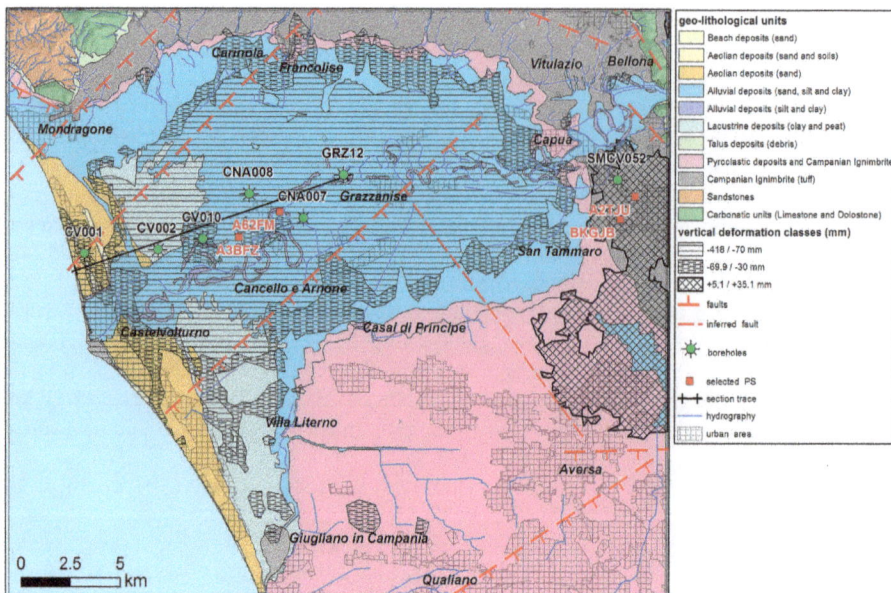

Figure 9. Correlation between cumulated amounts of vertical ground displacement classes, mapped in classes with wide thresholds to better highlight the main trends, and geology maps.

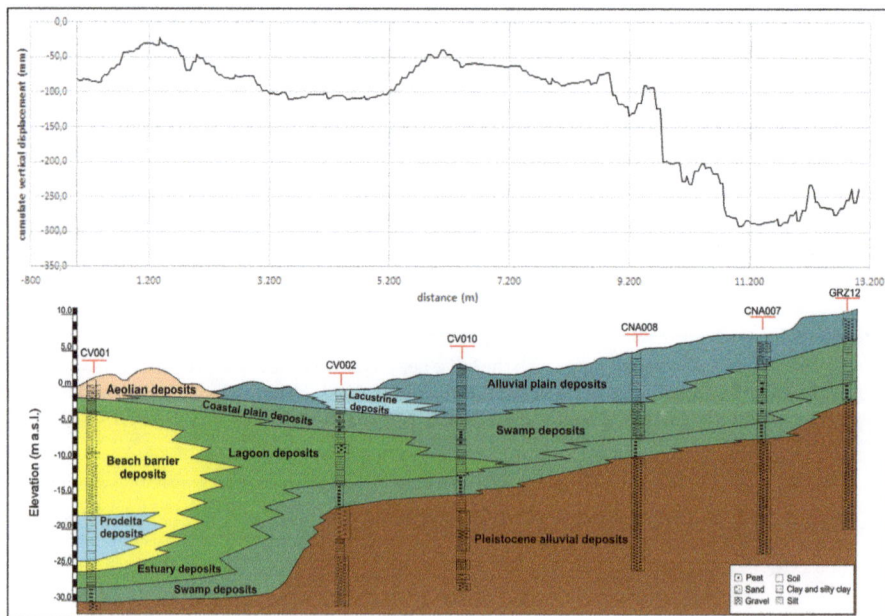

Figure 10. Correlation between stratigraphy and vertical velocity of displacement detected by Synthetic Aperture Radar (SAR) data.

The enhancement of the natural subsidence process has resulted from the uncontrolled groundwater exploitation, especially in the urban areas of the inner sector of the plain and along the coast with the highest values of subsidence (Figure 5).

In the north-eastern sector of the study area, the Campanian Ignimbrite tuff represents the main relative bedrock, which is locally covered by thin pyroclastic layers or alluvial and colluvial deposits (Figures 9 and 11). These tuffaceous units are formed mainly by consolidated rocky materials, which are substantially unaffected by the compaction processes due to secondary consolidation occurring under the lithostatic load of thick alluvial deposits. This geological setting clearly explains why the subsidence processes do not affect areas where the Campanian Ignimbrite and associated volcaniclastic cover units occur at shallow depths beneath the ground surface.

Unlike the central area of the Volturno Plain, the eastern sector of the study area is characterized by a slight uplift (values from +10 to +40 mm during 1992–2000) (Figure 9). The uplift appears to be continuous during the analyzed time intervals (1992–2000, 2003–2007 and 2003–2010; Figure 5). When analyzing the displacement time series (Figure 7a), we observed a change in uplift rates between the two decades of observation, as already detected from the maps of Figure 5. Minor oscillations are also related to the annual episodes of rainfall and groundwater variations. Actually, the uplifting sector is bounded to the south by the lateral termination of an E–W trending normal fault, which is known as the "Cancello fault" [51]. This tectonic element is located south of Marcianise (Figure 6) and has a vertical slip rate of approximately 1 mm/year from 130 ky to the present day, which has been estimated on the basis of morpho-tectonic analysis [51]. The area to the north of the fault is characterized by an average uplift of 1–4 cm in 19 years, which corresponds to a vertical rate of 0.5–2.0 mm/year. This scenario seems to be compatible with the active tectonic uplift component controlled by the "Cancello" fault activity and its estimated vertical rates. The south-western structural boundary between uplifting and subsiding sectors is tentatively interpreted as a fault zone, located between Casal di Principe and San Tammaro (Figure 9).

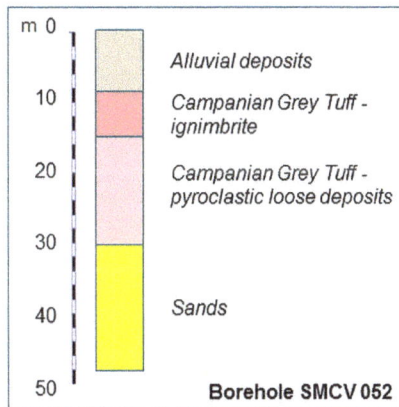

Figure 11. Example of a borehole stratigraphy inside the uplifting sector. For borehole location, see Figure 9.

6. Conclusions

The analysis and interpretation of three SAR datasets of both ascending and descending orbits acquired over the Campania coastal sectors during the period of June 1992–July 2010 provides new insights into the spatial variability of vertical ground deformation (subsidence/uplift) of the Volturno River coastal plain. These data are in agreement with the long-term tectonic subsidence process affecting the whole Tyrrhenian margin [45–48].

Ground deformation occurring in the Volturno plain was fully detectable by the SAR techniques. With reference to a period of about twenty years (1992–2010), we have documented a significant, continuous subsidence in some sectors of the alluvial and coastal plain and a moderate uplift in the eastern sector of the plain. Ground deformation patterns have been consistently imaged by the different SAR datasets, albeit with minor differences.

The integration of the three datasets has provided advantages with respect to the use of a single dataset:

1. By averaging and cumulating the different datasets, it was possible to minimize the contributions resulting from the different acquisition geometries (LOS angles) and SAR technique processing (PS-InSAR and PSP) of ERS-1/2—ENVISAT satellite data with respect to the RADARSAT satellite data;

2. By using multi-annual time series with different durations, it was possible to encompass a longer time period of observation, so that the final results of cumulated amount of vertical ground displacement is less affected by local and short-term changes.

According to our interpretation, the subsidence of the Volturno plain is regarded as a natural process, which is mainly due to the compaction under the lithostatic load of the alluvial/transitional sediments fill. The magnitude of the recorded subsidence has been found to be greater when the peat layers occur in the subsoil. Conversely, anthropic influences (e.g., water pumping and urbanization) are only considered as an additional factor, which locally enhances the subsidence processes. The detected uplift involving the eastern part of the study area is consistent with the tectonic activity already documented in the area [51].

The results of this study confirm the fundamental importance of using SAR data for a comprehensive understanding of rates and patterns of recent ground deformation over coastal plain regions within tectonically-active continental margins. They also provide a valuable tool for efficient territorial management addressed for coastal risk mitigation.

Acknowledgments: We thank three anonymous reviewers for their comments and suggestions that helped us to improve earlier versions of the manuscript. Patricia Sclafani is also acknowledged for the review of the English text.

Author Contributions: All the authors conceived and designed the research and interpreted the data; F.M. analyzed and processed the SAR data; M.S. provided a sequence stratigraphic framework of the delta system and the correlation between coastal marine and continental units; M.V. processed the stratigraphic data and elaborated the georeferenced geological map; D.R. analyzed the stratigraphic data and provided the geological profiles; F.M., D.R. and M.S. wrote the paper.

Conflicts of Interest: The authors declare no conflicts of interest.

References

1. Frihy, O.E.; El Sayed, E.E.; Deabes, E.A.; Gamai, I.H. Shelf sediments of Alexandria region, Egypt: Explorations and evaluation of offshore sand sources for beach nourishment and transport dispersion. *Mar. Georesour. Geotechnol.* **2010**, *28*, 250–274. [CrossRef]

2. Teatini, P.; Tosi, L.; Strozzi, T. Quantitative evidence that compaction of Holocene sediments drives the present land subsidence of the Po Delta, Italy. *J. Geophys. Res.* **2011**, *116*, 1–10. [CrossRef]

3. Teatini, P.; Tosi, L.; Strozzi, T.; Carbognin, L.; Cecconi, G.; Rosselli, R.; Libardo, S. Resolving land subsidence within the Venice Lagoon by persistent scatterer SAR interferometry. *Phys. Chem. Earth* **2012**, *40*, 72–79. [CrossRef]

4. Higgins, S.A. Review: Advances in delta-subsidence research using satellite methods. *Hydrogeol. J.* **2015**, *24*, 587–600. [CrossRef]

5. Warrick, R.A.; Provost, C.L.; Meier, M.F.; Oerlemans, J.; Woodworth, P.L. Changes in sea level, Climate Change. In *The Science of Climate Change*; Cambridge University Press: Cambridge, UK, 1996; pp. 359–405.

6. Xie, X.; Heller, P. Plate tectonics and basin subsidence history. *Geol. Soc. Am. Bull.* **2006**, *121*, 55–64. [CrossRef]

7. Long, A.J.; Waller, M.P.; Stupples, P. Driving mechanisms of coastal change: Peat compaction and the destruction of late Holocene coastal wetlands. *Mar. Geol.* **2005**, *225*, 63–84. [CrossRef]
8. Poland, J.F.; Davis, G.H. Land subsidence due to withdrawal of fluids. *Rev. Eng. Geol.* **1969**, *2*, 187–269.
9. Galloway, D.L.; Burbey, T.J. Review: Regional land subsidence accompanying groundwater extraction. *Hydrogeol. J.* **2011**, *19*, 1459–1486. [CrossRef]
10. Burgmann, R.; Rosen, P.A.; Fielding, E.J. Synthetic aperture radar interferometry to measure Earth's surface topography and its deformation. *Annu. Rev. Earth Planet. Sci.* **2000**, *28*, 169–209. [CrossRef]
11. Strozzi, T.; Wegmüller, U.; Tosi, L.; Bitelli, G.; Spreckels, V. Land subsidence monitoring with differential SAR interferometry. *Photogramm. Eng. Remote Sens.* **2001**, *67*, 1261–1270.
12. Herrera, G.; Fernández, J.A.; Tomás, R.; Cooksley Mulas, G.J. Advanced interpretation of subsidence in Murcia (SE Spain) using A-DInSAR data-modelling and validation. *Nat. Hazards Earth Syst. Sci.* **2009**, *9*, 647–661. [CrossRef]
13. Terranova, C.; Iuliano, S.; Matano, F.; Nardo, S.; Piscitelli, E.; Cascone, E.; D'Argenio, F.; Gelli, L.; Alfinito, M.; Luongo, G. The TELLUS Project: A satellite-based slow-moving landslides monitoring system in the urban areas of Campania Region. *Rend. Online Soc. Geol. Ital.* **2009**, *8*, 148–151.
14. Hanseen, R.F. *Radar Interferometry, Data Interpretation and Error Analysis*; Kluwer Academic Publishers: Dordrecht, The Netherlands, 2001; ISBN 978-0792369455.
15. Hooper, A.; Zebker, H.; Segall, P.; Kampes, B. A new method for measuring deformation on volcanoes and other natural terrains using InSAR persistent scatterers. *Geophys. Res. Lett.* **2004**, *31*, 1–5. [CrossRef]
16. Vilardo, G.; Ventura, G.; Terranova, C.; Matano, F.; Nardò, S. Ground deformation due to tectonic, hydrothermal, gravity, hydrogeological and anthropic processes in the Campania Region (Southern Italy) from Permanent Scatterers Synthetic Aperture Radar Interferometry. *Remote Sens. Environ.* **2009**, *113*, 197–212. [CrossRef]
17. Tofani, V.; Raspini, F.; Catani, F.; Casagli, N. Persistent Scatterer Interferometry (PSI) Technique for Landslide Characterization and Monitoring. *Remote Sens.* **2013**, *5*, 1045–1065. [CrossRef]
18. Gabriel, A.K.; Goldstein, R.M.; Zebker, H.A. Mapping small elevation changes over large areas: Differential radar interferometry. *J. Geophys. Res.* **1989**, *94*, 9183–9191. [CrossRef]
19. Costantini, M.; Falco, S.; Malvarosa, F.; Minati, F.; Trillo, F.; Vecchioli, F. Persistent Scatterer Pair Interferometry: Approach and Application to COSMO-SkyMed SAR Data. *IEEE J. Sel. Top. Appl. Earth Obs. Remote Sens.* **2014**, *7*, 2869–2879. [CrossRef]
20. Regione Campania—Settore Difesa del Suolo. Progetto TELLUS WebGIS (PSInSAR). 2009. Available online: http://webgis.difesa.suolo.regione.campania.it:8080/psinsar/map.phtml (accessed on 3 September 2017).
21. Regione Campania—Settore Difesa del Suolo. Progetto TELLUS web page. 2009. Available online: http://www.difesa.suolo.regione.campania.it/content/category/4/64/92/ (accessed on 3 September 2017).
22. Terranova, C.; Iuliano, S.; Matano, F.; Nardò, S.; Piscitelli, E. Relazione finale del Progetto TELLUS. PODIS Project of the Italian Ministry of Environment and of Protection of Territory and Sea—Campania Region. 2009. Available online: http://www.difesa.suolo.regione.campania.it/content/category/4/64/92/ (accessed on 3 September 2017).
23. EPRS-E. Not-Ordinary Plan of Environmental Remote Sensing Web Page. National Geoportal (NG) of the Italian Ministry of Environment and of Protection of Territory and Sea. 2015. Available online: http://www.pcn.minambiente.it/GN/en/projects/not-ordinary-plan-of-remote-sensing (accessed on 3 September 2017).
24. EPRS-E. Not-Ordinary Plan of Environmental Remote Sensing; ENVISAT Ascending e ENVISAT Descending Interferometric Products. National Geoportal (NG) of the Italian Ministry of Environment and of Protection of Territory and Sea. 2015. Available online: http://www.pcn.minambiente.it/viewer/ (accessed on 3 September 2017).
25. IREA-CNR InSAR WebGIS. Napoli Study Area. 2017. ENVISAT Dataset. Available online: http://webgis.irea.cnr.it/webgis.html (accessed on 3 September 2017).
26. Trasatti, E.; Casu, F.; Giunchi, C.; Pepe, S.; Solaro, G.; Tagliaventi, S.; Berardino, P.; Manzo, M.; Pepe, A.; Ricciardi, G.P.; et al. The 2004–2006 uplift episode at Campi Flegrei caldera (Italy): Constraints from SBAS-DInSAR ENVISAT data and Bayesian source inference. *Geophys. Res. Lett.* **2008**, *35*, 1–6. [CrossRef]
27. Ferretti, A.; Prati, C.; Rocca, F. Permanent Scatterers in SAR Interferometry. *IEEE Trans. Geosci. Remote Sens.* **2001**, *39*, 8–20. [CrossRef]

28. Costantini, M.; Falco, S.; Malvarosa, F.; Minati, F. A new method for identification and analysis of persistent scatterers in series of SAR images. In Proceedings of the 2008 IEEE International Geoscience and Remote Sensing Symposium (IGARSS), Boston, MA, USA, 7–11 July 2008; pp. 449–452.

29. Costantini, M.; Falco, S.; Malvarosa, F.; Minati, F.; Trillo, F. Method of Persistent Scatterer Pairs (PSP) and high resolution SAR interferometry. In Proceedings of the 2009 IEEE International Geoscience and Remote Sensing Symposium (IGARSS), Cape Town, South Africa, 12–17 July 2009; Volume 3, pp. 904–907.

30. Berardino, P.; Fornaro, G.; Lanari, R.; Sansosti, E. A new algorithm for surface deformation monitoring based on small baseline differential SAR interferograms. *IEEE Trans. Geosci. Remote Sens.* **2002**, *40*, 2375–2383. [CrossRef]

31. Lanari, R.; Mora, O.; Manunta, M.; Mallorquí, J.J.; Berardino, P.; Sansosti, E. A small baseline approach for investigating deformation on full resolution differential SAR interferograms. *IEEE Trans. Geosci. Remote Sens.* **2004**, *42*, 1377–1386. [CrossRef]

32. Lanari, R.; Casu, F.; Manzo, M.; Zeni, G.; Berardino, P.; Manunta, M. An overview of the Small Baseline subset algorithm: A DInSAR technique for surface deformation analysis. *Pure Appl. Geophys.* **2007**, *164*, 637–661. [CrossRef]

33. Vilardo, G.; Isaia, R.; Ventura, G.; De Martino, P.; Terranova, C. InSAR Permanent Scatterer analysis reveals fault re-activation during inflation and deflation episodes at Campi Flegrei caldera. *Remote Sens. Environ.* **2010**, *114*, 2373–2383. [CrossRef]

34. Iuliano, S.; Matano, F.; Caccavale, M.; Sacchi, M. Annual rates of ground deformation (1993–2010) at Campi Flegrei, Italy, revealed by Persistent Scatterer Pair (PSP)—SAR interferometry. *Int. J. Remote Sens.* **2015**, *36*, 6160–6191. [CrossRef]

35. Terranova, C.; Ventura, G.; Vilardo, G. Multiple causes of ground deformation in the Napoli metropolitan area (Italy) from integrated Persistent Scatterers Din-SAR, geological, hydrological, and urban infrastructure data. *Earth Sci. Rev.* **2015**, *146*, 105–119. [CrossRef]

36. Peduto, D.; Cascini, L.; Arena, L.; Ferlisi, S.; Fornaro, G.; Reale, D. A general framework and related procedures for multiscale analyses of DInSAR data in subsiding urban areas. *ISPRS J. Photogramm. Remote Sens.* **2015**, *105*, 186–210. [CrossRef]

37. Aucelli, C.P.P.; Di Paola, G.; Incontri, P.; Rizzo, A.; Vilardo, G.; Benassai, G.; Buonocore, B.; Pappone, G. Coastal inundation risk assessment due to subsidence and sea level rise in a Mediterranean alluvial plain (Volturno coastal plain e southern Italy). *Estuar. Coast. Shelf Sci.* **2016**, *198*, 597–609. [CrossRef]

38. Di Paola, G.; Alberico, I.; Aucelli, P.P.C.; Matano, F.; Rizzo, A.; Vilardo, G. Coastal subsidence detected by Synthetic Aperture Radar interferometry and its effects coupled with future sea-level rise: The case of the Sele Plain (Southern Italy). *J. Flood Risk Manag.* **2017**. [CrossRef]

39. Cocco, E.; Crimaco, L.; de Magistris, M.A. Dinamica ed evoluzione del litorale campano-laziale: Variazioni della linea di riva dall'epoca romana ad oggi nel tratto compreso tra la foce del Volturno e Torre S. In Proceedings of the "10° Congresso *AIOL*", Mondragone, Italy, 4–6 November 1992; pp. 543–555.

40. Donadio, C.; Vigliotti, M.; Valente, R.; Stanislao, C.; Ivaldi, R.; Ruberti, D. Anthropic vs. natural shoreline changes along the northern Campania coast, Italy. *J. Coast. Cons.* **2017**. [CrossRef]

41. Ruberti, D.; Vigliotti, M.; Di Mauro, A.; Chieffi, R.; Di Natale, M. Human influence over 150 years of coastal evolution in the Volturno delta system (southern Italy). *J. Coast. Cons.* **2017**. [CrossRef]

42. Ruberti, D.; Vigliotti, M. Land use and landscape pattern changes driven by land reclamation in a coastal area. The case of Volturno delta plain, Campania Region, southern Italy. *Environ. Earth Sci.* **2017**, *76*, 694.

43. ENEA. Analisi di specifiche situazioni di degrado della qualità delle acque in Campania. In *Riferimento ai Casi Che Maggiormente Incidono Negativamente Sulle Aree Costiere*; ENEA, Sezione Protezione-Idrogeologica: Roma, Italy, 2008.

44. Riccio, T. Analysis of Subsidence in Campania Plain. Ph.D. Thesis, cycle XXVIII, University of Campania "L. Vanvitelli", Aversa, Italy, 2016; p. 67.

45. Patacca, E.; Sartori, R.; Scandone, P. Tyrrhenian basin and Apenninic arcs: Kinematic relations since late Tortonian times. *Mem. Soc. Geol. Ital.* **1990**, *45*, 425–451.

46. Oldow, J.S.; D'Argenio, B.; Ferranti, L.; Pappone, G.; Marsella, E.; Sacchi, M. Large-scale longitudinal extension in the southern Apennines contractional belt, Italy. *Geology* **1993**, *21*, 1123–1126. [CrossRef]

47. Ferranti, L.; Oldow, J.S.; Sacchi, M. Pre-Quaternary orogen-parallel extension in the Southern Apennine belt, Italy. *Tectonophysics* **1996**, *260*, 325–347. [CrossRef]

48. Casciello, E.; Cesarano, M.; Pappone, G. Extensional detachment faulting on the Tyrrhenian margin of the southern Apennines contractional belt (Italy). *J. Geol. Soc. Lond.* **2006**, *163*, 617–629. [CrossRef]

49. Di Nocera, S.; Matano, F.; Pescatore, T.; Pinto, F.; Torre, M. Geological characteristics of the external sector of the Campania-Lucania Apennines in the CARG maps. *Rend. Online Soc. Geol. Ital.* **2011**, *12*, 39–43.

50. Matano, F.; Critelli, S.; Barone, M.; Muto, F.; Di Nocera, S. Stratigraphic and provenance evolution of the Southern Apennines foreland basin system during the Middle Miocene to Pliocene (Irpinia-Sannio successions, Italy). *Mar. Petr. Geol.* **2014**, *57*, 652–670. [CrossRef]

51. Cinque, A.; Ascione, A.; Caiazzo, C. Distribuzione spazio-temporale e caratterizzazione della fagliazione quaternaria in Appennino meridionale. In *Le Ricerche del GNDT nel Campo Della Pericolosità Sismica 1996–1999*; Galadini, F., Meletti, C., Rebez, A., Eds.; CNR-Gruppo Nazionale per la Difesa dai Terremoti: Roma, Italy, 2000; pp. 203–218.

52. Mariani, M.; Prato, R. I bacini neogenici costieri del margine tirrenico: Approccio sismico-stratigrafico. *Mem. Della Soc. Geol. Ital.* **1988**, *41*, 519–531.

53. Milia, A.; Torrente, M.M. Late Quaternary volcanism and transtensional tectonics in the Naples Bay, Campanian continental margin, Italy. *Miner. Petrol.* **2003**, *79*, 49–65. [CrossRef]

54. Milia, A.; Torrente, M.M.; Massa, B.; Iannace, P. Progressive changes in rifting directions in the Campania margin (Italy): New constrains for the Tyrrhenian Sea opening. *Glob. Planet. Chang.* **2013**, *109*, 3–17. [CrossRef]

55. Ortolani, F.; Aprile, F. Nuovi dati sulla struttura profonda della Piana Campana a sud-est del fiume Volturno. *Boll. Soc. Geol. Ital.* **1978**, *97*, 591–608.

56. Scandone, R.; Bellucci, F.; Lirer, L.; Rolandi, G. The structure of the Campanian Plain and the activity of the Neapolitain volcanoes (Italy). *J. Volcanol. Geotherm. Res.* **1991**, *48*, 1–31. [CrossRef]

57. Barberi, F.; Innocenti, F.; Lirer, L.; Munno, R.; Pescatore, T.; Santacroce, R. The Campanian Ignimbrite: A major prehistoric eruption in the Neapolitan area (Italy). *Bull. Volcanol.* **1978**, *41*, 1–22. [CrossRef]

58. Di Girolamo, P.; Ghiara, M.R.; Lirer, L.; Munno, R.; Rolandi, G.; Stanzione, D. Vulcanologia e Petrologia dei Campi Flegrei. *Boll. Soc. Geol.* **1984**, *103*, 349–413.

59. Deino, A.L.; Southon, I.; Terrasi, F.; Campajola, L.; Orsi, G. [14]C and [40]Ar/[39]Ar dating of the Campanian Ignimbrite. Phlegrean Fields, Italy. In Proceedings of the 8th International Conference on Geochronology, Cosmochronology and Isotope Geology (ICOG), Berkeley, CA, USA, 5–11 June 1994; Volume 3, p. 633.

60. De Vivo, B.; Rolandi, G.; Gans, P.B.; Calvert, A.; Bohrson, W.A.; Spera, F.J.; Belkin, H.E. New constraints on the pyroclastic eruptive history of the Campanian volcanic plain (Italy). *Miner. Petrol.* **2001**, *73*, 47–65. [CrossRef]

61. Deino, A.L.; Orsi, G.; de Vita, S.; Piochi, M. The age of the Neapolitan Yellow Tuff caldera-forming eruption (Campi Flegrei caldera, Italy) assessed by Ar-40/Ar-39 dating method. *J. Volcanol. Geotherm. Res.* **2004**, *133*, 157–170. [CrossRef]

62. Amorosi, A.; Pacifico, A.; Rossi, V.; Ruberti, D. Late Quaternary incision and deposition in an active volcanic setting: The Volturno valley fill, southern Italy. *Sediment. Geol.* **2012**, *282*, 307–320. [CrossRef]

63. Sacchi, M.; Molisso, F.; Pacifico, A.; Vigliotti, M.; Sabbarese, C.; Ruberti, D. Late-Holocene to recent evolution of Lake Patria, South Italy: An example of a coastal lagoon within a Mediterranean delta system. *Glob. Planet. Chang.* **2014**, *117*, 9–27. [CrossRef]

64. Pappone, G.; Alberico, I.; Amato, V.; Aucelli, P.P.C.; Di Paola, G. Recent evolution and the present-day conditions of the Campanian Coastal plains (South Italy): The case history of the Sele River Coastal plain. *WIT Trans. Ecol. Environ.* **2011**, *149*, 15–27.

65. Scorpio, V.; Aucelli, P.P.C.; Giano, S.I.; Pisano, L.; Robustelli, G.; Rosskopf, C.M.; Schiattarella, M. River channel adjustments in Southern Italy over the past 150 years and implications for channel recovery. *Geomorphology* **2015**, *251*, 77–90. [CrossRef]

66. Barra, D.; Romano, P.; Santo, A.; Campaiola, L.; Roca, V.; Tuniz, C. The Versilian transgression in the Volturno river plain (Campania, Southern Italy): Palaeoenvironmental history and chronological data. *Il Quaternario* **1996**, *9*, 445–458.

67. Amorosi, A.; Molisso, F.; Pacifico, A.; Rossi, V.; Ruberti, D.; Sacchi, M.; Vigliotti, M. The Holocene evolution of the Volturno River coastal plain (southern Italy). *J. Mediter. Earth Sci.* **2013**, *5*, 7–11.

68. Romano, P.; Santo, A.; Voltaggio, M. L'evoluzione morfologica della pianura del fiume Volturno (Campania) durante il tardo Quaternario. *Il Quaternario* **1994**, *7*, 41–56.

69. Lambeck, K.; Antonioli, F.; Anzidei, M.; Ferranti, L.; Leoni, G.; Scicchitano, G.; Silenzi, S. Sea level change along the Italian coast during the Holocene and projections for the future. *Quat. Int.* **2011**, *232*, 250–257. [CrossRef]
70. Putignano, M.L.; Ruberti, D.; Tescione, M.; Vigliotti, M. Evoluzione recente di un territorio di pianura a forte sviluppo urbano: La Piana Campana nell'area di Caserta. *Boll. Soc. Geol. Ital.* **2007**, *126*, 11–24.
71. Lu, Z.; Kwoun, O.; Rykus, R. Interferometric syntetic aperture radar (InSAR): Its past, present and future. *Photogramm. Eng. Remote Sens.* **2007**, *73*, 217–221.
72. Rott, H. Advances in interferometric synthetic aperture radar (InSAR) in earth system science. *Prog. Phys. Geogr.* **2009**, *33*, 769–791. [CrossRef]
73. Vasco, D.W.; Rucci, A.; Ferretti, A.; Novali, F.; Bissell, R.C.; Ringrose, P.S.; Mathieson, A.S.; Wright, I.W. Satellite-based measurements of surface deformation reveal fluid flow associated with the geological storage of carbon dioxide. *Geophys. Res. Lett.* **2010**, *37*, L03303. [CrossRef]
74. Colesanti, C.; Locatelli, R.; Novali, F. Ground deformation monitoring exploiting SAR permanent scatterers. *IEEE IGARSS* **2002**, *2*, 1219–1221.
75. Colesanti, C.; Ferretti, A.; Locatelli, R.; Novali, F.; Savio, G. Permanent Scatterers: Precision assessment and multi-platform analysis. *IEEE IGARSS* **2003**, *2*, 1193–1195.
76. Colesanti, C.; Wasowski, J. Investigating landslides with spaceborne synthetic aperture radar (SAR) interferometry. *Eng. Geol.* **2006**, *88*, 173–199. [CrossRef]
77. Ferretti, A.; Savio, G.; Barzaghi, R.; Borghi, A.; Musazzi, S.; Novali, F.; Prati, C.; Rocca, F. Submillimeter accuracy of InSAR time series: Experimental validation. *IEEE Trans. Geosci. Remote Sens.* **2007**, *45*, 1142–1153. [CrossRef]
78. Massironi, M.; Zampieri, D.; Bianchi, M.; Schiavo, A.; Franceschini, A. Use of PSInSAR™ data to infer active tectonics: Clues on the differential uplift across the Giudicarie belt (Central-Eastern Alps, Italy). *Tectonophysics* **2009**, *476*, 297–303. [CrossRef]
79. Franke, R. Scattered data interpolation: Test of some methods. *Math. Comput.* **1982**, *33*, 181–200.
80. Mueller, T.G.; Pusuluri, N.B.; Mathias, K.K.; Cornelius, P.L.; Barnhisel, R.I.; Shearer, S.A. Map quality for ordinary kriging and inverse distance weighted interpolation. *Soil Sci. Soc. Am. J.* **2004**, *68*, 2042–2047. [CrossRef]
81. Lundgren, P.; Casu, F.; Manzo, M.; Pepe, A.; Berardino, P.; Sansosti, E.; Lanari, R. Gravity and magma induced spreading of Mount Etna volcano revealed by satellite radar interferometry. *Geophys. Res. Lett.* **2004**, *31*, L04602. [CrossRef]
82. Manzo, M.; Ricciardi, G.P.; Casu, F.; Ventura, G.; Zeni, G.; Borgström, S.; Berardino, P.; Del Gaudio, C.; Lanari, R. Surface deformation analysis in the Ischia island (Italy) based on spaceborne radar interferometry. *J. Volcanol. Geotherm. Res.* **2006**, *151*, 399–416. [CrossRef]
83. Meckel, T.A.; Brink, U.S.; Williams, S.J. Sediment compaction rates and subsidence in deltaic plains: Numerical constraints and stratigraphic influences. *Basin Res.* **2007**, *19*, 19–31. [CrossRef]
84. Meckel, T.A.; Brink, U.S.; Williams, S.J. Current subsidence rates due to compaction of Holocene sediments in southern Louisiana. *Geophys. Res. Lett.* **2006**, *33*, L11403. [CrossRef]
85. Shi, C.; Zhang, D.; You, L.; Li, B.; Zhang, Z.; Zhang, O. Land subsidence as a result of sediment consolidation in the Yellow River Delta. *J. Coast. Res.* **2007**, *23*, 173–181.
86. Tornqvist, T.E.; Wallace, D.J.; Storms, J.E.A.; Wallinga, J.; Van Dam, R.L.; Blaauw, M.; Derksen, M.S.; Klerks, C.J.W.; Meijneken, C.; Snijders, E.M.A. Mississippi Delta subsidence primarily caused by compaction of Holocene strata. *Nat. Geosci.* **2008**, *1*, 173–176. [CrossRef]
87. Van Asselen, S. The contribution of peat compaction to total basin subsidence: Implications for the provision of accommodation space in organic-rich deltas. *Basin Res.* **2011**, *23*, 239–255. [CrossRef]

geosciences

MDPI

Article

Exploitation of Satellite A-DInSAR Time Series for Detection, Characterization and Modelling of Land Subsidence

Roberta Bonì [1,*], Claudia Meisina [1], Francesca Cigna [2], Gerardo Herrera [3,4,5], Davide Notti [1], Stephanie Bricker [2], Harry McCormack [6], Roberto Tomás [3,4,7], Marta Béjar-Pizarro [3,4], Joaquín Mulas [3,4] and Pablo Ezquerro [3]

[1] Department of Earth and Environmental Sciences, University of Pavia, Via Ferrata 1, 27100 Pavia, Italy; claudia.meisina@unipv.it (C.M.); davidenotti@gmail.com (D.N.)
[2] British Geological Survey, Natural Environment Research Council, Nicker Hill, Keyworth, Nottinghamshire NG12 5GG, UK; francesca.cigna@gmail.com (F.C.); step@bgs.ac.uk (S.B.)
[3] Geohazards InSAR Laboratory and Modeling Group, Instituto Geológico y Minero de España (IGME), C/. Alenza 1, 28003 Madrid, Spain; g.herrera@igme.es (G.H.); roberto.tomas@ua.es (R.T.); m.bejar@igme.es (M.B.-P.); j.mulas@igme.es (J.M.); p.ezquerro.martin@gmail.com (P.E.)
[4] Unidad Asociada de Investigación IGME-UA de Movimientos del Terreno Mediante Interferometría Radar (UNIRAD), Universidad de Alicante, P.O. Box 99, 03080 Alicante, Spain
[5] Earth Observation and Geohazards Expert Group (EOEG), EuroGeoSurveys, the Geological Surveys of Europe, 36–38, Rue Joseph II, 1000 Brussels, Belgium
[6] CGG, NPA Satellite Mapping, Crockham Park, Edenbridge Kent TN8 6SR, UK; Harry.McCormack@cgg.com
[7] Departamento de Ingeniería Civil, Escuela Politécnica Superior, Universidad de Alicante. P.O. Box 99, 03080 Alicante, Spain
* Correspondence: roberta.boni01@universitadipavia.it; Tel.: +39-0382-985-842

Academic Editors: Ruiliang Pu and Jesus Martinez-Frias
Received: 28 February 2017; Accepted: 6 April 2017; Published: 11 April 2017

Abstract: In the last two decades, advanced differential interferometric synthetic aperture radar (A-DInSAR) techniques have experienced significant developments, which are mainly related to (i) the progress of satellite SAR data acquired by new missions, such as COSMO-SkyMed and ESA's Sentinel-1 constellations; and (ii) the development of novel processing algorithms. The improvements in A-DInSAR ground deformation time series need appropriate methodologies to analyse extremely large datasets which consist of huge amounts of measuring points and associated deformation histories with high temporal resolution. This work demonstrates A-DInSAR time series exploitation as valuable tool to support different problems in engineering geology such as detection, characterization and modelling of land subsidence mechanisms. The capabilities and suitability of A-DInSAR time series from an end-user point of view are presented and discussed through the analysis carried out for three test sites in Europe: the Oltrepo Pavese (Po Plain in Italy), the Alto Guadalentín (Spain) and the London Basin (United Kingdom). Principal component analysis has been performed for the datasets available for the three case histories, in order to extract the great potential contained in the A-DInSAR time series.

Keywords: A-DInSAR time series; land subsidence; groundwater level change; principal component analysis (PCA)

---

1. Introduction

Land subsidence represents the main response to superficial and deep deformations induced by multiple natural and anthropic phenomena (i.e., vadose zone processes, such as swelling/shrinkage of clay soils, soil consolidation, aquifer compaction, solid and fluid extraction, and load-induced compaction) which take place at different spatio-temporal scales. The impacts of land subsidence can be infrastructural, economic, environmental, and social [1,2], including impacts on the natural heritage, natural surficial drainage system, agricultural activities, building foundations and transportation network, alteration of irrigation network, reduction of aquifer storage, ground failure, and it enhances the risk of floods, affecting the human life and activities. The increasing factors of risk are mainly due to rapid urban development, relatively young alluvium soils, and weak mitigation and adaptation strategies [3]. Therefore, in the pre-mitigation investigation phase, the identification of land subsidence areas and the understanding of driving factors is fundamental in order to adopt suitable land use planning and sustainable management of the available resources. Then, land subsidence investigations are essential to delineate the magnitude and type of deformation related to the temporal evolution of surface displacements (i.e., linear or non-linear), the spatial extension of the affected areas and the mechanism of land subsidence.

When land subsidence evidence is the result of superimposed processes, it is difficult to discern, and to map the different processes and to evaluate the associated triggering factors. Therefore, the complexity of the mechanism recognition requires a multidisciplinary approach including the expertise of engineering geologists, hydrogeologists, and geotechnical engineers. Until now, the scientific community carried out different strategies and a combination of different methods, including field measurements, remote sensing tools, and integrated approaches, to solve the complexity of the problem in many areas of the world [4–8].

Recent advanced ground deformation investigations make use of satellite synthetic aperture radar (SAR) data to examine the mechanisms of land subsidence around the world. In particular, these investigations exploit advanced differential interferometric synthetic aperture radar (A-DInSAR) techniques, which are based on the processing of multiple interferograms derived from a large set (at least 20 images) of SAR images. These techniques allow the retrieval of displacement time series of measuring points over wide areas at millimeter resolution [9–15] and have already been successfully applied to monitoring the evolution of different processes.

In the last two decades, A-DInSAR techniques have experienced major developments, which are mainly related to (i) the progress of satellite SAR data acquired by new missions, such as COSMO-SkyMed and ESA's Sentinel-1 constellations; and (ii) the development of novel processing algorithms. The improvements in the displacement time series (TS) obtained by A-DInSAR techniques need appropriate methodologies to analyze extremely large datasets which consist of very large amounts of measuring points and associated deformation histories with high temporal resolution.

This work contributes to address this crucial aspect, by exploiting the great potential contained in A-DInSAR time series. In this article, the relevant contributions of A-DInSAR time series are analyzed distinguishing the main advantages and limitations for three topics that commonly concern land subsidence investigations:

1. Detection of the magnitude and distribution of land subsidence;
2. Characterization of the mechanisms involved in land subsidence; and
3. Modelling of land subsidence due to groundwater level changes.

In particular, the paper provides insights into A-DInSAR time series capabilities and suitability as valuable supporting tools for these topics through interesting case histories of moderate rates of displacement in flat areas (Oltrepo Pavese, Po Plain, in Italy) and of land subsidence due to groundwater level change (Alto Guadalentín Basin, in Spain and the London Basin, in the United Kingdom). A-DInSAR time series analysis has emerged as a fundamental tool not only to monitor, but also to detect, ground motion areas. An innovative approach based on A-DInSAR time series analysis is presented to detect ground motion areas with significant deformational behaviors, such as linear, non-linear, and seasonal movements. Moreover, displacement time series acquired by multi-temporal SAR sensors allow the extraction of a historical archive of the temporal evolution of land movements. The collection of multi-sensor displacement time series proved to be crucial for the characterization of land subsidence mechanisms, such as transient and inelastic aquifer deformation otherwise undetectable by considering the average velocity. Finally, changes, such as acceleration and deceleration in the displacement rates, were simulated by calibrating a 1-D model using A-DInSAR time series.

2. Data and Methods

In the framework of the Panel on Land Subsidence of the U.S. National Research Council (NRC) (1991) two information needs were recognized [16]:

1. Earth-science data and information on the magnitude and distribution of land subsidence in order to recognize and to assess future problems; and
2. Research on subsidence processes and engineering methods in order to prevent damage.

According to these information needs, the support of A-DInSAR time series for land subsidence investigations is presented here. A multidisciplinary approach was adopted by combining geological, geotechnical, hydrogeological, and A-DInSAR data to gain insight into land subsidence detection, characterization, and modelling through three representative test cases: Oltrepo Pavese, in Italy, Alto Guadalentín Basin, in Spain, and the London Basin, in the United Kingdom (Figure 1).

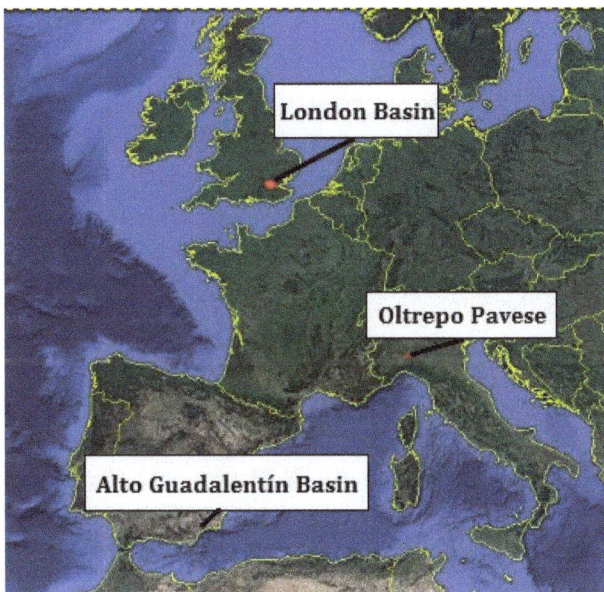

Figure 1. Location of the study areas.

Regarding the detection of ground motion areas, a novel methodology was implemented in the plain area of the Oltrepo Pavese (Italy) in order to distinguish different components of motion. The characterization of the mechanisms of land subsidence were analyzed in detail in the Alto Guadalentín Basin in Spain, where the highest subsidence rates measured in Europe (>10 cm/year) were recognized as a direct consequence of long-term aquifer exploitation [17]. Finally, the London Basin was chosen as a study area to use the A-DInSAR time series to characterize the hydrological properties of the deposits of the basin and to integrate the A-DInSAR time series in the modelling of ground deformation. The prediction of the ground motion expected in response to the groundwater level variations was performed in the London Basin thanks to the availability of detailed time-series data from the observation borehole network provided by the Environment Agency (around ~200 piezometers), to reconstruct the historical groundwater level changes across the basin for the 1990s and 2000s.

Additional details about the data for each test case are summarized in Table 1.

Table 1. Test case characteristics and available A-DInSAR data.

Test Case	SAR Data	Sensor	Processing Technique	Time Span	Type of Land Subsidence	Area (km²)
Oltrepo Pavese (Italy)	ERS-1/2	C-band	SqueeSAR™	1992–2000	Natural and anthropic causes	440
	RADARSAT-1	C-band	SqueeSAR™	2003–2010		
Alto Guadalentín Basin (Spain)	ERS-1/2	C-band	StaMPS	1992–2000	Anthropogenic subsidence due to groundwater extraction	277
	ENVISAT	C-band	StaMPS	2003–2007		
	ALOS-PALSAR	L-band	SPN	2007–2010		
	COSMO-SkyMed	X-band	SPN	2011–2012		
London Basin (United Kingdom)	ERS-1/2	C-band	IPTA	1992–2000	Natural and anthropic causes	1360
	ENVISAT	C-band	IPTA	2002–2010		

2.1. The Oltrepo Pavese Test Case

The Oltrepo Pavese (Italy) is the plain sector of the Po River (Figure 2) covering an area of approximately 440 km². The Oltrepo Pavese is a representative site of similar geological contexts in the Po Plain, where geohazards, due to natural and anthropic factors, were previously recognized [18]. Severely damaged structures were also observed in this area due to volume changes of clayey soils (shrinkage and swelling during drying and wet periods, respectively). Regarding the land use coverage of Oltrepo Pavese, discontinuous urban fabrics are the most prevalent land use class and 68% of the monitored area is covered by non-irrigated arable land.

The plain of the Oltrepo Pavese is constituted by alluvial quaternary deposits, originated from the action of Apennine streams and of the Po River [19,20]. These deposits overlay Miocene-Pliocene marine substratum, composed of sandy-marls, sandstones, conglomerates, gypsy-marls, and calcareous-marls. Three main geomorphological units were previously distinguished [20,21] in the quaternary deposits (1) Post-würmian alluvial deposits; (2) Würmian-holocenic alluvial deposits; (3) Pre-würmian alluvial deposits.

Post-würmian alluvial deposits are the most recent sediments that are mainly localized close to the Po River. These deposits are composed of sand, sandy silt, and silt, and they contain a shallow phreatic aquifer.

Würmian-holocenic alluvial deposits are made up of alternating sand and gravel, with interbedded clays or argillaceous silt. These deposits contain a shallow phreatic aquifer and deeper aquifers, of both phreatic and confined types. The aquifer geometry is controlled by the deeply buried structures in the tertiary marine basement, consisting of a series of folds and fold-faults. These alluvial deposits are extensively covered by clayey-silty deposits, which act as a seal, limiting water infiltration [22].

Pre-würmian alluvial deposits, located in the southern part of the study area, consist of older fluvial terraces, and are composed of gravel and sand with a silty matrix. In the Pre-würmian unit, the depth of the groundwater level in this unit reaches values up to 20 m depth.

Regarding the Quaternary sediment thickness, a decrease from west to east is observed in the study area. The minimum thickness and the outcrop of the marine substratum is evident in correspondence with the Stradella thrust, where neotectonic activity was previously described [23].

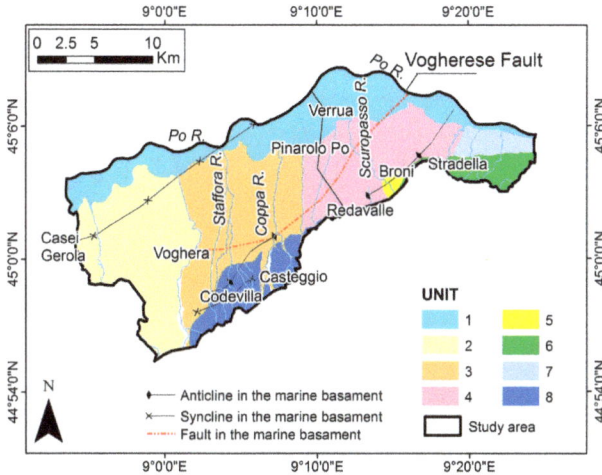

Figure 2. Engineering geological map of the plain area of the Oltrepo Pavese, modified from [24].

The geotechnical properties of the Quaternary deposits in the first 15 m of depth of the Oltrepo Pavese plain are summarized in Figure 2 [24]. Six geotechnical classes of non-cohesive soils (from I1 to I6) and four geotechnical classes of cohesive soils (from C1 to C4) have been distinguished. Therefore, taking into account the geotechnical classification of these soils, eight engineering geological units were introduced as representative of homogeneous geotechnical profiles (Figure 2). Unit 1 represents the Post-würmian alluvial deposit of the Po River, and is mainly constituted of non-cohesive soils (I3). Conversely, units 2, 3, 4, and 7 are Würmian-holocenic alluvial deposits, characterized by cohesive soils. Unit 5 is composed of the alluvial fan of the Scuropasso River, and units 6 and 8 by the Pre-würmian terraced deposits.

The A-DInSAR dataset available for the Oltrepo Pavese test case is composed of SAR images acquired by sensors operating in the C-band (wavelength: 5.6 cm; frequency: 5.3 GHz) onboard the ERS-1, ERS-2, and RADARSAT-1 (RSAT) satellites. The scenes were acquired in the ascending mode, covering the time intervals from 9 July 1992 to 2 August 2000, and from 24 March 2003 to 5 May 2010. The dataset acquired in the descending mode covers the time intervals from 3 April 1992 to 7 January 2001 and from 28 April 2003 to 5 January 2009. The ERS-1/2 images were acquired with a nominal repeat cycle of 35 days, while the RADARSAT images were acquired with a revisiting period of 24 days. These scenes were processed with the SqueeSAR™ technique by Tele-Rilevamento Europa (TRE srl, Milano, Italy). The algorithm allows the extraction of movement measurements, not only from traditional persistent scatterers (PS), such as anthropic structures or rocks, but also from distributed scatterers (DS), such as sparse vegetated areas [10]. This permitted to have a high density of A-DInSAR data over non-urban areas. ERS-1/2 descending and RADARSAT-1 ascending data were exploited to monitor the time intervals, from 1992 to 2000, and from 2003 to 2010 (Figure 3), in order to analyze the datasets with higher spatial and temporal resolutions.

Various geohazard-mapping methodologies were previously implemented by the use of A-DInSAR techniques [25–30]. The problem of detection of land subsidence using A-DInSAR data is compounded by the large-scale analysis of an enormous amount of measuring points. Often, the strategies to detect and to map the land subsidence areas are based on the spatial distribution of the average velocities and on statistic approaches to automatically detect clusters with significant movements. Some difficulties were observed to distinguish ground deformation due to different processes, or in detecting shallow deformations caused by seasonal processes.

In this work, the procedure proposed by [31] is exploited for the identification of land subsidence areas and to disentangle the contribution of different processes to the spatial and temporal distribution of displacement estimated through multisensor A-DInSAR time series. The first step of the performed activity was the decomposition of the vertical and E-W components of motion and the displacement time series (TS) accuracy assessment. Then, different statistical tests, such as the principal component analysis (PCA) and an automated time series classification, were applied. Thereafter, areas with significant movement, so-called "ground motion areas", were mapped. These areas correspond to a cluster of a minimum 3 of PS, with a maximum distance of 50 m, characterized by the same trends (i.e., linear, non-linear, and seasonal). Furthermore, the methodology to detect ground motion areas using the average velocity [25] was implemented in the same area using ERS-1/2 and Radarsat data to assess the results.

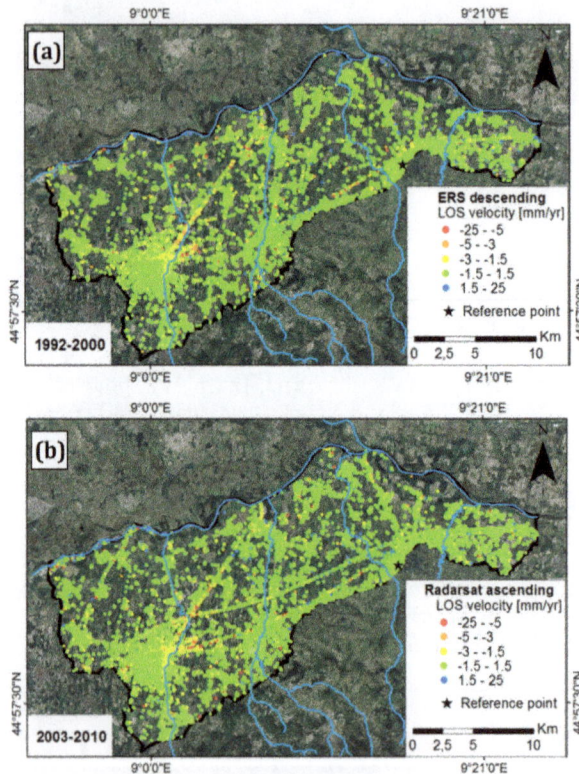

Figure 3. (a) Line of sight (LOS) velocity measured by the use of ERS-1/2 descending data (time interval from 1992 to 2000) and (b) RADARSAT ascending data (time interval from 2003 to 2010), modified from [31]. The reference points are also reported.

2.2. The Alto Guadalentín Test Case

The Guadalentín Basin is located in Southeast Spain, and has an extension of 277 km^2 (Figure 4). The area was chosen because the highest subsidence rates measured in Europe (>10 cm/year) have been recorded here, as a direct consequence of long-term aquifer exploitation. Urban sites are located in the area, but agriculture is the prevalent land use.

The Guadalentín Basin is underlain by Neogene-Quaternary sediments transported by the Guadalentín River along an intramontane depression located in the eastern part of the Baetic Cordillera, which is an ENE-WSW-oriented alpine orogenic belt resulting from the ongoing convergence of the African and Iberian plates [32]. The Guadalentín is a tributary river of the Segura River which geographically divides the Bajo from the Alto Guadalentín sub-basins, where Lorca City is located (Figure 4). The basin is mainly composed of Quaternary alluvial fan systems that overlap with Tertiary deposits. The Tertiary deposits are composed of conglomerate and calcarenite that outcrop at the border of the basin. The main active fault system of the study area, the NE-SW-oriented Alhama de Murcia Fault (AMF) [33] is represented in Figure 4. The deposits of the basin overlap the Paleozoic metamorphic complexes [34].

The Alto Guadalentín aquifer system is characterized by Plio-Quaternary detrital and alluvial material, including clays, sands, and conglomerates with clay and/or silt matrices; Miocene detritical with conglomerate and sand deposits; and local Triassic carbonate rocks (Figure 4). The lower impermeable limit of the aquifer is composed of Mesozoic marl, and marl with intercalated sand and conglomerates. The aquifer geometry is controlled by the horst and graben structures of the impermeable limit [35]. Since 1960, agricultural advance has led to the exploitation of the aquifer system, which resulted in the aquifer being declared temporarily overexploited in 1987 [36]. Historically, the piezometric level was close to the surface and, as artesian wells were exploited, groundwater drawdown became apparent in 1972 [35].

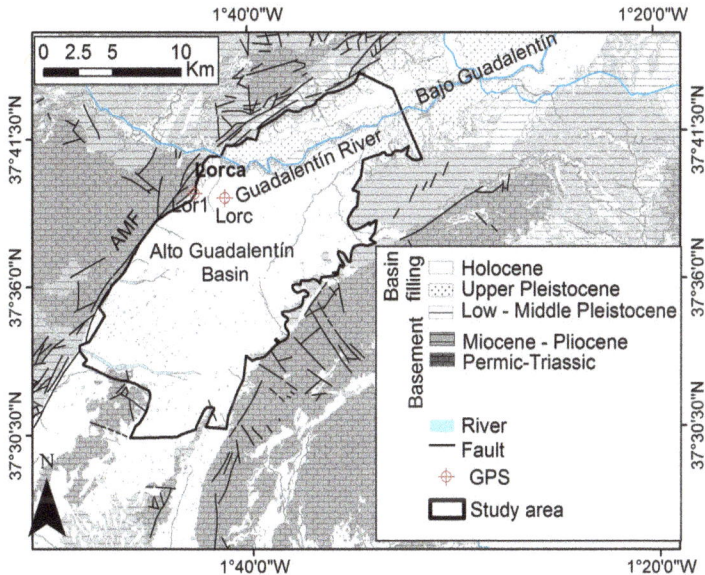

Figure 4. Geological setting of the Alto Guadalentín Basin modified from [37]. The localization of the Global Positioning System (GPS) stations, called Lor1 and Lorc, is also reported.

The A-DInSAR datasets available for the Alto Guadalentín are C-band (ERS-1, ERS-2, and ENVISAT), L-band (ALOS PALSAR), and X-band SAR data (COSMO-SkyMed). ERS-1/2 and ENVISAT scenes were acquired in the descending mode, covering the time intervals from 22 June 1992, to 21 December 2000, and from 15 March 2003 to 15 March 2007. ALOS PALSAR and COSMO-SkyMed scenes were acquired in the ascending mode, covering the time intervals from 19 January 2007 to 14 June 2010, and from 17 May 2011 to 14 October 2012. The SAR images acquired by ERS-1, ERS-2, and ENVISAT sensors were processed using the small baseline approach [13,38]. For each dataset, we first used Delft Object-oriented Radar Interferometric Software (DORIS) [11] to produce the interferograms and then we performed the time series analysis using Stanford Method for Persistent Scatterers

(StaMPS) [15]. Finally, results from both datasets were merged to determine the temporal evolution of displacement over the complete ERS-ENVISAT period. ALOS PALSAR and COSMO-SkyMed dataset were processed using SPN software [12,14]. The use of the A-DInSAR dataset acquired by different sensors using different incidence angles clearly affects the capability to measure the vertical component of the displacement. More precisely, ERS-1/2 and ENVISAT have a 23° incidence angle, which allows the estimation of 92% of vertical displacements, while ALOS and COSMO-SkyMed satellites only detect 83% and 75% of vertical displacements, respectively. As a result, the identification of these geometrical distortions due to the different acquisition angles of the various sensors is essential to perform A-DInSAR analyses. Consequently, LOS velocities have been projected along the vertical direction for each dataset in order to homogenize the datasets (Figure 5). Furthermore, local comparisons were performed with Global Positioning System (GPS) data available for two continuous stations (see the location in Figure 4) located in the study area, demonstrating the high consistency of the vertical motion measurements between the two different surveying techniques. An average absolute error of 4.6 ± 4 mm for the ALOS data and of 4.8 ± 3.5 mm for the COSMO-SkyMed data confirmed the reliability of the projection along the vertical direction [39].

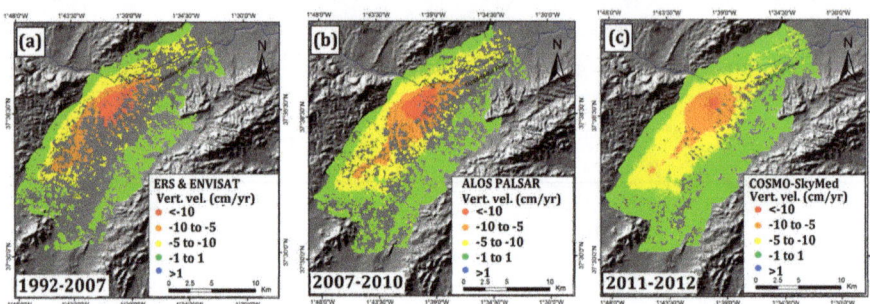

Figure 5. (**a**) Vertical velocity (Vert. vel.) measured by the use of ERS and ENVISAT data (time interval from 1992 to 2007); (**b**) ALOS PALSAR (time interval from 2007 to 2010); and (**c**) COSMO-SkyMed (time interval from 2011 to 2012).

The land subsidence triggered by groundwater exploitation depends on three main factors [4]: the areal and vertical distribution of subsidence-prone materials, their current state of stress, and their stress history. Previous works highlight that the same aquifer system can exhibit different deformation responses, according to the stratigraphic characteristic and the amount of groundwater pumped. Indeed, the complexity of the deformation responses to the applied stress is due to the different composition of the hydrostratigraphic units and the different changing patterns of groundwater levels that the units have experienced [40]. Principal component analysis was performed using the ERS and ENVISAT, ALOS, and COSMO-SkyMed data to analyze the spatio-temporal deformation pattern of each dataset. Then, A-DInSAR and piezometers time series have been compared to give insights about the evolution of the aquifer deformation behavior across the basin.

2.3. The London Basin Test Case

The study area located in the London Basin (Southern England) covers an extension of about 1360 km^2 (Figure 6). This test site was chosen to model land subsidence due to groundwater level change by applying a 1D model, thanks to the large availability of geological, hydrogeological, and geotechnical data. The London Basin is a densely-urbanized area, where, London, the capital and largest city in the United Kingdom, and the most populated in Europe, is located. The basin overlies the Palaeozoic basement, which is bounded to the south by the Variscan Front [41]. The basin is underlain by Paleogene deposits that overlie the Chalk Group [42,43]. These deposits comprise the

Thanet Formation, mainly composed of a fine sand; the Lambeth Group, consisting of vertically- and laterally-variable sequences, mainly of clay with silty and sandy horizons; the Harwich Formation, a silty, sandy clay with gravel beds, and; the London Clay Formation, a dense, fissured clay [44,45]. Quaternary deposits, primarily river terrace deposits, are associated with the River Thames and artificially modified ground. The city of London lies within a graben bounded by the Northern Boundary fault to the north and the Wimbledon-Streatham fault and the Greenwich fault to the south.

The Chalk Group is the principal aquifer in the London Basin supporting both public water and industrial groundwater use supply [46,47]. In the central area of the basin, the chalk aquifer is confined by the overlying Palaeogene formations. The chalk aquifer is recharged on the interfluves—in the Chiltern Hills to the north and North Downs to the south, where the Chalk outcrops and the aquifer becomes unconfined. The unconfined aquifer is characterized by larger seasonal water table variations, where fracture lineaments represent the main river drainage systems [48]. The lithological variations and the faulting on the chalk aquifer have a key control on the variability of the aquifer properties across the Thames Basin [49]. Historic overexploitation of the Chalk aquifer up to the 1960s led to widespread lowering of groundwater levels, reaching a depth of up to 90 m below sea-level in the central London Basin [46]. Thus, the Chalk aquifer became unconfined, leading to under-drainage and desaturation of the London Clay [41,46]. In the mid-1960s, Chalk groundwater level rapidly recovered at a rate of up to 3 m/year [46]. Therefore, an action plan was developed by the London Underground, Thames Water and the Environment Agency (EA) (i.e., the GARDIT, General Aquifer Research Development and Investigation Team) strategy to control groundwater recovery and re-saturation of the London Clay, that could potentially have negative impacts on the foundations of structures and infrastructure in the basin. Thus, the EA implemented an observation borehole network within the basin in order to monitor and manage changes in groundwater levels. Furthermore, an artificial recharge scheme was licensed in North London, NLARS (North London Artificial Recharge Scheme; [46]) to support the groundwater levels control [49,50].

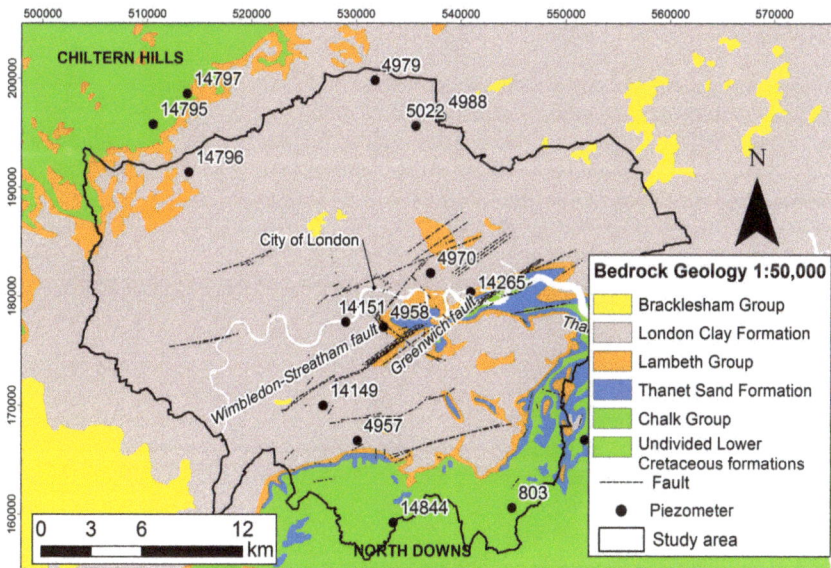

Figure 6. Geological setting of London Basin, modified from [51]. Geological materials © NERC 2016. All rights reserved. British National Grid. Projection: Transverse Mercator. Datum: OSGB 1936. The localization of the piezometers is also reported.

The A-DInSAR datasets available for the London Basin are ERS-1 and ERS-2 SAR scenes acquired in ascending and ENVISAT scenes acquired in descending mode. The first dataset covers the time interval from 19 June 1992 to 31 July 2000 and the second one from 13 December 2002 to 17 September 2010. The available scenes were processed using the GAMMA SAR and Interferometry software (Gamma Remote Sensing, Bern, Switzerland) and, in particular, the Interferometric Point Target Analysis (IPTA) algorithm [52]. Considering the horizontal movements negligible in the study area, LOS displacements and velocities were projected for both datasets along the vertical direction (Figure 7). Ninety-five percent of the PS targets within the study area show an uncertainty of the average LOS velocity in the range between 0.09 and 1.09 mm/year in the ERS-1/2 dataset, and between 0.17 and 1.13 mm/year in the ENVISAT dataset [53].

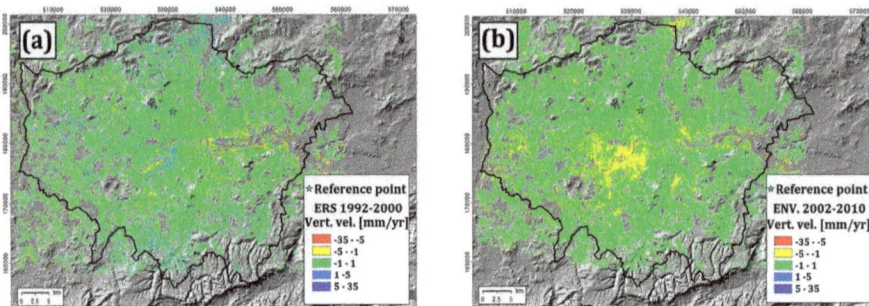

Figure 7. (**a**) Vertical velocity (Vert. vel.) measured by the use of ERS (time interval from 1992 to 2000) and (**b**) ENVISAT (ENV.) data (time interval from 2002 to 2010), overlapped onto shaded relief of NEXTMap® DTM at 50 m resolution; modified from [53]. Reference points for each dataset are also reported. ERS-1/2 and ENVISAT PSI data © CGG NPA Satellite Mapping. NEXTMap® Britain © 2003, Intermap Technologies Inc., All rights reserved.

Principal component analysis was also performed across the London Basin, using ERS-1/2 and ENVISAT data, to identify spatio-temporal deformation patterns.

Furthermore, A-DInSAR time series were also exploited to simulate the ground motion response due to the changes in the groundwater level. Many authors have exploited A-DInSAR data for the calibration and validation of models in which the land subsidence is due to groundwater level variations [54–57]. In this study, A-DInSAR time series were integrated in a one-dimensional model (1-D) to simulate the ground motion in response to changes in hydraulic head in the Chalk aquifer. The 1-D model is based on the inversion of the equation introduced by [58] to calculate the storage coefficient (S):

$$S = \Delta d/\Delta h \tag{1}$$

where Δd is the vertical displacement as estimated by A-DInSAR time series Δh. The assumption of one-dimensional consolidation is motivated by the hypothesis that the ground deformation is only vertical for this area, justified by the fact that the horizontal displacements are believed to not be significant [51]. The 1-D model assumes that the aquifer pore pressure instantaneously equilibrates with piezometric level changes in the aquifer and any time-lag between the piezometer level variations and the compaction of the geological layers is not accounted for. This assumption is justified by the inspection of the ground motion and groundwater records across the basin [51]. The simulation of the ground motion was conducted by using the following equation [59,60]:

$$\Delta d = S \times \Delta h \tag{2}$$

The simulation was performed at nine piezometers representative of semi-confined and confined aquifer condition using the average displacement time series within the buffer areas with a radius of 500 m from each piezometer. In the first step, the geological sequence was compared with the groundwater level variation to classify the aquifer state according to its confined condition (i.e., confined, semi-confined, unconfined) and the geological interval over which the piezometric head varied. Then, the aquifer storage coefficient (S) was calculated by using Equation (1) in the time interval where a good fit between piezometer data and average displacement time series was evident.

Finally, A-DInSAR time series were compared with the simulated ground motion in order to estimate the absolute average error of the model. Furthermore, the capability to use the A-DInSAR time series as the input of the model was assessed in order to simulate the hydraulic head changes. The inverted model was performed at piezometer 4988 (see the location in Figure 6) using the following equation:

$$\Delta h = \Delta d / S \tag{3}$$

3. Results

3.1. Time Series Support for Detection of Land Subsidence: The Oltrepo Pavese Test Case

The outcomes of the A-DInSAR time series analysis acquired by ERS-1/2 and RADARSAT-1 sensors show that the Oltrepo Pavese is affected by three deformational behaviors with linear, non-linear, and seasonal trends. The cumulated displacements observed in the 1992–2000 period is higher than those detected in the period 2003–2010, and a decrease of deformation was observed.

The town of Voghera and the railway of Voghera–Pavia (Figure 8) are located in the most affected area by land subsidence with linear movements (principal component 1, PC1), with average LOS velocity in the range from −2 to −3.7 mm/year, in the period 1992–2000, and in the range of −2 to −4.8 mm/year in range of −2 to −4.8 mm/year, in the period 2003–2010. Areas of moderate uplift were observed via principal component 2 (PC2) of the ERS-1/2 and RADARSAT-1 data, in the sector from Stradella to Monteacuto (Figure 8), where geomorphic evidence of an active emergent thrust was previously observed [61]. Ground motion areas characterized by seasonal movements (principal component 3, PC3) are mainly located in the southern sector of Voghera and in proximity to Lungavilla, Codevilla, Casei Gerola, and Broni. The seasonal deformation behaviors in the Oltrepo Pavese are mainly due to seasonal fluctuations of the groundwater level and swelling-shrinkage of clayey soils [31]. Ground motion areas delineated using the PC3 of ERS-1/2 and RADARSAT-1 data in proximity of Codevilla (see location in Figure 8) match with an area where a high density of damaged buildings was recorded [62]. This site is located over Engineering Geological Unit 8, where shallow alluvial deposits of the first 8 m of depth exhibit swelling potentials from medium to very high [62]. The seasonal component of motion detected in the TS is directly correlated with the effective rainfall detected at the nearest pluviometric station, located in Voghera (Figure 8).

Changes in trends in the displacement time series were observed at the end of 1999 and 2008 in the areas of Broni, Voghera, Lungavilla, and Casei Gerola, and, in most of the cases, the non-linear trend are superimposed on seasonal ground motion areas identified using PC3.

Even though the average velocity may be useful to detect physical processes characterized by linear trends, the same parameter seems not to be efficient in detecting ground motion areas affected by non-linear and seasonal movements. The challenge of exploitation of the time series to detect the ground motion areas is to take into account different deformational behaviors. Hence, the reliability of the results clearly depends on the quality of the time series. Consequently, post-processing checks of the time series are fundamental to remove problems due to phase unwrapping, regional unreal trends, and anomalous displacement detected on certain dates [31]. Anomalous displacements were recorded on 9 March 1997 and 16 July 2000 by the ERS-1/2 descending datasets, and on 10 December 2008, for the RADARSAT-1 ascending dataset in the Oltrepo Pavese. The anomalous displacement identified in the RADARSAT-1 dataset might be related to the snowfall that occurred on the day of the SAR acquisition. Thus, these SAR scenes were not included in the following analyses.

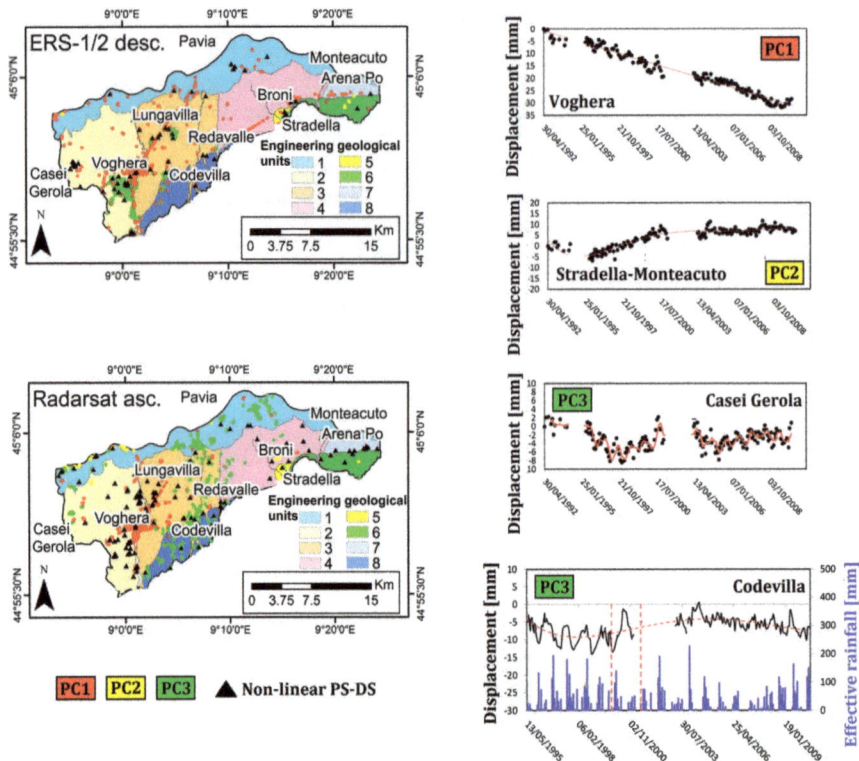

Figure 8. Ground motion areas detected using PC1, PC2, and PC3, and non-linear PS-DS of the ERS-1/2 descending and Radarsat ascending data overlapped on the engineering geological units. Time series (TS) of the deformational behavior detected at Voghera, Stradella–Monteacuto, Casei Gerola, and Codevilla are also reported. For Codevilla, TS were compared with effective rainfall measured at the Voghera weather station.

Furthermore, the same A-DInSAR datasets were analyzed using the methodology based on the average velocity [25] in order to validate the areas detected using the PC1 and to assess the results obtained using these different approaches. The distribution of the so-called "anomalous areas" is represented in Figure 9. Eighty-four percent of the anomalous areas detected in the period from 1992 to 2000 and from 2003 to 2010 coincide with ground motion areas detected using the PC1 of the ERS-1/2 and Radarsat data, respectively. Three and seven percent, respectively, of the anomalous areas detected in the period from 1992 to 2000 and from 2003 to 2010 coincide with ground motion areas detected via PC2. Three percent of the anomalous areas corresponds to ground motion areas detected via PC3 using ERS-1/2 and Rasarsat data. Therefore, 10% and 6% of the anomalous areas are not detected using PCA-based methodology. However, the extension of these areas reaches values of 0.2–0.3 km^2, resulting in localized movements. Otherwise, the approach based on the average velocity allows the detection of 35% and 69% of the ground motion areas detected via PC1 of the ERS-1/2 and Radarsat data. Seventy-two and sixty-seven percent, respectively, of the ground motion areas detected via PC2 can be identified using the average velocity of ERS and Radarsat data. Finally, 51% and 11% of the ground motion areas detected via PC3 of the ERS and Radarsat data, respectively, can be mapped using the average velocity. Overall, the results demonstrate that, when using the PCA-based methodology, most of anomalous areas based on the average velocity can be detected and, as expected, ground motion areas mainly due to seasonal movements can be detected using the time series information.

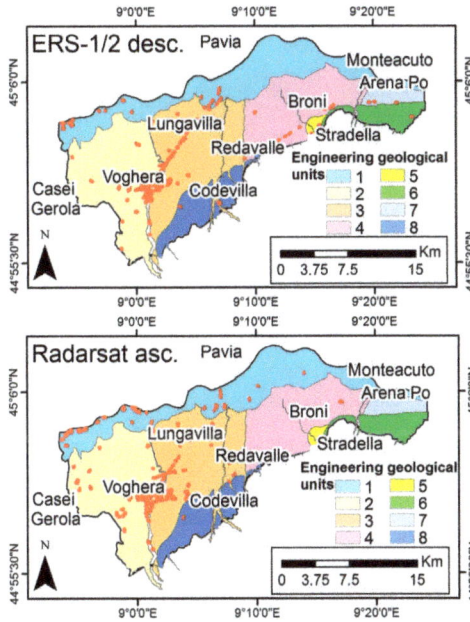

Figure 9. Anomalous areas detected using the average velocity [25] overlapped on the engineering geological units.

3.2. Time Series Support for the Characterization of Land Subsidence Mechanisms: The Alto Guadalentín Test Case

To better understand the land subsidence in the Alto Guadalentín Basin, the time series of multi-temporal A-DInSAR data were analyzed. First, principal component analysis (PCA) was performed for each dataset. Figure 10 shows that the different datasets are characterized by only one principal component (PC1) that explains the percentage of variance higher than 99%. The spatial distribution of principal component (PC) scores, which gives insight into the correlation between the measuring point and the trend of PC1, shows that the area affected by this kind of trend remains the same over the whole monitored period. The PC1 trend is well represented by the eigenvector value, and the land subsidence trend is evident for each dataset.

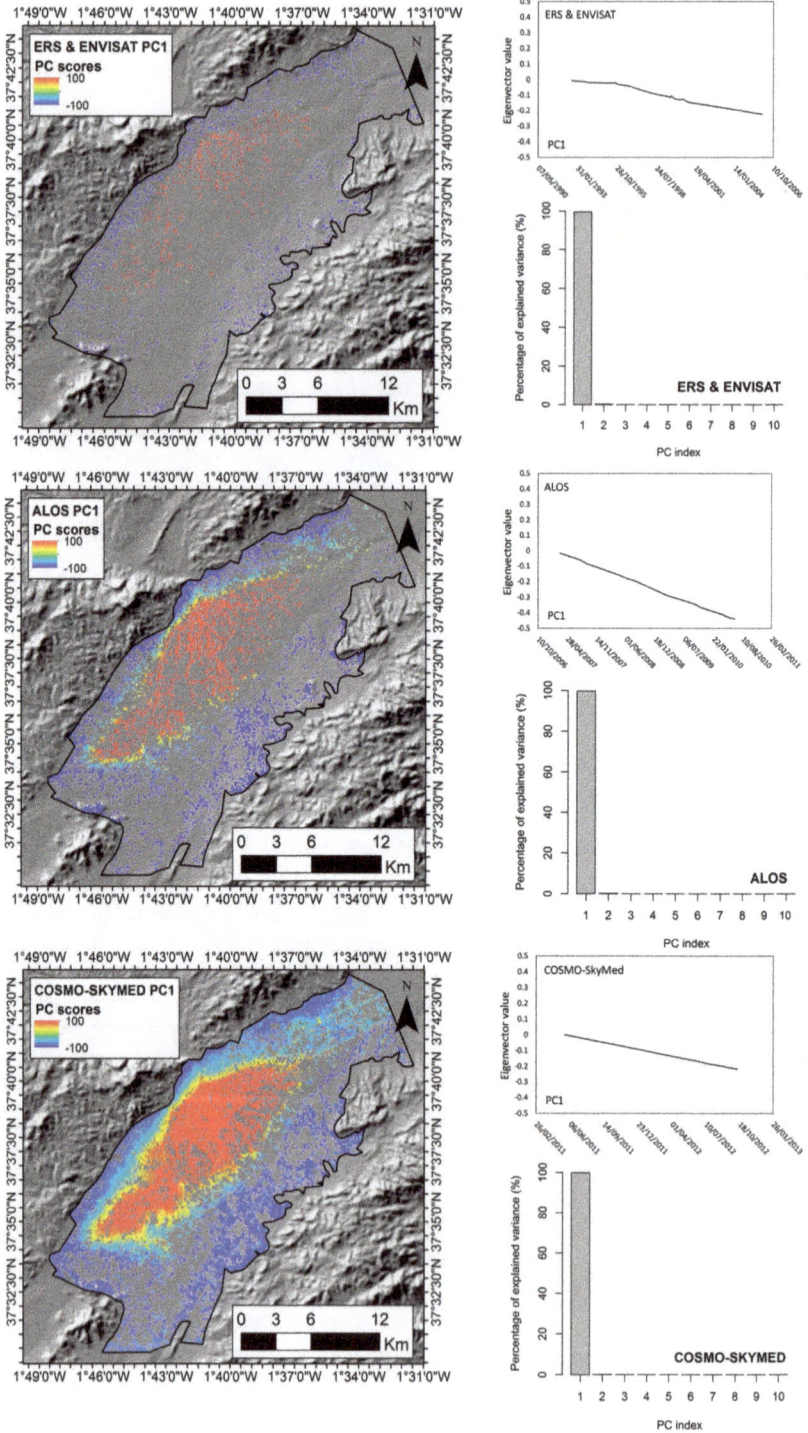

Figure 10. Principal component (PC) score maps of ERS and ENVISAT, ALOS, and COSMO-SkyMed data. For each dataset, the eigenvector value and the percentage of explained variance are also reported.

Furthermore, the cumulative displacement obtained by the use of the ERS-ENVISAT, ALOS PALSAR, and COSMO-SkyMed sensors in the period from 1992 to 2012 were computed (Figure 11). The maximum cumulative land subsidence reached up to 250 cm, with an average subsidence of 180 cm over an extension of 14.8 km^2, located in the central sector of the Alto Guadalentín Basin (Figure 11). The areal extent of the subsidence exhibits a SW–NE elliptical shape parallel to the valley direction, showing an apparent deceleration trend over time. Cross-comparisons were performed between the groundwater level variations, the lithological data, and the displacement time series for five representative sites. Close to the piezometer P1, located in the southern zone, in which the highest groundwater drawdown (162 m) from 1992 to 2007 was measured, the cumulated land subsidence reached 13 cm in the period 1992–2012 (Figure 11). This evidence is linked to the absence of soft soils in the southern border of the basin [39]. In the western part of the basin, where the piezometers P2, P4a, and P4b are located, the soft soil thickness is moderate (from 18 to 21 m). Groundwater drawdowns between 31 and 107 m in the period 1992–2007 correspond to cumulative subsidence in the range of 38–48 cm during the period 1992–2012. Piezometer P2 gives insight into a steady lowering of the piezometric level from 1989 to 2013 (Figure 11) that is correlated with the steady subsidence rate. Otherwise, piezometers P4a–b, which are jointly analyzed due to their proximity and characteristics, exhibit a high groundwater level variability that is not in agreement with the estimated subsidence linear rate. In the eastern part of the basin, piezometers P3a–b display the piezometric level variations are in agreement with subsidence accelerations and decelerations. Thus, cross-comparisons between A-DInSAR time series and piezometric measurements have allowed the recognition of transient aquifer compaction due to piezometric lowering in the period 1992–2007 (Figure 11), which was followed by decelerations of the subsidence when piezometric recovery was observed until 2012. In the central part of the basin where the highest thicknesses of soft soils are found (approximately 190 m), the time series of piezometer P5 reveals a groundwater level recovery from 2007 to 2013 that does not correspond with measured subsidence steady rates (Figure 11). An inelastic, unrecoverable, and permanent deformation of this aquifer was observed, as previously reported [39]. This land subsidence mechanism was triggered by the groundwater exploitation beginning in the early 1970s.

It is worth noting that the A-DInSAR time series reported in Figure 11 were generated considering a linear displacement velocity in the temporal gaps (11 months) between the ALOS PALSAR and COSMO-SkyMed datasets.

3.3. Time Series Support for Modelling of Land Subsidence: The London Basin Test Case

Principal component analysis (PCA) was performed using ERS-1/2 and ENVISAT data, in order to find the spatio-temporal deformation pattern across the London Basin. Figure 12 shows that both datasets are characterized by two principal components of ground motion. The first component represents land subsidence and the second one is an uplift trend. ERS-1/2 dataset shows that 91.6% of the variance is explained by the first component of motion and 4.4% by the second one. The variance is explained be the first and second components of motion (82.5% and 5.4%, respectively) using the ENVISAT data. Such comparisons were performed between the distribution of PC1 and PC2 with the areas of observed natural and anthropogenic geohazards mapped using the PanGeo procedure [26]. These areas were identified combining the same A-DInSAR datasets with geological data [53]. The distribution of PC1 detected using ERS-1/2 and ENVISAT data matches with natural ground movements, such as those due to compressible ground, and anthropogenic ground instability, such as that due to underground construction. The distribution of the PC2 detected using ERS-1/2 and ENVISAT data corresponds to areas where groundwater rise was previously observed [53] and areas in which causes of uplift are unknown [53], whereas geohazard areas due to tectonic uplift movements characterized by low confidence levels [53] do not show evident correlation with the PC2 trend.

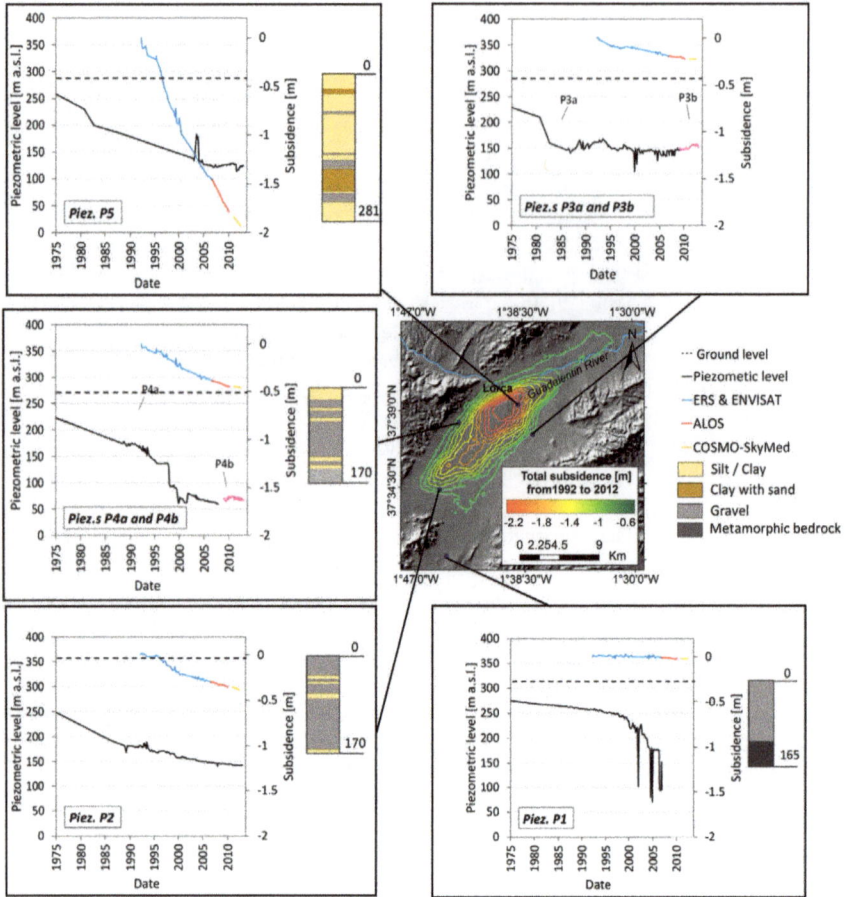

Figure 11. Cumulated displacement from 1992 to 2012 and cross-comparison of the piezometric level variations from 1975 to 2012 (black and pink lines) at the different piezometers (P1, P2, P3a, P3b, P4a, P4b, and P5) with the displacement time series detected by the satellite sensors (blue, red, and yellow lines). For some piezometers the lithological column is reported. The dotted line is the ground level. Piez. stands for piezometer.

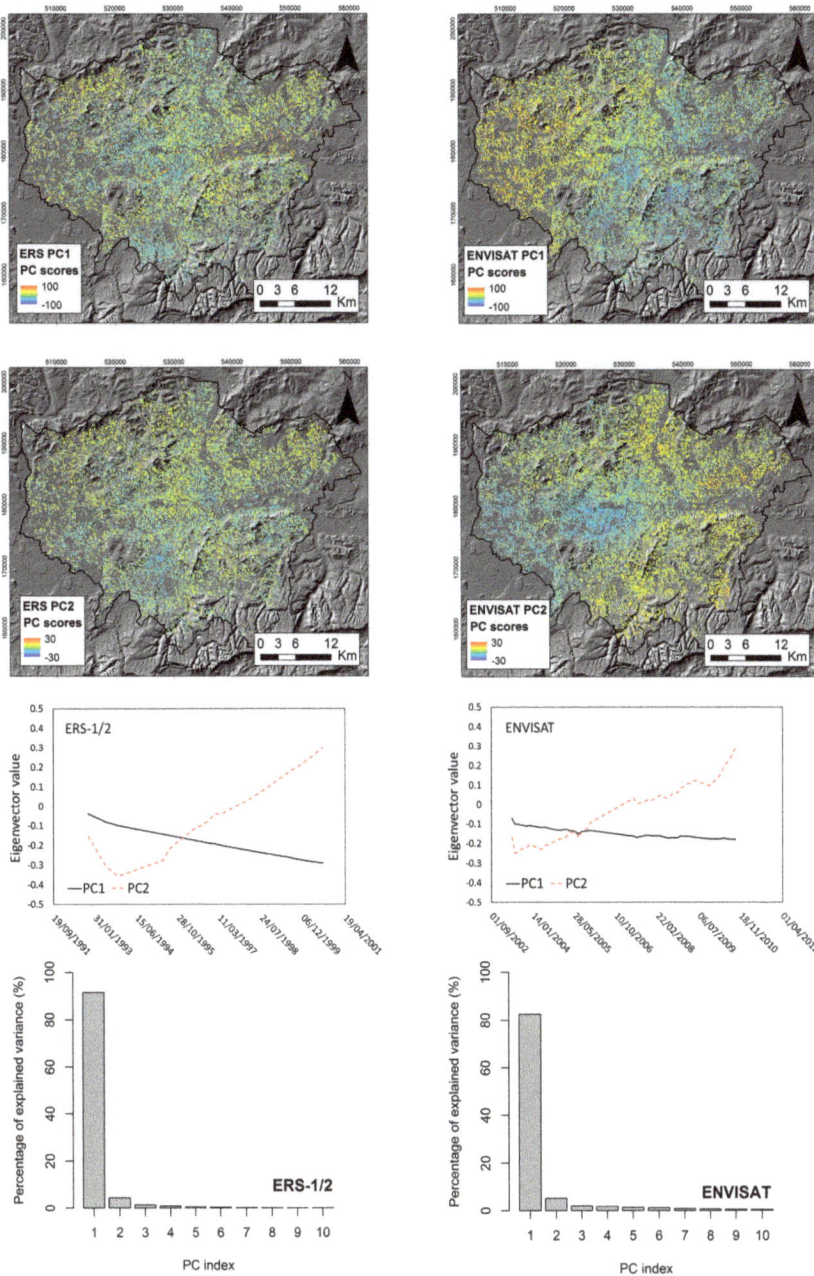

Figure 12. Principal component (PC) score maps of the ERS-1/2 and ENVISAT data using the first and the second component of the movements (PC1, PC2). For each dataset, the eigenvector value and the percentage of the explained variance are also reported.

A-DInSAR time series were used to simulate the acceleration and deceleration of the displacement in response to the changes in the piezometric level across the London Basin. Figure 13 shows the simulated displacements at four piezometers in order to represent the main responses to hydraulic head changes. These are obtained by using the groundwater level data and the estimated storage coefficients as inputs to Equation (2).

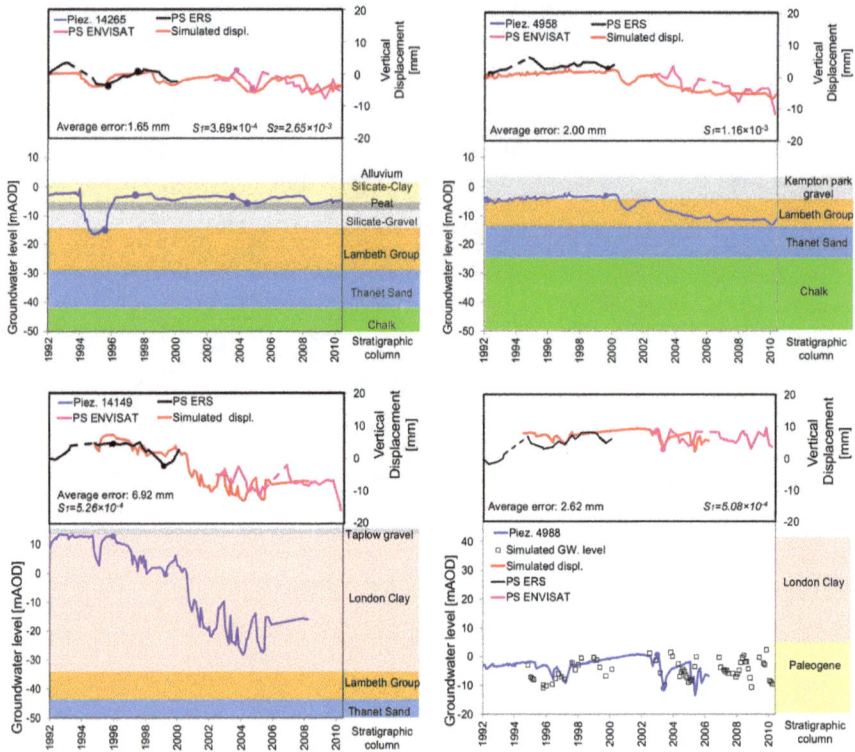

Figure 13. Comparison of A-DInSAR time series for 1992–2000 (PS ERS) and 2002–2010 (PS ENVISAT) and simulated displacements (mm) with the groundwater level variations (m) (modified from [51]). In addition, the stratigraphic column is represented. The dots represent the start and the end of the calibration period for the simulation. Piezometer localization is shown in Figure 6. For piezometer 4988, the simulated groundwater (GW) level is also reported. Groundwater level data © Environment Agency copyright and/or database rights 2015. All rights reserved.

In the North London Basin, when the Chalk is confined by the London Clay, the hydraulic head change of 11.50 m coincides with approximately 5.85 mm of surface displacement (Figure 13; piezometer 4988). In the Central London Basin, temporary transition of the water table across the units was detected (Figure 13; piezometer 14265). In this regard, two different storage coefficients were evaluated for the different aquifer conditions in the periods 01 February 1996 to 12 December 1997 and 14 January 2004 to 13 December 2004 for the confined and semi-confined conditions, respectively. The semi-confined conditions in the Lambeth group were observed at piezometer 4958. The hydraulic head change of 2.45 m coincides with approximately 2.85 mm of surface displacement. In the South London Basin, the aquifer is confined by thicker deposits of the London Clay and lower values of the storage coefficient were observed. A hydraulic head change of 13.16 m produces around 6.92 mm of surface displacement (Figure 13; piezometer 14149). The absolute average difference error between the simulated and A-DInSAR time series was also estimated.

The use of A-DInSAR time series for the modelling has allowed finding a correlation between the derived storage coefficient and the aquifer condition. Indeed, where the Chalk is confined by the London Clay the storage coefficient is of the order of 1×10^{-4}, whilst where overlain by the Lambeth Group and semi-confined conditions are expected to exist, the storage is higher, typically 1×10^{-3}, revealing the additional storage provided by sand-rich horizons in the Lambeth Group.

It is noteworthy how another benefit provided by the use of the simulated displacements is the filling of the temporal gap (approximately two years) between the ERS and ENVISAT data. The ENVISAT ground motion time series in Figure 13 were indeed adjusted for the position of the ground on 13 December 2002 to match with the modelled ground motion time series based on the ERS data. A vertical displacement of −5 mm was estimated for piezometer 14149 and −2 mm for piezometer 14265.

Furthermore, Equation (3) was applied in order to simulate the hydraulic head changes based on the observed ground motion at piezometer 4988 (Figure 13). The results show that the A-DInSAR time series could be used to simulate groundwater level changes following this innovative approach. It is worth noting that the relative error between the simulated hydraulic head changes and the measured groundwater level variations using time series acquired by the ERS and ENVISAT sensors is 25% of the hydraulic head change. The frequency of measurements between the displacement and piezometric time series is about 4–6 measurements per year versus 12 measurements per year. Therefore, the difference in the temporal sampling of these measurements could be influencing the error of the model.

4. Conclusions

This work aimed at exploiting the support of A-DInSAR time series for different facets of land subsidence mechanism assessment, starting from detection, up to the characterization and 1-D modelling. The main advantages and limitations of the A-DInSAR time series from an end-user point of view are discussed through three well-studied test sites: the Oltrepo Pavese (Po Plain, in Italy), the Alto Guadalentín (Spain), and the London Basin (UK). In particular, time series analysis was carried out in the Oltrepo Pavese for the detection of the ground motion areas. The time series support for the characterization and the modelling of land subsidence mechanisms was assessed, respectively, in the Alto Guadalentín Basin and in the London Basin.

Although various methodologies were implemented to map the ground motion areas using the average velocity detected by A-DInSAR data [25–30], the use of the time series for the detection of ground motion areas is still not a common practice in the scientific community. The main advantage of A-DInSAR time series for the detection of the ground motion areas is the unprecedented mapping of different deformational behaviors, such as in the case of the Oltrepo Pavese. By exploiting the A-DInSAR time series, the distribution of the areas affected by linear, non-linear, and seasonal movements was assessed. Furthermore, two approaches were applied to detect ground motion areas in order to give insight into the obtained results using the time series and the average velocity. Long-term movements detected using principal component analysis (PCA) are consistent with the results obtained using the average velocity. Otherwise, additional seasonal trends are observed using only PCA.

For the test site of the Alto Guadalentín Basin, PCA analysis was performed using the ERS and ENVISAT, ALOS, and COSMO-SkyMed data. The different datasets show that the Alto Guadalentín Basin is characterized by one component of motion that displays a land subsidence trend. Furthermore, the integrated analysis of A-DInSAR time series acquired by multi-temporal sensors allowed the characterization of the long-term subsidence. The collection of A-DInSAR time series acquired by multi-temporal sensors provided valuable information about the typology of aquifer deformation. Indeed, the use in conjunction of multi-sensor A-DInSAR data prove to be fundamental for the reconstruction of long-term evolution of land subsidence due to groundwater management, as in the case of the history of the Alto Guadalentín aquifer exploitation, confirming the potentials of the back-monitoring approach. Furthermore, the analysis of time series carried out for the Alto Guadalentín Basin has allowed the detection of transient aquifer deformation and inelastic aquifer deformation, not otherwise detectable by exploiting the average velocity. Further consideration to be taken into account for time series acquired by different sensors is how to fill the temporal gap between the datasets. With regard to this issue, the linear velocity was assumed in the temporal gaps (11 months) of the ALOS PALSAR and COSMO-SkyMed datasets. The results are consistent with the measurements obtained by two GPS stations.

Finally, PCA analysis was also performed for the ERS-1/2 and ENVISAT data available for the London Basin. Two components of ground motion were recognized, being related to land subsidence and uplift trends. PCA results are consistent with the areas delineated using the PanGeo procedure. Only geohazard areas due to tectonic uplift movements with low confidence levels show no evident correlation with PC2.

Furthermore, changes, such as acceleration and deceleration in the displacement rates in the London Basin, were simulated by calibrating a 1-D model using A-DInSAR time series. The modelling also has the benefit of filling the temporal gap between ERS and ENVISAT time series (about two years). A-DInSAR time series were also used as input for an inverse model in order to simulate the hydraulic head. This approach could be used to exploit A-DInSAR time series for groundwater level change monitoring in areas where no observation boreholes are available. Thus, we believe that new opportunities and perspectives will emerge for the detection and interpretation of subsiding phenomena using A-DInSAR time series provided by improved acquisition frequency, achievable through new high-resolution SAR satellite sensors, such as COSMO-SkyMed and Sentinel-1 constellations, thanks to the shorter revisiting time.

More widely, the findings of this work confirmed that time series represent an added value for a variety of activities in the framework of land subsidence mitigation and management. These observations have significant implications for similar studies in other geological contexts.

Acknowledgments: Groundwater level data for the London Basin were provided by the Environment Agency (Licence No. A2719; non-commercial use), and the ERS and ENVISAT PS data via the EC FP7 PanGeo project. Francesca Cigna and Stephanie Bricker publish with the permission of the Executive Director of BGS. Part of this work was supported by the Spanish Ministry of Economy and Competitiveness and EU FEDER funds under projects TIN2014-55413- C2-2-P and ESP2013-47780-C2-2-R.

Author Contributions: Roberta Bonì carried out this study and wrote this paper in the framework of her Ph.D. project at the University of Pavia, and during her research visits to the Geological Survey of Spain (IGME) in January–March 2015, and the British Geological Survey (BGS) in June–September 2015. Claudia Meisina provided guidance and support throughout the research process for the three case histories. Francesca Cigna supervised Roberta Bonì during her visit at BGS and revised the manuscript. Davide Notti supported the analysis of the A-DInSAR time series for the Oltrepo Pavese and Alto Guadalentín Basin. Stephanie Bricker provided support for the characterization of hydrogeological properties of the Chalk aquifer. Harry McCormack processed ERS-1/2 and ENVISAT data for the London Basin. Gerardo Herrera supervised Roberta Bonì during her visit to IGME. Roberto Tomás provided support to understand the land subsidence mechanism in the Alto Guadalentín Basin. Marta Béjar-Pizarro contributed analyses of the SAR data for the Alto Guadalentín Basin. Joaquin Mulas supported the analysis of the geological data about the Alto Guadalentín Basin. Pablo Ezquerro supported the analysis of the land subsidence in the Alto Guadalentín Basin. All authors co-wrote and reviewed the manuscript.

Conflicts of Interest: The authors declare no conflict of interest.

References

1. Holzer, T.L.; Galloway, D.L. Impacts of land subsidence caused by withdrawal of underground fluids in the United States. *Rev. Eng. Geol.* **2005**, *16*, 87–99. [CrossRef]
2. Abidin, H.Z.; Andreas, H.; Gumilar, I.; Sidiq, T.P.; Gamal, M. Environmental Impacts of Land Subsidence in Urban Areas of Indonesia. In Proceedings of the FIG Working Week 2015, TS 3—Positioning and Measurement, Sofia, Bulgaria, 17–21 May 2015.
3. Syvitski, J.; Higgins, S. Going under: The world's sinking deltas. *New Sci.* **2012**, *216*, 40–43. [CrossRef]

4. Galloway, D.; Jones, D.R.; Ingebritsen, S.E. *Land Subsidence in the United States*; US Geological Survey: Reston, VA, USA, 1999; p. 177.

5. Cigna, F.; Osmanoğlu, B.; Cabral-Cano, E.; Dixon, T.H.; Ávila-Olivera, J.A.; Gardŭno-Monroy, V.H.; DeMets, C.; Wdowinski, S. Monitoring land subsidence and its induced geological hazard with Synthetic Aperture Radar Interferometry: A case study in Morelia, Mexico. *Remote Sens. Environ.* **2012**, *117*, 146–161. [CrossRef]

6. Bozzano, F.; Esposito, C.; Franchi, S.; Mazzanti, P.; Perissin, D.; Rocca, A.; Romano, E. Understanding the subsidence process of a quaternary plain by combining geological and hydrogeological modelling with satellite InSAR data: The Acque Albule Plain case study. *Remote Sens. Environ.* **2015**, *168*, 219–238. [CrossRef]

7. Jones, C.E.; An, K.; Blom, R.G.; Kent, J.D.; Ivins, E.R.; Bekaert, D. Anthropogenic and geologic influences on subsidence in the vicinity of New Orleans, Louisiana. *J. Geophys. Res. Solid Earth* **2016**, *121*, 3867–3887. [CrossRef]

8. Herrera, G.; Fernández, J.A.; Tomás, R.; Cooksley, G.; Mulas, J. Advanced interpretation of subsidence in Murcia (SE Spain) using A-DInSAR data-modelling and validation. *Nat. Hazards Earth Syst. Sci.* **2009**, *9*, 647. [CrossRef]

9. Crosetto, M.; Monserrat, O.; Cuevas-González, M.; Devanthéry, N.; Crippa, B. Persistent scatterer interferometry: A review. *ISPRS J. Photogramm. Remote Sens.* **2016**, *115*, 78–89. [CrossRef]

10. Ferretti, A.; Fumagalli, A.; Novali, F.; Prati, C.; Rocca, F.; Rucci, A. A new algorithm for processing interferometric data-stacks: SqueeSAR. *IEEE Trans. Geosci. Remote Sens.* **2011**, *49*, 3460–3470. [CrossRef]

11. Kampes, B.M.; Hanssen, R.F.; Perski, Z. Radar interferometry with public domain tools. In Proceedings of the FRINGE 2003 Workshop, Frascati, Italy, 1–5 December 2003.

12. Arnaud, A.; Adam, N.; Hanssen, R.; Inglada, J.; Duro, J.; Closa, J.; Eineder, M. ASAR ERS Interferometric phase continuity. In Proceedings of the International Geoscience and Remote Sensing Symposium, Toulouse, France, 21–25 July 2003.

13. Berardino, P.; Fornaro, G.; Lanari, R.; Sansosti, E. A new algorithm for surface deformation monitoring based on small baseline differential SAR interferograms. *IEEE Trans. Geosci. Remote Sens.* **2002**, *40*, 2375–2383. [CrossRef]

14. Duro, J.; Closa, J.; Biescas, E.; Crosetto, M.; Arnaud, A. High Resolution Differential Interferometry using time series of ERS and ENVISAT SAR data. In Proceedings of the 6th Geomatic Week Conference, Barcelona, Spain, February 2005; Available online: https://pdfs.semanticscholar.org/d4bc/b3461ddb06da0704815bb40d815d780c8eb8.pdf (accessed on 10 April 2017).

15. Hooper, A. A multi-temporal InSAR method incorporating both persistent scatterer and small baseline approaches. *Geophys. Res. Lett.* **2008**, *35*, L16302. [CrossRef]

16. US National Research Council. *Mitigating Losses from Land Subsidence in the United States*; National Academy Press: Washington, DC, USA, 1991.

17. González, P.J.; Fernández, J. Drought-driven transient aquifer compaction imaged using multitemporal satellite radar interferometry. *Geology* **2011**, *39*, 551–554. [CrossRef]

18. Meisina, C.; Zucca, F.; Fossati, D.; Ceriani, M.; Allievi, J. Ground deformation monitoring by using the permanent scatterers technique: The example of the Oltrepo Pavese (Lombardia, Italy). *Eng. Geol.* **2006**, *88*, 240–259. [CrossRef]

19. Brambilla, G. Prime considerazioni cronologico-ambientali sulle filliti del Miocene superiore di Portalbera (Pavia-Italia settentrionale). In *Nuove Ricerche Archeologiche in Provincia di Pavia, Proceedings of the Convegno di Casteggio, Casteggio, Italy, 14 October 1990*; Civico Museo Archeologico di Casteggio e dell'Oltrepò Pavese: Casteggio, Italy, 1992; pp. 109–113. (In Italian)

20. Pellegrini, L.; Vercesi, P.L. Considerazioni morfotettoniche sulla zona a sud del Po tra Voghera (PV) e Sarmato (PC). *Atti Tic. Sci. Terra* **1995**, *38*, 95–118. (In Italian)

21. Cavanna, F.; Marchetti, G.; Vercesi, P.L. *Idrogeomorfologia e Insediamenti a Rischio Ambientale. Il Caso Della Pianura Dell'Oltrepò Pavese e del Relativo Margine Collinare*; Ricerche & Risultati, Valorizzazione dei progetti di ricerca 1994/97; Fondazione Lombardia Ambiente; Isabel Litografia: Gessate, Italy, 1998; pp. 14–72. (In Italian)

22. Pilla, G.; Sacchi, E.; Ciancetti, G. Studio idrogeologico, idrochimico ed isotopico delle acque sotterranee del settore di pianura dell'Oltrepò Pavese (Pianura lombarda meridionale). *G. Geol. Appl.* **2007**, *5*, 59–74. (In Italian)

23. Boni, A.; Boni, P.; Peloso, G.F.; Gervasoni, S. Dati sulla neotettonica del foglio di Pavia (59) e di parte dei fogli voghera (71) ed alessandria (70). *CNRPF Geodin. Pubbl.* **1980**, *356*, 1199–1223. (In Italian)

24. Meisina, C. Engineering geological mapping for urban areas of the Oltrepo Pavese plain (Northern Italy). In Proceedings of the 10th Congress of the International Association for Engineering Geology and the Environment (IAEG), Nottingham, UK, 6–10 September 2006.

25. Meisina, C.; Zucca, F.; Notti, D.; Colombo, A.; Cucchi, A.; Savio, G.; Giannico, C.; Bianchi, M. Geological interpretation of PSInSAR data at regional scale. *Sensors* **2008**, *8*, 7469–7492. [CrossRef] [PubMed]

26. Bateson, L.; Cuevas, M.; Crosetto, M.; Cigna, F.; Schijf, M.; Evans, H. PANGEO: Enabling Access to Geological Information in Support of GMES: Deliverable 3.5 Production Manual, version 1; 2012. Available online: http://nora.nerc.ac.uk/19289/ (accessed on 10 April 2017).

27. Bianchini, S.; Cigna, F.; Righini, G.; Proietti, C.; Casagli, N. Landslide hotspot mapping by means of persistent scatterer interferometry. *Environ. Earth Sci.* **2012**, *67*, 1155–1172. [CrossRef]

28. Lu, P.; Casagli, N.; Catani, F.; Tofani, V. Persistent Scatterers Interferometry Hotspot and Cluster Analysis (PSI-HCA) for detection of extremely slow-moving landslides. *Int. J. Remote Sens.* **2012**, *33*, 466–489. [CrossRef]

29. Peduto, D.; Cascini, L.; Arena, L.; Ferlisi, S.; Fornaro, G.; Reale, D. A general framework and related procedures for multiscale analyses of DInSAR data in subsiding urban areas. *ISPRS J. Photogramm. Remote Sens.* **2015**, *105*, 186–210. [CrossRef]

30. Di Martire, D.; Novellino, A.; Ramondini, M.; Calcaterra, D. A-differential synthetic aperture radar interferometry analysis of a Deep Seated Gravitational Slope Deformation occurring at Bisaccia (Italy). *Sci. Total Environ.* **2016**, *550*, 556–573. [CrossRef] [PubMed]

31. Bonì, R.; Pilla, G.; Meisina, C. Methodology for Detection and Interpretation of Land subsidence Areas with the A-DInSAR Time Series Analysis. *Remote Sens.* **2016**, *8*, 686. [CrossRef]

32. Bourgois, J.; Mauffret, A.; Ammar, A.; Demnati, A. Multichannel seismic data imaging of inversion tectonics of the Alboran Ridge (Western Mediterranean Sea). *Geo-Mar. Lett.* **1992**, *12*, 117–122. [CrossRef]

33. Martınez-Dıaz, J.J. Stress field variation related to fault interaction in a reverse oblique-slip fault: The Alhama deMurcia fault, Betic Cordillera, Spain. *Tectonophysics* **2002**, *356*, 291–305. [CrossRef]

34. Instituto Geológico y Minero de España (IGME). *Mapa Geologico de España, 1:50.000, Sheet Lorca (953)*; Servicio de Publicaciones Ministerio de Industria: Madrid, Spain, 1981. (In Spanish)

35. Cerón, J.C.; Pulido-Bosch, A. Groundwater problems resulting from CO_2 pollution and overexploitation in Alto Guadalentín aquifer (Murcia, Spain). *Environ. Geol.* **1996**, *28*, 223–228.

36. Confederación Hidrográfica del Segura (CHS). Plan especial ante situaciones de alerta y eventual sequia en la cuenca del Segura: Confederacion hidrografica del segura. Technical Report. 1996. (In Spanish). Avaliable online: https://www.chsegura.es/chs/cuenca/sequias/pes/eeapes.html (accessed on 11 April 2017).

37. Martín, V.J.M.; Espinosa, G.J.S.; Pérez, R.A. *Mapa geológico de España: E. 1:50,000*; Servicio de Publicaciones, Ministerio de Industria y Energía, Instituto geológico y minero de España (IGME): Madrid, Spain, 1973.

38. González, P.J.; Fernández, J. Error estimation in multitemporal InSAR deformation time series, with application to Lanzarote, Canary Islands. *J. Geophys. Res.* **2011**, *116*, B10404. [CrossRef]

39. Bonì, R.; Herrera, G.; Meisina, C.; Notti, D.; Béjar-Pizarro, M.; Zucca, F.; González, P.J.; Palano, M.; Tomás, R.; Fernández, J.; et al. Twenty-year advanced DInSAR analysis of severe land subsidence: The Alto Guadalentín Basin (Spain) case study. *Eng. Geol.* **2015**, *198*, 40–52. [CrossRef]

40. Zhang, Y.; Xue, Y.-Q.; Wu, J.-C.; Ye, S.-J.; Wei, Z.-W.; Li, Q.-F.; Yu, J. Characteristics of aquifer system deformation in the Southern Yangtse Delta, China. *Eng. Geol.* **2007**, *90*, 160–173. [CrossRef]

41. Royse, K.R.; de Freitas, M.; Burgess, W.G.; Cosgrove, J.; Ghail, R.C.; Gibbard, P.; King, C.; Lawrence, U.; Mortimore, R.N.; Owen, H.G.; et al. Geology of London, UK. *Proc. Geol. Assoc.* **2012**, *123*, 22–45. [CrossRef]

42. Ford, J.R.; Mathers, S.J.; Royse, K.R.; Aldiss, D.T.; Morgan, D.J.R. Geological 3D modelling: Scientific discovery and enhanced understanding of the subsurface, with examples from the UK. *Z. Dtsch. Ges. Geowiss.* **2010**, *161*, 205–218. [CrossRef]

43. Mathers, S.J.; Burke, H.F.; Terrington, R.L.; Thorpe, S.; Dearden, R.A.; Williamson, J.P.; Ford, J.R. A geological model of London and the Thames Valley, southeast England. *Proc. Geol. Assoc.* **2014**, *125*, 373–382. [CrossRef]

44. Ellison, R.A.; Woods, M.A.; Allen, D.J.; Forster, A.; Pharaoh, T.C.; King, C. *Geology of London: Special Memoir for 1:50,000 Geological Sheets 256 (North London), 257 (Romford), 270 (South London), and 271 (Dartford) (England and Wales)*; British Geological Survey: Nottingham, UK, 2004.

45. Sumbler, M.G. *British Regional Geology: London and the Thames Valley*, 4th ed.; HMSO for the British Geological Survey: London, UK, 1996.
46. Jones, M.A.; Hughes, A.G.; Jackson, C.R.; Van Wonderen, J.J. Groundwater resource modelling for public water supply management in London. *Geol. Soc. Lond. Spec. Publ.* **2012**, *364*, 99–111. [CrossRef]
47. Fry, V.A. Lessons from London: Regulation of open-loop ground source heat pumps in central London. *Q. J. Eng. Geol. Hydrogeol.* **2009**, *42*, 325–334. [CrossRef]
48. Bloomfield, J.P.; Bricker, S.H.; Newell, A.J. Some relationships between lithology, basin form and hydrology: A case study from the Thames basin, UK. *Hydrol. Process.* **2011**, *25*, 2518–2530. [CrossRef]
49. O'Shea, M.J.; Sage, R. Aquifer recharge: An operational drought-management strategy in North London. *Water Environ. J.* **1999**, *13*, 400–405. [CrossRef]
50. O'Shea, M.J.; Baxter, K.M.; Charalambous, A.N. The hydrogeology of the Enfield-Haringey artificial recharge scheme, north London. *Q. J. Eng. Geol. Hydrogeol.* **1995**, *28* (Suppl. 2), S115–S129. [CrossRef]
51. Bonì, R.; Cigna, F.; Bricker, S.; Meisina, C.; McCormack, H. Characterisation of hydraulic head changes and aquifer properties in the London Basin using Persistent Scatterer Interferometry land subsidence data. *J. Hydrol.* **2016**, *540*, 835–849. [CrossRef]
52. Werner, C.; Wegmüller, U.; Wiesmann, A.; Strozzi, T. Interferometric point target analysis with JERS-1 L-band SAR data. In Proceedings of the IEEE International Geoscience and Remote Sensing Symposium, IGARSS 2003, Toulouse, France, 21–25 July 2003; Volume 7, pp. 4359–4361.
53. Cigna, F.; Jordan, H.; Bateson, L.; McCormack, H.; Roberts, C. Natural and anthropogenic geohazards in greater London observed from geological and ERS-1/2 and ENVISAT persistent scatterers land subsidence data: Results from the EC FP7-SPACE PanGeo Project. *Pure Appl. Geophys.* **2015**, *172*, 2965–2995. [CrossRef]
54. Chaussard, E.; Bürgmann, R.; Shirzaei, M.; Fielding, E.J.; Baker, B. Predictability of hydraulic head changes and characterization of aquifer-system and fault properties from InSAR-derived ground deformation. *J. Geophys. Res. Solid Earth* **2014**, *119*, 6572–6590. [CrossRef]
55. Hoffmann, J.; Galloway, D.L.; Zebker, H.A. Inverse modelling of interbed storage parameters using land subsidence observations, Antelope Valley, California. *Water Resour. Res.* **2003**, *39*. [CrossRef]
56. Calderhead, A.I.; Therrien, R.; Rivera, A.; Martel, R.; Garfias, J. Simulating pumping-induced regional land subsidence with the use of InSAR and field data in the Toluca Valley, Mexico. *Adv. Water Resour.* **2011**, *34*, 83–97. [CrossRef]
57. Teatini, P.; Castelletto, N.; Ferronato, M.; Gambolati, G.; Janna, C.; Cairo, E.; Marzorati, D.; Colombo, D.; Ferretti, A.; Bagliani, A.; et al. Geomechanical response to seasonal gas storage in depleted reservoirs: A case study in the Po River basin, Italy. *J. Geophys. Res. Earth Surf.* **2011**, *116*. [CrossRef]
58. Hoffmann, J.; Zebker, H.A.; Galloway, D.L.; Amelung, F. Seasonal subsidence and rebound in Las Vegas Valley, Nevada, observed by Synthetic Aperture Radar Interferometry. *Water Resour. Res.* **2001**, *37*, 1551–1566. [CrossRef]
59. Tomás, R.; Herrera, G.; Delgado, J.; Lopez-Sanchez, J.M.; Mallorquí, J.J.; Mulas, J. A ground subsidence study based on DInSAR data: Calibration of soil parameters and subsidence prediction in Murcia City (Spain). *Eng. Geol.* **2010**, *111*, 19–30. [CrossRef]
60. Ezquerro, P.; Herrera, G.; Marchamalo, M.; Tomás, R.; Béjar-Pizarro, M.; Martínez, R. A quasi-elastic aquifer deformational behavior: Madrid aquifer case study. *J. Hydrol.* **2014**, *519*, 1192–1204. [CrossRef]
61. Benedetti, L.C.; Tapponnier, P.; Gaudemer, Y.; Manighetti, I.; van der Woerd, J. Geomorphic evidence for an emergent active thrust along the edge of the Po Plain: The Broni-Stradella fault. *J. Geophys. Res. Solid Earth* **2003**, *108*. [CrossRef]
62. Meisina, C. Light buildings on swelling/shrinking soils: Case histories from Oltrepo Pavese (north-western Italy). *Int. Conf. Probl. Soils* **2003**, *2*, 28–30.

geosciences

MDPI

Article

Assessing the Feasibility of a National InSAR Ground Deformation Map of Great Britain with Sentinel-1

Alessandro Novellino [1,*], Francesca Cigna [2], Mouna Brahmi [3], Andrew Sowter [4], Luke Bateson [1] and Stuart Marsh [5]

[1] British Geological Survey, Nicker Hill, Keyworth NG12 5GG, UK; alessn@bgs.ac.uk
[2] Formerly at British Geological Survey, Nicker Hill, Keyworth NG12 5GG, UK; francesca.cigna@gmail.com
[3] Department of Earth Sciences, Environment and Resources, Federico II University of Naples, Naples 80138, Italy; mouna.brahmi@unina.it
[4] Geomatic Ventures Limited, Nottingham Geospatial Building, Nottingham NG7 2TU, UK; andrew.sowter@geomaticventures.com
[5] University of Nottingham, Nottingham Geospatial Building, Nottingham NG7 2TU, UK; stuart.marsh@nottingham.ac.uk
* Correspondence: alessn@bgs.ac.uk; Tel.: +44-0-115-936-3407

Academic Editors: Ruiliang Pu and Jesus Martinez-Frias
Received: 27 February 2017; Accepted: 27 March 2017; Published: 30 March 2017

Abstract: This work assesses the feasibility of national ground deformation monitoring of Great Britain using synthetic aperture radar (SAR) imagery acquired by Copernicus' Sentinel-1 constellation and interferometric SAR (InSAR) analyses. As of December 2016, the assessment reveals that, since May 2015, more than 250 interferometric wide (IW) swath products have been acquired on average every month by the constellation at regular revisit cycles for the entirety of Great Britain. A simulation of radar distortions (layover, foreshortening, and shadow) confirms that topographic constraints have a limited effect on SAR visibility of the landmass and, despite the predominance of rural land cover types, there is potential for over 22,000,000 intermittent small baseline subset (ISBAS) monitoring targets for each acquisition geometry (ascending and descending) using a set of IW image frames covering the entire landmass. Finally, InSAR results derived through ISBAS processing of the Doncaster area with an increasing amount of Sentinel-1 IW scenes reveal a consistent decrease of standard deviation of InSAR velocities from 6 mm/year to ≤2 mm/year. Such results can be integrated with geological and geohazard susceptibility data and provide key information to inform the government, other institutions and the public on the stability of the landmass.

Keywords: InSAR; Sentinel-1; geohazards

1. Introduction

Satellite interferometric synthetic aperture radar (InSAR) data have proven effective and valuable in the analysis of geohazards based on multi-temporal radar images and have played a key role in recent years representing a cost- and time-efficient support and/or alternative to geological and geotechnical investigations [1].

Starting from the end of the eighties and beginning of the nineties [2,3], onward, the potential of satellite InSAR for cm or sub-cm level surface deformation mapping over a large area was revealed. Those works represented a milestone for the breakthrough of InSAR methods which proliferated during the following decades with a steady improvement and validation of several and different techniques (such as persistent scatterer interferometry—PSI, and small baseline subset—SBAS; [4]) through a large amount of scientific literature. This was especially thanks to international projects, such as Terrafirma [5], funded by the European Space Agency (ESA), and the EC-FP7 PanGeo project [6], and thanks to new SAR missions, such as ENVISAT in 2002, TerraSAR-X in 2007 and COSMO-SkyMed the following year. The proportional increase in the number of citations per year compared to the corresponding number of papers can be related to the increasing efforts in InSAR applications rather than the improvement or development of InSAR techniques (Figure 1).

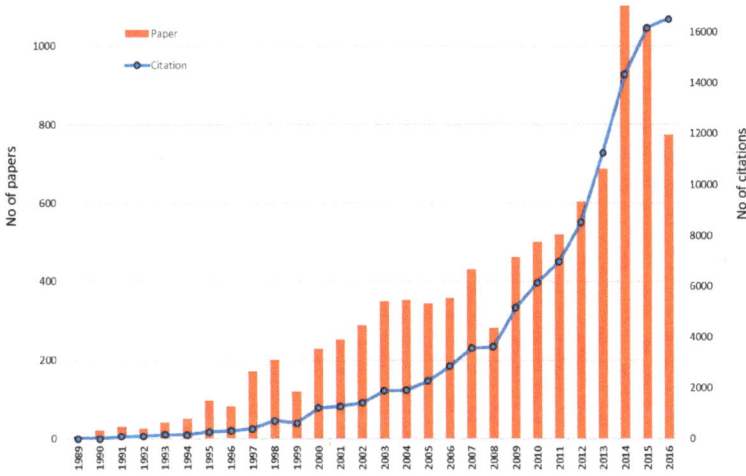

Figure 1. Number of articles, proceedings papers, and reviews found as of December 2016 in the ISI Web of Knowledge under the terms "InSAR" and "Interferometry" in the areas of remote sensing applied to Earth sciences; data from Thomson Reuters.

Italy is the only country worldwide where an InSAR database has been assembled successfully from the processing of ~15,000 ERS and ENVISAT SAR scenes to cover an entire nation (>300,000 km^2), and is now being extended with X-band COSMO-SkyMed data. The processing is part of a project funded by the Italian Ministry for the Environment, Territory and Sea in the framework of the Extraordinary Plan of Environmental Remote Sensing and essentially driven by the country's susceptibility to hydro-geomorphological risk [7].

In April 2014, with the launch of the first satellite of ESA's Sentinel-1 mission, Sentinel-1A [8], a new challenge has been set for InSAR applications: to provide, on a national scale, near real-time disaster monitoring, assessment, and response for geological processes [9]. In this paper, we aim to assess the feasibility of InSAR techniques to monitor the landmass of Great Britain to extend the work done by [10] who simulated SAR geometric distortions and predicted persistent scatterer densities for ERS and ENVISAT C-band SAR and InSAR applications in Great Britain. The target region encompasses an area of 243,610 km^2 made up of a main landmass and minor islands and islets. Since not all of the different geohazards affecting Great Britain produce ground displacement (e.g., flooding, storms, and wildfires), our assessment is limited to ground instability phenomena, such as natural compaction of recent river deposits [11] and loessic silt (largely confined to Southeast Essex and Northern Kent), shrink-swell terrains [12] found mainly in the southeast of the country, landsliding [13], land subsidence/uplift due to groundwater abstraction/recharge [14], glacial isostatic adjustment [15], dissolution of soluble rocks [16] principally in Cheshire, Pennines, and Wales, ground motions induced by human activities [17], and excavations and mine-water management in active and abandoned mining areas [18]. The United Kingdom is a country with limited direct experience of natural disasters and the British Geological Survey (BGS) archive confirms that just eleven people are known to have died as the results of British earthquakes since 1580 [19]. Moreover, despite an estimated 32,000 natural sinkholes, none has been responsible for a single death. Landslides are responsible of less than one death per year, apart from the Aberfan disaster of 1966 that claimed over 144 victims [20]. However, despite the sources of geohazards associated with ground deformation being heterogeneously distributed across Great Britain, most of the landmass can be potentially affected by geohazards, which caused 1–2 billion euros of damage per year in the UK in 1998–2009 [21].

Considering the latest set of 30-year averages data from the Met Office [22] for 1981–2010, climatic variables, such as rainfall (>1150 mm/year) and cloud coverage, suggest that active remote sensing technologies (e.g., radar sensors) are particularly suitable for monitoring the country.

The framework of this work follows the four conditions that must be fulfilled for accomplishing a nationwide InSAR analysis: we analyse the Sentinel-1 archive image availability in Section 2.1, the effect of topography on the SAR images in Section 2.2, the effect of land use on radar targets distribution in Section 2.3, the feasibility of Sentinel-1 in terms of the kinematic, temporal, and spatial extension of the investigated phenomenon in Section 2.4 and the effect of different stacks of Sentinel-1 images on InSAR results over Doncaster in Section 2.5. The future perspectives relating to the implementation of the Sentinel-1 nationwide monitoring are finally discussed with respect to the benefits for geohazards susceptibility and risk assessment.

2. Materials, Methods, and Discussion

2.1. Sentinel-1 Archive Data

ESA is developing the Sentinels, a new family of six missions, each based on a constellation of two satellites to fulfil revisit and coverage requirements, providing robust datasets for Copernicus Services. The Sentinel-1 constellation ensures the continuity of the ERS and ENVISAT missions and has been specifically designed to provide systematic and regular SAR observations with high temporal (six-day revisit time) and spatial (~250 km swath and ~165 km azimuth width) coverage. The Terrain Observation by Progressive scans SAR (TOPSAR) of the Interferometric Wide Swath (IWS) is the main acquisition mode over land [23], in addition to Stripmap and Extra Wide Swath (EWS) products. Sentinel-1 SAR products are freely available for both research and commercial applications, thus providing the scientific community, as well as public and private companies with more opportunities to explore.

Currently, Sentinel-1 products are available through the following platforms:

- Sentinels Scientific Data Hub (https://scihub.copernicus.eu/) operated by ESA, the most completed and updated (within 24 hours for systematically archived data) archive where SAR data are identified according to relative orbit and slice number. In 2015, alone, more than 1000 out of the 15,243 users registered from the UK [24].
- Alaska Satellite Facility [25], as part of an agreement between the U.S. State Department and the European Commission, where products are identified from their path and frame.
- Sentinel Data Access Service [26] operated by the Satellite Applications Catapult in association with the UK Space Agency where SLC IWS SAR data are identified according to the Absolute Orbit number.

Starting from October 2014, when just two IW SLC images were available over Great Britain (Figure 2), and tracking the status of the archive, we found that there were a total ~11,200 SLC Sentinel-1A images as of the end of December 2016. The analysis shows that the average number of scenes acquired per month for each acquisition geometry (ascending and descending) increased from a few tens in 2014 to over 100–200 in 2015 and up to ~400 at the end of 2016. The scenes are almost equally distributed among the ascending and descending orbits, as opposed to the observed preponderance of ERS and ENVISAT descending frames with respect to ascending ones from 1991 to 2010 [10]. The current acquisition rate has been achieved after the completion of the commissioning phase of the second satellite of the constellation, Sentinel-1B (launched in April 2016 and orbiting Earth 180° apart from its twin), when further products have been made available from mid-September 2016 and the number of scenes acquired per month over the landmass has increased to ~400. A short gap occurs between 16 June 2016 and 28 June 2016 due to the persistence of the onboard anomaly of the SAR antenna [27].

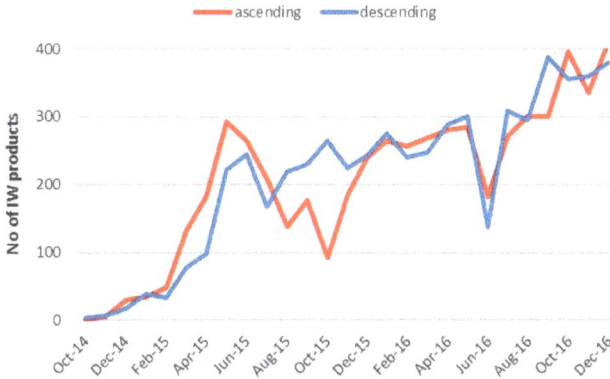

Figure 2. Number of Sentinel-1 IW products available over Great Britain from the Sentinel Data Hub as of December 2016.

The Sentinel-1A IW coverage of Great Britain is achieved by 46 different ascending frames (from 158 to 200) and 41 different descending frames (from 392 to 428) distributed along a total of six paths in each geometry (Figure 3). Among the possible path-frame combinations, a full coverage of Great Britain can be achieved with just 15 IW frames.

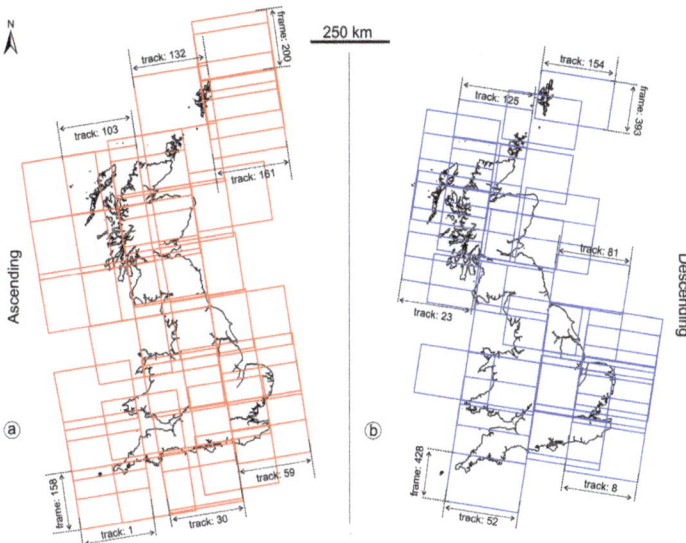

Figure 3. Satellite SAR image availability over Great Britain: spatial distribution of Sentinel-1A frames for the (**a**) ascending and (**b**) descending modes. Along the same path a step of 5 in frame ID is required to cover the next or previous adjacent area.

However, it becomes apparent that the opportunity to use always the same frames to cover the same area has been affected by some inconsistency in the Sentinel-1 acquisition as shown in the example of Figure 4. In this case, the Sentinel-1A ascending stack along path 30 has been selected to cover the municipality of Doncaster. Since 6 January 2016 there has been a regular succession of the acquisitions of frames 169 and 174 with frame 173 until the end of March 2016. The latter covers exactly the same area of frame 172 which has been acquired just twice until now: on 24/11/2014 and 25/11/2016. The heterogeneity of the archive is increased during spring and summer as frames 170, 174, and 175 prevail until the beginning of October 2016, when frame 171 (from Sentinel-1B) cover Doncaster (Figure 4).

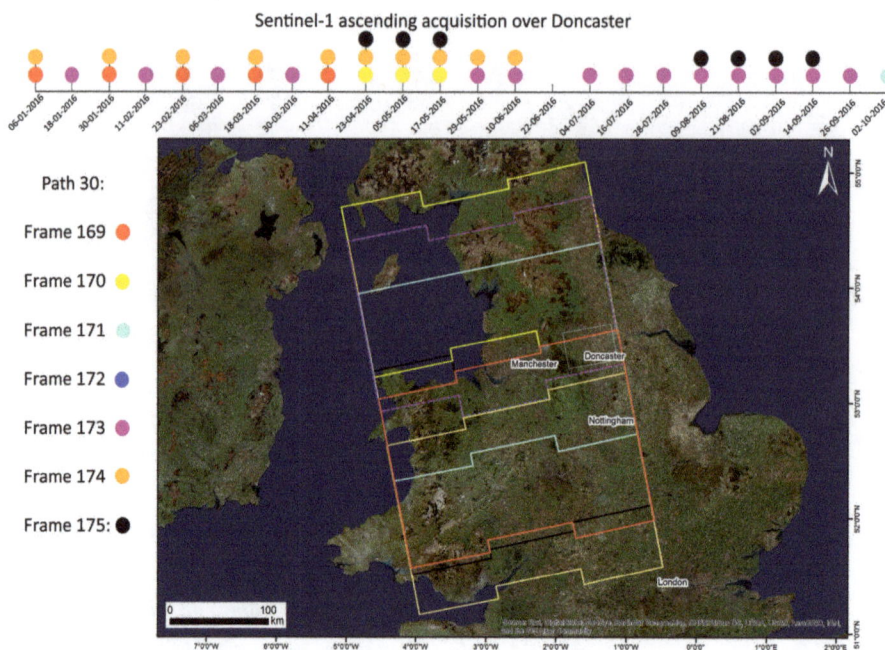

Figure 4. Sentinel-1 IW image availability over path 30 for Doncaster during the period between January and October 2016.

2.2. Topographic Constraints

The visibility of the landmass to the radar satellite sensor depends on the orientation of the employed satellite line-of-sight (LOS) and radar acquisition geometry with respect to that of the imaged terrain. Visibility varies within different portions of the same Sentinel-1 IW scene depending on local terrain slope (β) and aspect (α), local incident angle (θ), which varies between 29.1° and 46° from the near to the far range and is ~37.5° at scene centre, and the satellite heading angle (γ), which is 9.1° on average at the latitudes of Great Britain. We extract the terrain slope and aspect from the Shuttle Radar Topography Mission (SRTM) digital surface model (DSM) at 90 m resolution which—in terms of spatial resolution—is the closest one to the Sentinel-1 ground range resolution after multi-looking using the factors of 22 and 5 in range and azimuth, respectively, as adopted by [28].

Depending on their effects, range geometrical distortions can be classified as foreshortening, layover (Lv), and shadow (Sh) (e.g., [29,30]). Their presence hampers the application of both interferometric and non-interferometric techniques (e.g., [31]). A priori identification of these distortions is, therefore, essential to plan InSAR analyses, especially over the foreslope of hilly and mountainous regions where the assessment of the most suitable LOS geometry ensures that the target area is visible prior to the acquisition of long SAR imagery stacks. To identify Sentinel-1 IW geometrical distortions in the ascending and descending acquisition modes in Great Britain, we calculated the R-Index (RI) [10,32]:

$$RI = \sin[\theta - \beta \times \sin(A)] \qquad (1)$$

where A is the aspect correction factor, which is computed as $A = \alpha - \gamma$ for descending mode data, and $A = \alpha + \gamma + 180°$ for ascending mode data. Essentially RI, represents the ratio between the pixel size in slant and ground range geometry and its values range between -1 and $+1$. Areas of foreshortening show RI above 0 and up to $\sin(\theta)$. The latter represents the specific threshold to discriminate foreshortening from good visibility regions, and is identified as RI_0 [10]. Radar layover affects slopes facing the sensor, where $\theta < \beta$ and, therefore, RI takes on negative values between 0 and -1, whilst shadow occurs when slopes facing away from the sensor are characterised by $\beta \geq 90° - \theta$. Since passive layover (Lv) and shadow (Sh) cannot be discriminated on backslopes by using the sole R-Index, Lv and Sh are separately calculated by using the formula from [32], and eventually masked out to obtain the visibility map by multiplying the RI by a reclassified binary map where pixels take on values of 0 and 1 where Lv or Sh occur and do not, respectively, as follows:

$$Visibility = RI \times Lv \times Sh \qquad (2)$$

The total areas of shadow within Great Britain were found to be as small as 9.6 km^2 in ascending, and 8.8 km^2 in descending, mode due to the presence of very few areas characterised by slopes steeper than 52.5° (Figure 5). Total areas of layover (including active and passive) are 8805 km^2 in ascending and 7939 km^2 in descending mode (Figure 5). Of these, 10.5 km^2 in ascending and 11.2 km^2 in descending mode are active layover regions, which were determined as a subset of total layover, by considering DSM pixels showing R between -1 and 0. Total extension of layover and shadow distortions for the landmass is ~1.0% in each imaging mode (Figure 6). The results of the analysis show that, for almost the entire territory, the layover effects that are observed for the descending geometry over slopes facing E can be compensated by employing the ascending geometry, and vice versa, for W-facing slopes (Figure 5). Areas resulting in layover or shadow distortions in both geometries, corresponding to ~185 km^2, cannot be monitored using SAR imagery with either orbital mode, but our simulation shows that fortunately these do not occur anywhere in Great Britain but are limited to Northwest Highlands and Grampian Mountains. This allows to include in the InSAR studies also challenging phenomena like coastal landslides whose analyses otherwise are strongly compromised by geometrical distorsions. Areas of foreshortening were determined by considering an R-index of 0 to +0.6 (the RI_0 of S-1), and extended to 75,070 km^2 (~32.7% of the landmass) in ascending, and 74,994 km^2 (~32.6%) in descending mode. The absence of any type of distortions is observed over around 153,273 km^2 (~67.2%) in ascending and 155,054 km^2 (~67.5%) in descending mode (Figure 5). Differently from the results of the simulation for the ERS and ENVISAT acquisition modes [10] the larger θ of Sentinel-1 (~37.5° instead of ~23.3°), reduces the overall size of the areas affected by SAR foreshortening and shadow effects in both geometries over the entire territory.

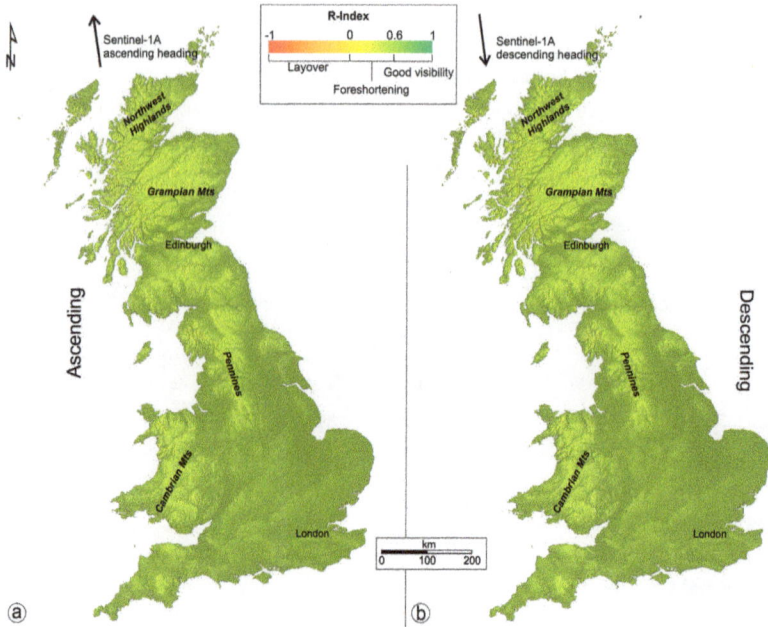

Figure 5. Identification of image distortions to the Sentinel-1 IW ascending (**a**) and descending (**b**) modes over Great Britain.

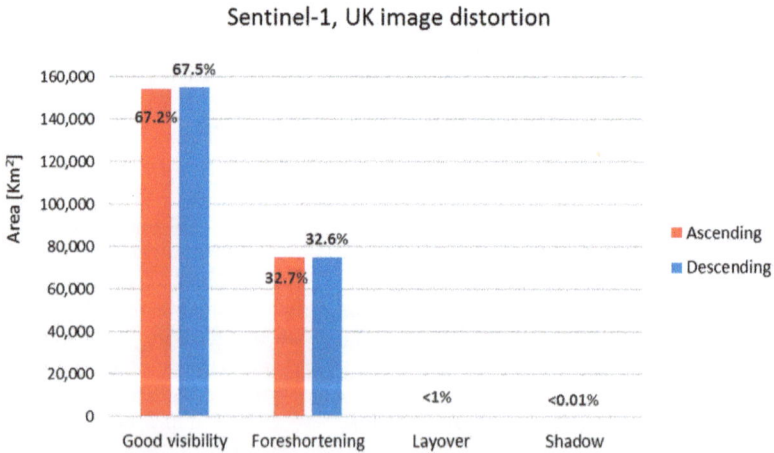

Figure 6. Total areas of simulated Sentinel-1 IW geometric distortions (shadow and layover include both active and passive distortions) across Great Britain, based on the 90 m resolution SRTM DSM.

2.3. Land Cover Constraints

Land cover represents a critical factor reducing InSAR capabilities, in terms of accuracy of the results, due to temporal decorrelation effects. In order to remove or at least minimize such effects, two main families of methods were developed over the last decade [4]: the PSInSARTM [33] and SBAS [34]. With PS, temporal decorrelation effects are reduced by considering targets with low amplitude dispersion, while SBAS family relies on forming interferograms only between images separated by a short time interval and with a small difference in look and squint angle. However, the predominance of non-irrigated arable land and pastures, covering ~130,000 km^2 across the landmass, exerts significant control on the potential for InSAR methods to identify scatterers in Great Britain, with particularly critical effects on rural and grassland regions where only a few radar targets per square kilometre can be extracted and monitored via InSAR processing of C-band data [10]. Novel InSAR methods, such as SqueeSAR™ [35] and ISBAS [36,37], aim to increase the coverage of results in non-urban regions by considering, respectively, distributed scatterers (e.g., debris, non-cultivated land, or low vegetation cover), or intermittently coherent surfaces. We assess ISBAS density using the freely-available CORINE land cover data for the reference year 2012 (CLC2012) and at 100 m resolution, following the empirical approach illustrated in [38].

Therefore, for each single land cover class the number of ISBAS targets N falling inside the land cover class is calculated. This is implemented by counting the number of geocoded ISBAS Nk (of the reference SAR data) for each single subarea k of each land cover class (i) and then summing up the results (N = ΣNk). Then, the entire area of the i-th class within each site A was calculated. Next, the ISBAS density p for each class i (ISBAS targets per km^2) was derived:

$$p_i = N_i / A_i \qquad (3)$$

The procedure described above was implemented in a geographical information system (GIS) and executed for a number of calibration sites. In particular, the p_i was estimated from a total of 64,844 km^2 including six subareas (East Anglia, East Midlands, Hampshire, Lancashire, Yorkshire, and Southern Scotland) where 36 out of the 39 land cover types of Great Britain occur and 5,155,619 ISBAS pixels have been identified (Table 1). The exploitation of existing ISBAS datasets over different areas has been the input data for the quantitative calibration of countrywide land cover data towards extending the pre-survey feasibility assessment to the entire landmass and land cover types mapped across the country (Figure 7).

Table 1. Sentinel-1A datasets used to calibrate the CORINE land cover map CLC2012 for Great Britain. Asc., Ascending; Desc., Descending.

Area	Orbit	No. of Scenes	Dates	ISBAS Point	Surface Area (km^2)
East Anglia	Asc	34	25 December 1014–29 June 2016	778,455	6730
East Midlands	Asc	39	12 March 2015–14 September 2016	402,829	3439
Hampshire	Asc	31	12 March 2015–17 May 2016	263,570	2510
Lancashire	Desc	36	8 May 2015–11 September 2016	353,703	3550
Yorkshire	Desc	36	7 March 2015—9 September 2016	447,564	4596
Southern Scotland	Asc	24	12 March 2015–18 January 2016	2,909,498	44,017

By using the same sensor with the same wavelength over the two acquisition geometries, normalization of absolute ISBAS densities has not been required. Furthermore, climate conditions are similar in the seven calibration sites (Met Office, [22]), therefore, no initial climate classification and compensation were performed.

Standing at the ISBAS average density, ISBAS standard deviation and extension of each land cover types (Figure 8), ranging from the 122 ISBAS/km^2 over continuous urban fabric to just 9 ISBAS/km^2 over estuaries, and accounting for the topographic constraints, ~22,000,000 ISBAS points can be acquired, corresponding to an area of ~170,000 km^2, namely 70% of the Great Britain landmass. No ISBAS points were detected over burnt areas, costal lagoons, permanently irrigated land, sea and ocean, water bodies, and water courses.

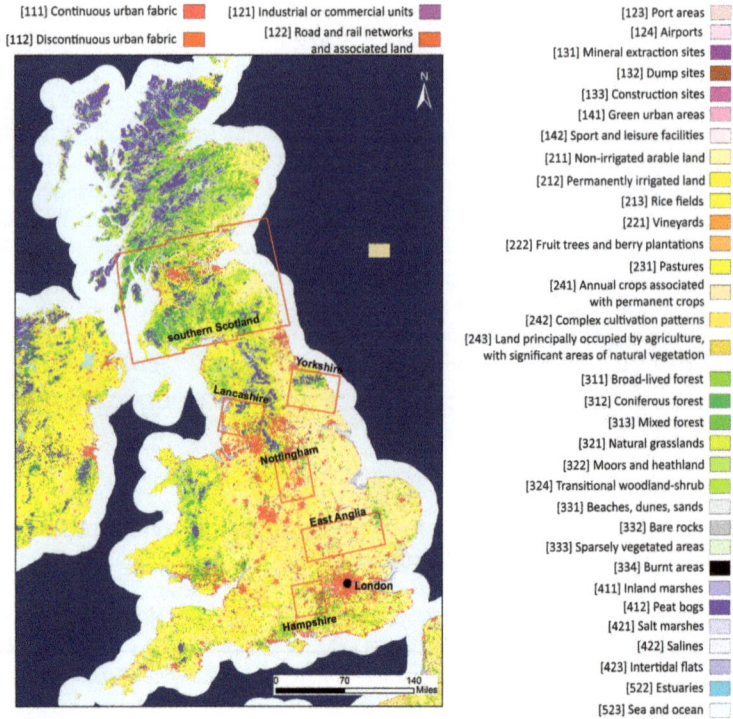

Figure 7. CORINE land cover map 2012 (CLC2012) of Great Britain, along with the footprints of the six test sites where the ISBAS average density has been extracted.

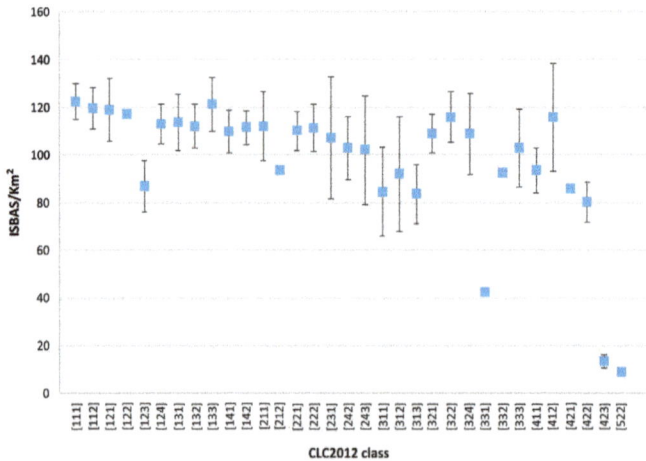

Figure 8. Average ISBAS densities estimated using the six calibration sites for the 33 CLC2012 classes, according to the code used in Figure 7, mapped in Great Britain where at least one ISBAS point occurred in the calibration sites. Therefore, burnt areas, coastal lagoons, permanently irrigated land, water bodies, and water courses classes have been excluded.

2.4. Kinematic and Geometrical Constraints to Monitor Geohazards

Even if a sufficient number of Sentinel-1 data over Great Britain is already available (see Section 2.1), geometrical distortions represent a spatially-limited constraint (see Section 2.2) and land cover always allows a sufficient density of ISBAS radar targets (see Section 2.3), three systematic conditions can still negatively affect the analysis:

1. The extension of the investigated phenomena must be over ~0.008 km^2 which refers to the size of the smallest possible feature that can be detected by Sentinel-1 data accounting for the spatial resolution of its IW mode. Apart from dolines and sinkholes particular landslide typology (especially falls and topples), UK geohazards have an extension at the kilometric scale.
2. Rather than providing 3-D components, ISBAS measurements are along the LOS direction and, therefore, insensitive to horizontal displacements in the along-track direction [39]. To decompose the InSAR LOS signal into its vertical and horizontal components we, therefore, need sufficient acquisitions of both ascending and descending SAR data [40]. In the absence of both geometries, quantification of the fraction of maximum motions measurable along the satellite LOS can be quantified by means of the R-Index for areas of good visibility or foreshortening. In this regard, most of the UK geohazards do not prevent InSAR analysis by being associated with predominant vertical deformation (see Section 1) and, specifically, this is true for dolines, sinkholes, collapsible ground, compressible ground, and shrink-swell terrains.
3. Finally, a third limitation is represented by the incapability for resolving deformation $\geq \lambda/4$ between two SAR acquisitions [41] that constrains the measurable displacement velocity to ≤ 86 cm/year, or 1.4 cm for Sentinel-1 scenes six days apart. Such values are usually large enough to completely assess and monitor strain rates associated with rock and soil viscous, viscoelastic, or creep behaviour induced by human or natural-changes in stress conditions. At the same time, events partially detectable by SAR platforms represent an unlikely occurrence in the UK, temporally in the case of interseismic elastic deformation where an earthquake of ≥ 3.7 ML statistically hits every year, and spatially in the case of fall and topple landslides, whose recurrence is mainly limited to Western Scotland [13].

An example of the S-1 kinematic and geometrical constraints is given in Figure 8 which focuses on a sinkhole of 12 m in diameter which appeared twice (14 August 2015 and 15 September 2015) in Mancunian Way, Manchester and determined the partial collapse of the carriageway within diamicton deposits (Figure 9). The descending interferogram spanning across the date of the event has not been able to detect the phenomenon due to the (1) and (3) conditions, despite its vertical component of movement.

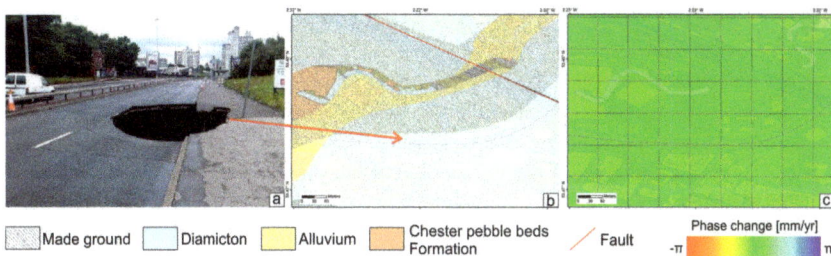

Figure 9. The sinkhole on Mancunian Way (**a**) Image copyright: MEN SYNDICATION). The 10k Geological map of superficial deposits (**b**) and the wrapped interferogram between 22 July 2015 and 15 August 2015 (**c**). The black square refers to the pixel size and geolocation error is half the size of the pixel. Contains Ordnance Data © Crown Copyright and database rights 2017.

2.5. Test Site

The Doncaster area (see Figure 4) has been selected to validate the simulated distortion maps and the CLC2012 map and inspect the influence of an increasing number of SAR images on the LOS velocities. By using the ISBAS approach, a multithreaded solution for the Punnet software was adopted where 28 threads enhance the process model with, parallel, flows of execution within a process.

Starting from the first available acquisition, on 12-3-2015, an initial stack of 20 Sentinel-1A scenes in ascending mode along satellite track 30 has been considered and then gradually incremented by adding the five temporally successive Sentinel-1A and also -1B images in order to evaluate the effect of the number of SAR images on the standard deviation of LOS velocities (σ). Therefore, the 20 images stack encompasses the time span from the first acquisition until 29-5-2016, the 25 images stack the time span until 9-8-2016, the 30 images stack the time span until 2-10-2016, the 35 images stack the time span until 1-11-2016 and the 40 images stack the time span until 1-12-2016. In all the processing the 18-1-2016 acquisition has been set as the master, the scenes were multi-looked by a factor of 5 × 22 in azimuth and range, respectively, and interferograms were generated within a maximum baseline of 200 m and one year in space and time, respectively. The derivation of linear velocities for each coherent point (with coherence above 0.25) was then performed using a reference point at the HOOB GNSS station (latitude: 53.4829° N; longitude: −1.3797° E) located northeast of Doncaster [42]. Coherent pixels cover 80% of the 1750 km^2 extended area with an average density of 51 ISBAS/km^2 (Figure 10). With an average RI-ascending of 0.57, shadowing and layover are not present, while foreshortening, with an extension of 1517.8 km^2, represents 87% of the area.

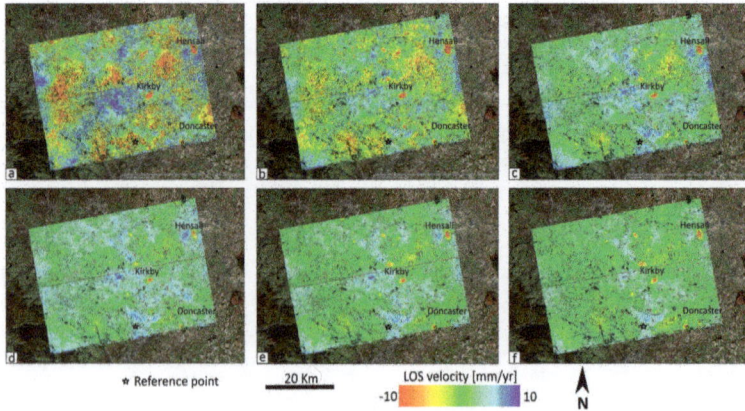

Figure 10. Sentinel-1A LOS velocities for the Doncaster area with 15 (**a**), 20 (**b**), 25 (**c**), 30 (**d**), 35 (**e**) and 40 (**f**) Sentinel-1 images.

The increased number of interferograms for each successive processing led to highlight the two main areas of subsidence north of Doncaster, namely Kirkby and Hensall, through the reduction of tropospheric noise and the consequent decrease in σ which assumes a progressively lower values compared to the average LOS velocity. The ISBAS standard deviation ranges from 5.9 mm/year with 105 interferograms which is 58 times that of the average LOS velocity, to 1.9 with 741 interferograms, which represent less than the average LOS velocity (Figure 11). The potential cause of deformation affecting those sites cannot be related to groundwater extraction from the Sherwood Sandstone Group whose depth here is usually ≤5 m here, due to the stable piezometric levels of the Carlton Hanger Lane [43]. In this perspective, further research should focus on analyzing the 3D displacement vector and InSAR time series to identify a possible link to the former underground mining activities in the area.

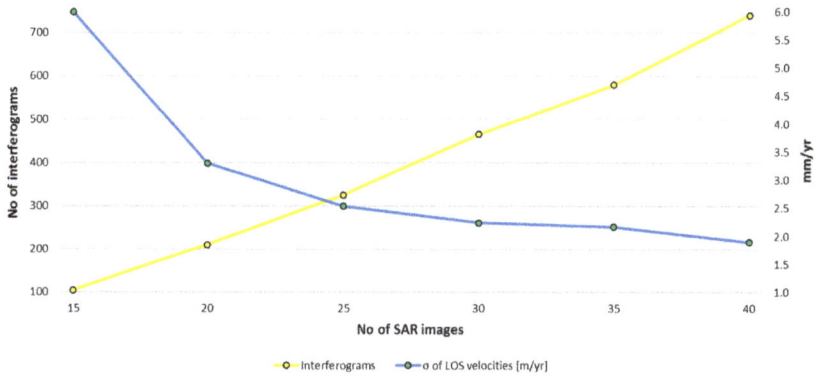

Figure 11. The inverse relationship between the number of interferograms and the LOS standard deviation for the Doncaster area.

3. Conclusions and Future Perspectives

In this paper, the feasibility of the Sentinel-1 satellite to provide InSAR-derived ground deformation map for Great Britain has been analysed in relation to the availability of archived stacks of imagery suitable for interferometric applications, terrain visibility for the two radar acquisition geometries, the density of good radar reflectors for the various land cover types, and the kinematic and geometrical constraints of the investigable phenomena. We showed that topography exerts limited control on the visibility of the British landmass, with only ~1.0% of the total territory potentially affected by radar shadow and layover to the Sentinel-1 IW modes in each acquisition mode. Calibration of the CLC2012 using seven external ISBAS datasets allowed the characterisation of the land cover classes in terms of the average of potential pixel density that ISBAS could provide for each category. Despite the predominance of non-urban land cover across the landmass, the Sentinel-1 shorter repeat-cycles combined with an improved processing technique like ISBAS addresses the need for more dense networks of monitoring targets across rural and partly-vegetated regions with an average density of 90 ISBAS/km² and a potentiality of ~22,000,000 pixels for each acquisition mode covering Great Britain. Furthermore, all of the Great Britain landmass, as of 2016, has enough IW acquisitions ($\geq$30 on average) to guarantee satisfactory results in terms of standard deviation of the LOS velocity as shown in the Doncaster test site.

The implementation of the ground motion study of the whole of Great Britain with interferometric measurements aim to:

- Provide information at regional and national scales to identify ground displacements that occur over hundreds or thousands of square kilometres and characterise areas prone to risk because they are affected by critical geohazards. With such information decision-makers (e.g., UK Oil and Gas Authority, Coal Authority, Environmental Agency, or Health and Safety Executive) can analyse and evaluate different scenarios and plan specific actions based on homogeneous and reliable measurements. Furthermore, InSAR data can be updated regularly as more Sentinel-1 images are acquired, and can provide new ways to design early warning systems covering entire nations, where satellite information can highlight areas where in situ sensors and continuous monitoring tools should be installed or field checks should be optimized.
- Realize an important database to encourage the sharing of methodologies, data, and results. In this regard the manual data analysis and interpretation of ~22 M ISBAS points makes the process cumbersome, operator-dependent, and time-consuming, while semi-automated methodologies can be particularly advantageous for analysing and classifying remote sensing data at the regional scale.

Acknowledgments: A. Novellino, F. Cigna, and L. Bateson published with the permission of the Executive Director of BGS.

Author Contributions: A. Novellino and F. Cigna conceived and designed the experiments; A. Novellino and A. Sowter performed the experiments; all the authors analyzed the data; A. Novellino, F. Cigna and S. Marsh wrote the paper.

Conflicts of Interest: The authors declare no conflict of interest.

References

1. Simons, M.; Rosen, P.A. Interferometric Synthetic Aperture Radar Geodesy. *Treatise Geophys.* **2007**, *3*, 391–446.
2. Gabriel, A.K.; Goldstein, R.M.; Zebker, H.A. Mapping small elevation changes over large areas: Differential radar interferometry. *J. Geophys. Res.* **1989**, *94*, 9183–9191. [CrossRef]
3. Massonnet, D.; Rossi, M.; Carmona, C.; Adragna, F.; Peltzer, G.; Feigl, K.; Rabaute, T. The displacement field of the Landers earthquake mapped by radar interferometry. *Nature* **1993**, *364*, 138–142. [CrossRef]
4. Hooper, A. A multi-temporal InSAR method incorporating both persistent scatterer and small baseline approaches. *Geophys. Res. Lett.* **2008**, *35*, L16302. [CrossRef]
5. Terrafirma Project. Available online: http://www.terrafirma.eu.com/ (accessed on 28 March 2017).
6. Pangeo Project. Available online: http://www.pangeoproject.eu/ (accessed on 28 March 2017).
7. Salvati, P.; Bianchi, C.; Rossi, M.; Guzzetti, F. Societal landslide and flood risk in Italy. *Nat. Hazards Earth Syst. Sci.* **2010**, *10*, 465–483. [CrossRef]
8. ESA—Sentinel Mission. Available online: https://sentinel.esa.int/web/sentinel/home (accessed on 28 March 2017).
9. Milillo, P.; Riel, B.; Minchew, B.; Yun, S.; Simons, M.; Lungrden, P. On the Synergistic Use of SAR Constellations' Data Exploitation for Earth Science and Natural Hazard Response. *IEEE J STARS* **2015**, *9*, 1095–1100. [CrossRef]
10. Cigna, F.; Bateson, L.; Jordan, C.J.; Dashwood, C. Simulating SAR geometric distortions and predicting Persistent Scatterer densities for ERS-1/2 and ENVISAT C-band SAR and InSAR applications: Nationwide feasibility assessment to monitor the landmass of Great Britain with SAR imagery. *Rem. Sens. Env.* **2014**, *152*, 441–466. [CrossRef]
11. Aldiss, D.; Burke, H.; Chacksfield, B.; Bingley, R.; Teferle, F.N.; Williams, S.; Blackman, D.; Burren, R.; Press, N. Geological interpretation of current subsidence and uplift in the London area, UK, as shown by high precision satellite-based surveying. *Proc. Geol. Assoc.* **2014**, *125*, 1–13. [CrossRef]
12. Jones, L.D.; Terrington, R. Modelling volume change potential in the London Clay. *Q. J. Eng. Geol. Hydrogeol.* **2011**, *44*, 1–15. [CrossRef]
13. Pennington, C.; Freeborough, K.; Dashwood, C.; Dijkstra, T.; Lawrie, K. The National Landslide Database of Great Britain: Acquisition, communication and the role of social media. *Geomorphology* **2015**, *249*, 44–51. [CrossRef]
14. Boni, R.; Cigna, F.; Bricker, S.; Meisina, C.; McCormack, H. Characterisation of hydraulic head changes and aquifer properties in the London Basin using Persistent Scatterer Interferometry ground motion data. *J. Hydrol.* **2016**, *540*, 835–849. [CrossRef]
15. Bradley, S.L.; Milne, G.A.; Teferle, F.N.; Bingley, R.M.; Orliac, E.J. Glacial isostatic adjustment of the British Isles: New constraints from GPS measurements of crustal motion. *Geophys. J. Int.* **2009**, *178*, 14–22. [CrossRef]
16. Cooper, A.H.; Farrant, A.R.; Price, S.J. The use of karst geomorphology for planning, hazard avoidance and development in Great Britain. *Geomorphology* **2011**, *134*, 118–131. [CrossRef]
17. Wilson, M.P.; Davies, R.J.; Foulger, G.R.; Julian, B.R.; Styles, P.; Gluyas, J.G.; Almond, S. Anthropogenic earthquakes in the UK: A national baseline prior to shale exploitation. *Mar. Petrol. Geol.* **2015**, *30*, 1–17. [CrossRef]
18. Bateson, L.; Cigna, F.; Boon, D.; Sowter, A. The application of the intermittent SBAS (ISBAS) InSAR method to the south Wales coalfield, UK. *Int. J. Appl. Earth Obs. Geoinform.* **2015**, *34*, 249–257. [CrossRef]
19. British Geological Survey. Have Earthquakes Caused Deaths in Britain? Available online: http://www.earthquakes.bgs.ac.uk/education/faqs/faq14.html (accessed on 28 March 2017).
20. Gibson, A.D.; Culshaw, M.G.; Dashwood, C.; Pennington, C.V.L. Landslide management in the UK—The problem of managing hazards in a 'low-risk' environment. *Landslides* **2013**, *10*, 599–610. [CrossRef]

21. European Environment Agency. Mapping the Impacts of Natural Hazards and Technological Accidents in Europe—An Overview of the Last Decade. EEA Technical Report No 13/2010. 2011. Available online: http://www.eea.europa.eu/publications/mapping-the-impacts-of-natural (accessed on 28 March 2017).

22. Met Office—UK Climate. Available online: http://www.metoffice.gov.uk/public/weather/climate/ (accessed on 28 March 2017).

23. De Zan, F.; Guarnieri, A.M. TOPSAR: Terrain observation by progressive scans. *IEEE Trans. Geosci. Remote Sens.* **2006**, *44*, 2352–2360. [CrossRef]

24. European Space Agency. SPA-COPE-ENG-RP-066: Sentinels Data Access Annual Report (3/10/2014–30/11/2015), 2016. Available online: https://sentinel.esa.int/web/sentinel/user-guides/document-library/-/asset_publisher/xlslt4309D5h/content/sentinels-data-access-annual-report (accessed on 28 March 2017).

25. Alaska Satellite Facility. Available online: https://www.asf.alaska.edu/ (accessed on 28 March 2017).

26. Sentinel Data Access Service. Available online: http://sedas.satapps.org (accessed on 28 March 2017).

27. Copernicus—Open Access Hub. Available online: https://scihub.copernicus.eu/news/ (access on 13 October 2016).

28. ESA. Sentinel-1 User Handbook 2013. Available online: https://sentinel.esa.int/ (access on 13 October 2016).

29. Kropatsch, W.G.; Strobl, D. The generation of SAR layover and shadow maps from digital elevation models. *IEEE Trans. Geosci. Remote Sens.* **1990**, *28*, 98–107. [CrossRef]

30. Colesanti, C.; Wasowski, J. Investigating landslides with space-borne Synthetic Aperture Radar (SAR) interferometry. *Eng. Geol.* **2006**, *88*, 173–199. [CrossRef]

31. Cigna, F.; Bianchini, S.; Casagli, N. How to assess landslide activity and intensity with Persistent Scatterer Interferometry (PSI): The PSI-based matrix approach. *Landslides* **2013**, *10*, 267–283. [CrossRef]

32. Notti, D.; Herrera, G.; Bianchini, S.; Meisina, C.; García-Davalillo, J.C.; Zucca, F. A methodology for improving landslide PSI data analysis. *Int. J. Remote Sens.* **2014**, *35*, 2186–2214.

33. Ferretti, A.; Prati, C.; Rocca, F. Permanent scatterers in SAR interferometry. *IEEE Trans. Geosci. Remote Sens.* **2001**, *39*, 8–20. [CrossRef]

34. Berardino, P.; Fornaro, G.; Lanari, R.; Sansosti, E. A new Algorithm for Surface Deformation Monitoring based on Small Baseline Differential SAR Interferograms. *IEEE Trans. Geosci. Remote Sens.* **2002**, *40*, 2375–2383. [CrossRef]

35. Ferretti, A.; Fumagalli, A.; Novali, F.; Prati, C.; Rocca, F.; Rucci, A. A new algorithm for processing interferometric data-stacks: SqueeSAR. *IEEE Trans. Geosci. Remote Sens.* **2011**, *49*, 3460–3470. [CrossRef]

36. Gee, D.; Sowter, A.; Novellino, A.; Marsh, S.; Gluyas, J. Monitoring land motion due to natural gas extraction: Validation of the Intermittent SBAS (ISBAS) DInSAR algorithm over gas fields of North Holland, the Netherlands. *Mar. Petrol. Geol.* **2016**, *77*, 1338–1354. [CrossRef]

37. Sowter, A.; Amat, M.B.C.; Cigna, F.; Marsh, S.; Athab, A.; Alshammari, L. Mexico City land subsidence in 2014–2015 with Sentinel-1 IW TOPS: Results using the Intermittent SBAS (ISBAS) technique. *Int. J. Appl. Earth Obs.* **2016**, *52*, 230–240. [CrossRef]

38. Plank, S.; Singer, J.; Thuro, K. Assessment of number and distribution of persistent scatterers prior to radar acquisition using open access land cover and topographical data. *ISPRS J. Photogramm.* **2013**, *85*, 132–147. [CrossRef]

39. Hu, J.; Li, Z.W.; Ding, X.L.; Zhu, J.J.; Zhang, L.; Sun, Q. Resolving three-dimensional surface displacements from InSAR measurements: A review. *Earth Sci. Rev.* **2014**, *133*, 1–17. [CrossRef]

40. Wright, T.J.; Parsons, B.E.; Lu, Z. Toward mapping surface deformation in three dimensions using InSAR. *Geophys. Res. Lett.* **2004**, *31*. [CrossRef]

41. Rosen, P.A.; Hensley, S.; Joughin, I.R.; Li, F.K.; Madsen, S.N.; Rodriguez, E.; Goldstein, R.M. Synthetic aperture radar interferometry. *Proc. IEEE* **2000**, *88*, 333–382. [CrossRef]

42. Natural Environment Research Council (NERC). British Isles Continuous GNSS Facility (BIGF). Available online: http://www.bigf.ac.uk/ (access on 25 March 2017).

43. Environment Agency, Scottish Environment Protection Agency and Irish Office of Public Works. Shoothill GaugeMap. Available online: http://www.gaugemap.co.uk/# (access on 24 March 2017).

geosciences

MDPI

Article

Assessment of Landslide Pre-Failure Monitoring and Forecasting Using Satellite SAR Interferometry

Serena Moretto [1,2,*], Francesca Bozzano [1,2], Carlo Esposito [1], Paolo Mazzanti [1,2] and Alfredo Rocca [2]

[1] Department of Earth Sciences, Sapienza University of Rome, Rome 00185, Italy; francesca.bozzano@uniroma1.it (F.B.); carlo.esposito@uniroma1.it (C.E.); paolo.mazzanti@uniroma1.it (P.M.)
[2] NHAZCA S.r.l., Spin-off of Sapienza, University of Rome, Rome 00185, Italy; alfredo.rocca@nhazca.com
*   Correspondence: serena.moretto@uniroma1.it; Tel.: +39-32-7127-8508

Academic Editors: Francesca Cigna and Jesus Martinez-Frias
Received: 28 February 2017; Accepted: 9 May 2017; Published: 12 May 2017

Abstract: In this work, the ability of advanced satellite interferometry to monitor pre-failure landslide behaviours and the potential application of this technique to Failure Forecasting Methods (FFMs) are analysed. Several limits affect the ability of the technique to monitor a landslide process, especially during the pre-failure phase (tertiary creep). In this study, two of the major limitations affecting the technique have been explored: (1) the low data sampling frequency and (2) the phase ambiguity constraints. We explored the time series of displacements for 56 monitored landslides inferred from the scientific literature and from different in situ and remote monitoring instruments (i.e., extensometers, inclinometers, distometers, Ground Base InSAR, and total station). Furthermore, four different forecasting techniques have been applied to the monitoring data of the selected landslides. To analyse the reliability of the FFMs based on the InSAR satellite data, the 56 time series have been sampled based on different satellite features, simulating the satellite revisit time and the phase ambiguity constraints. Our analysis shows that the satellite InSAR technique could be successful in monitoring the landslide's tertiary creep phase and, in some cases, for forecasting the corresponding time of failure using FFMs. However, the low data sampling frequency of the present satellite systems do not capture the necessary detail for the application of FFMs in actual risk management problems or for early warning purposes.

Keywords: A-DInSAR; InSAR; landslide; failure forecasting methods; landslide forecasting

MSC: 86A60

1. Introduction

The scientific community started to explore landslide behaviour during the first half of the twentieth century [1,2]. In 1950, Terzaghi recognized the connection between landslide evolution and creep theory, laying the foundation for landslide prediction. Landslide forecasting remains a current issue and a significant challenge in natural hazard risk mitigation. Since the beginning of the 1960s, many authors have been studying this topic [3–14]. Saito and Uezawa initiated this field of research, and the first successful landslide prediction was performed by Saito in 1965 for the Dosan Line landslide [4]. They defined semi-empirical methods to estimate the timing of slope failure using displacement monitoring data acquired via different techniques.

Subsequently, different semi-empirical methods for landslide prediction were developed. In 1985, a new equation based on laboratory experiments was derived by Fukuzono [6] and later validated by Voight [8]:

$$\Omega'' = A\Omega'^{\alpha}, \tag{1}$$

This equation describes the relationship between velocity (Ω') and acceleration (Ω'') of the surface displacement during the pre-failure phase under constant stress and temperature conditions using two empirical constants A and α.

Based on Equation (1), failure forecasting methods (FFMs) that estimate the time of failure (TF) were developed by Cornelius and Voight [11].

The methods are based on the creep theory [2], which describes the time-dependent deformational behaviour of the slope under constant stress conditions. Specifically, the final creep phase known as tertiary creep (TC), characterized by an acceleration of the deformation affecting a slope, is modelled using semi-empirical functions to predict the TF. Pre-failure monitoring data are the basic requirement for the application of FFMs. Many successful predictions are described in the literature [15–21] that were developed with data acquired from different monitoring techniques such as extensometers, inclinometers, GB-InSAR, etc.

In this paper, the potential of satellite interferometry to monitor the pre-failure landslide behaviour for the application of FFMs is assessed.

The advanced techniques of satellite SAR image processing (A-DInSAR Advanced Differential Synthetic Aperture Radar Interferometry, [22–25]) can provide the time series of displacements for stable radar reflectors in the investigation area. The processing algorithms developed during the last few decades have made A-DInSAR a powerful technique for landslide risk management. It currently allows for landslide detection, mapping, characterization, and monitoring at different scales, both for large area and site specific analysis. However, the present features of the technique influence its suitability for landslide forecasting and early warning purposes, even if, in some cases, satellite InSAR can be successfully applied to disaster risk reduction by identifying unstable slope indicators [26].

The limitations that affect the capability of the technique to monitor a landslide process during the pre-failure phase can be summarized as follows [27]: (i) radar distortions; (ii) presence of backscatter; (iii) availability of data; (vi) areal extension of the landslide process; and (v) deformational behaviour (length of tertiary creep and rate of deformations). In this study, the site-specific feasibility of satellite InSAR monitoring has been deliberately neglected, and the focus is on the constraints related to the slope's deformational behaviour. Two major limitations of the technique for pre-failure landslide monitoring were accounted for: (i) the long satellite revisit time (RT, i.e., the sampling period of the data acquisition) and (ii) the phase ambiguity constraints.

The satellite RT is one of the major limitations affecting the A-DInSAR technique for forecasting purposes because it reduces the applicability of the technique to landslide processes with a TC long enough to be captured by the satellites. The phase ambiguity constraint influences the ability of the technique to monitor rapid displacements, which characterize the pre-failure stage. Furthermore, there is a strong dependence between RT and phase ambiguity. Specifically, if no prior information is available, it is not possible to achieve A-DInSAR results if the displacements are greater than $\lambda/4$ (where λ is the sensor's wavelength) between two consecutive SAR acquisitions and two adjacent pixels [28,29].

Considering these limitations, in this paper the potential of satellite interferometry for landslide forecasting has been explored by back-analysing 56 case histories.

2. Materials and Methods

Is it possible to monitor a landslide during TC with satellite interferometry? Is it possible to apply landslide forecasting methods using the InSAR time series of displacement?

In this study, we try to answer these questions by exploring two of the primary limitations affecting satellite InSAR: (i) the low satellite data sampling frequency and (ii) the ambiguity phase constraints.

To assess the satellite InSAR potential for landslide pre-failure monitoring and forecasting, the following approach has been adopted:

1. A database collection composed of 56 past landslides, including pre-failure displacement data inferred from the scientific literature;
2. The digitization of the pre-failure datasets from graphs;
3. The sampling of 56 time series of displacements based on different satellites' revisit time to reproduce the satellite data acquisition;
4. Simulation of the phase ambiguity constraint based on the different satellites' features (wavelength and revisit time);
5. The application of four FFMs on three different datasets: (i) digitized datasets, which represent the actual monitoring datasets; (ii) sampled datasets (based on the revisit time constraint); and (iii) simulated datasets (considering the phase ambiguity limit);
6. A comparison between the forecast results obtained using actual monitoring data and simulated satellite data.

The pre-failure monitoring data of 56 past landslides have been explored by observing their evolution from the beginning of the acceleration phase to the occurrence of failure.

The collected monitoring data show different characteristics; they were acquired using different monitoring instruments (i.e., extensometers, inclinometers, distometers, GB-InSAR, and total station) and were displayed in the instruments using different formats, including displacement vs. time, velocity vs. time, and inverse velocity vs. time. Furthermore, the data was displayed in different units such as millimetres, centimetres, metres, mm/day, mm/hour, and cm/day. To homogenize the information and to compare the results, all the datasets are reported in mm/day.

The 56 landslide datasets collected (Table 1) show different features in terms of: triggering factor, mechanism, volume, material, and deformational behaviour.

Table 1. Landslide database collected from the scientific literature.

Landslide	Reference	ID	Tertiary Creep (TC) Length (days)	Average Velocity (mm/day)
Jitsukiyama	[30]	1	0.01	0.04 (mm/s)
Moriwaki Experiment	[31]	2	0.02	0.04 (mm/min)
Unnamed Mufundirawa	[32]	3	0.1	0.87 (mm/s)
Letlhakane Mine	[33]	4	0.2	68.60 (mm/h)
Dosan Line	[34]	5	1.2	664.79
La Saxe	[35]	6	1.6	4541.68
Betz Post 2 mc	[15]	7	3.2	63.42
Asamushi	[34]	8	3.4	163.76
Rock Dump	[36]	9	4.1	1011.21
Cavallerizzo	[37]	10	4.3	26.32
Roesgrenda B	[38]	11	6	171.93
Tuckabianna West	[39]	12	6.1	39.55
Afton	[40]	13	9.4	591.07
Excavation A	[40]	14	9.7	14.22
Preonzo	[41]	15	10.6	91.30
Silla Montecchi	[42]	16	10.8	80.31
Eskihisar Coal Mine	[43]	17	11.5	6.72
Ryan Call Slide1	[44]	18	12.8	1564.10
Crack Gauge	[45]	19	13	11.98
Ruahihi	[46]	20	14.1	54.58
Es.2 Rose & Hungr2007	[15]	21	15	52.16
Roesgrenda A	[38]	22	20.1	2.00

Table 1. Cont.

Landslide	Reference	ID	Tertiary Creep (TC) Length (days)	Average Velocity (mm/day)
Abbotsford	[46]	23	20.4	106.26
New Tredegar	[47]	24	22.5	36.22
Victorious East Pit	[40]	25	22.7	4.56
Haveluck	[40]	26	25.8	1.94
Bohemia	[48]	27	30.3	0.51
Val Pola	[49]	28	30.5	51.33
Luscar_51B2	[40]	29	31.6	76.73
Ooigawa Railroad	[34]	30	38.9	12.76
Telfer	[40]	31	39.4	2.73
Liberty Pit Mine	[15]	32	40.2	4.73
Selbourn	[50]	33	41.5	100.54
Betz Post 18mc	[15]	34	44	6.58
Lijiaxia	[51]	35	44.7	4.47
Takabayama	[52]	36	44.9	145.78
Kennekott#1	[53]	37	46.9	57.19
Chuquicamata Mine2	[52]	38	47.6	224.24
Ryan Call Slide2	[44]	39	53.7	21.35
Vajont	[13]	40	54.6	30.41
Luscar 50A2	[40]	41	60.9	1272.11
Hogart	[54]	42	77.8	63.85
Roberts	[55]	43	105	57.46
Nevis Bluff	[56]	44	110.4	2.98
Monte Beni	[16]	45	132.2	19.67
Xintan	[57]	46	138.9	459.60
Azimi 1988	[7]	47	143.8	3.39
Threatening Rock	[58]	48	153.8	2.49
Bomba	[59]	49	167.9	10.86
Smoky River	[40]	50	174.2	1.59
Scalate	[55]	51	183.7	0.76
Brada River	[60]	52	187.8	0.10
Excavation C	[40]	53	211.3	1.22
Delabole Quarry	[61]	54	380.7	2.89
Saleshan	[62]	55	631.3	1.88
Puigcercos Scarp	[63]	56	1013	0.39

For the objective of this work, the most important feature is the deformational behaviour of the slope before failure; specifically, the TC duration and the deformation rates occurring during the TC. The beginning of TC has been manually defined on the strain rate vs. time plot as the point at which the velocity begins to increase just before the final collapse. As shown in Table 1, the length of the TC period is extremely variable, ranging from several minutes to more than 1000 days. Additionally, the average velocity during the pre-failure phase varies substantially from 0.01 m/day to more than 4 m per day.

Monitoring data presented graphically in the scientific literature have been digitized using the MATLAB® code "grabit". An example of the digitized data for the Vajont landslide is shown in Figure 1. To simulate the satellite data acquisition, the monitoring datasets of the 56 landslides have been sampled in the velocity time space based on different satellites' RT values (Figure 2). The features of nine satellites in the past, present, and future have been considered (Table 2).

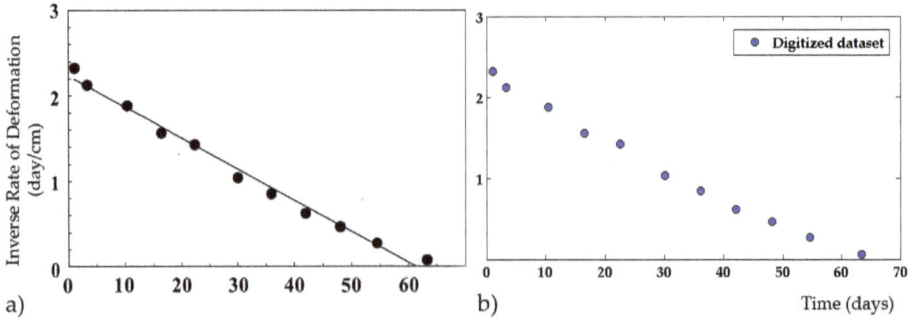

Figure 1. An example of the digitized monitoring data for the Vajont landslide (9 October 1963). (a) Monitoring data available in the literature ([12] after Muller [64] and Voight [8]) referencing the inverse rates (days/cm) of horizontal slope movement registered before the Mt. Toc catastrophic collapse. Velocity measurements were obtained from stations near the toe of the moving mass [12]; (b) Digitized data.

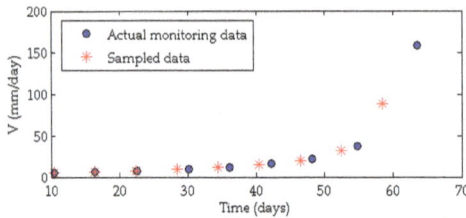

Figure 2. Example of sampled datasets (red stars) based on the Sentinel 1A + 1B revisit time of 6 days. Blue circles represent the pre-failure monitoring data of the Vajont landslide.

Table 2. Satellites considered in this study and their principal features (Revisit Time (RT) and wavelength (λ)).

Period	Satellite	Revisit Time (days)	Band	λ (mm)
	ALOS PALSAR 1	46	L	236
Past	J-ERS	44	L	235
	ERS 1/2-Envisat	35	C	56
		4	X	31
	COSMO-SkyMed	8	X	31
		16	X	31
Present	Sentinel-1A	12	C	56
	Sentinel 1A + 1B	6	C	56
	ALOS PALSAR 2	14	L	236
	TerraSAR-X	11	X	31
Future	Radarsat Constellation	4	C	56

Furthermore, according to the satellites' wavelengths and RT values, time series have been customized to consider the phase ambiguity limits. Because the phase is a periodic function and the wavelength number in the two travel paths is unknown, it follows that displacements greater than $\lambda/2$ between two consecutive acquisitions may remain unsolved or underestimated. Furthermore, for Advanced DInSAR (i.e., Persistent Scatterers Interferometry (PSI), Small Baseline Subset (SBAS), and similar techniques), the maximum amount of displacement that can be measured between two consecutive acquisitions is $\lambda/4$ [28]. To simulate this limit, we assumed that satellite InSAR is not able to monitor displacements greater than $\lambda/4$ between two consecutive acquisitions. For example, considering the Sentinel features, measurement points with displacements greater than 14 mm in 6 days have not been considered (Figure 3). Because the datasets are in the velocity–time space, the measurement points with a velocity greater than the velocity of ambiguity ($V_{ambiguity}$) have been removed, where $V_{ambiguity}$ is defined as:

$$V_{ambiguity} = (\lambda/4)/RT,\qquad(2)$$

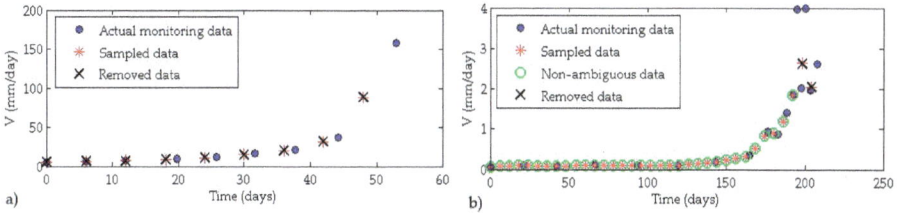

Figure 3. An example of the ambiguity phase simulation based on the Sentinel 1A+1B features (RT = 6 days and Vambiguity = 2.33 mm/day). Removed data (black X) represent the measurements with a velocity greater than the velocity of ambiguity, while the green circles represent data with V < Vambiguity. (**a**) Vajont's datasets; in this case, the deformations occurring during the pre-failure stage were too fast to be monitored with satellite interferometry; (**b**) Excavation C datasets. The raw data consist of surface displacement data acquired using a prism monitoring system.

Four FFMs have been used to obtain the TF estimation [65]:

i.   Inverse velocity method (INV, [6]);
ii.   Log rate vs. log acceleration technique (LOG, [11]);
iii. Linearized least-square technique (LSM, [11]);
iv.   Non-linear least-square technique (NL, [11]).

The INV method is a graphical technique based on the linear approximation of the inverse velocity data, while the other methods are numerical techniques derived via the manipulation of Equation (1). The INV method is the most utilized due to its simple implementation in actual emergency situations. It works well for linearly distributed data ($\alpha = 2$); however, the other methods allow data characterized by non-linear trends to be fit based on the estimation of the two constants A and α.

The four forecasting methods have been applied to all the landslide time series using the following datasets:

i.   the actual monitoring data, i.e., the digitized datasets;
ii.   the sampled time series, i.e., the sampled time series based on the satellite's revisit time;
iii. the simulated time series, which considers the phase ambiguity constraint.

Using this approach, it was possible to assess the suitability of the satellite InSAR technique for monitoring the landslide process during the tertiary creep phase and for the application of FFMs based on the revisit time and phase ambiguity constraints of different satellite sensors.

3. Results

The FFM results are reported in the following subsections according to the datasets used: (i) the actual monitoring data (Section 3.1); (ii) the sampled data (Section 3.2) and (iii) the simulated time series (Section 3.3).

3.1. Actual Monitoring Data

The FFM results for the actual monitoring data are compared to the results obtained for the sampled and simulated time series. The forecasting methods have been applied to 56 past landslides considering all the tertiary creep phases from the beginning to the last available datum. For each landslide, the actual time of the slope failure (actual failure) is known, allowing us to estimate the prediction error. The prediction error obtained for each landslide according to each forecasting method is shown in Figure 4. The error (E) has been defined as the difference in days between the actual failure and the computed time of failure (Tf):

$$E = \text{Actual Failure} - \text{Tf}, \tag{3}$$

Based on this equation, positive error values would represent safe predictions in a priori analysis [29]; in contrast, negative error values are related to unsafe predictions (Tf is delayed with respect to the actual failure).

Landslide ID values are reported on the x-axis, while the prediction error is on the y-axis. ID values have been assigned according to an increasing TC length. Additionally, a decrease in data sampling frequency is contained in our database toward the higher ID values.

Figure 4 shows that the prediction error increases towards the higher ID values. Predictions close to the actual failure have been obtained for the low ID values (from ID 1 to ID 21), i.e., based on the datasets with higher sampling frequency and a short TC phase.

Overall, the semi-empirical methods returned predictions with an error less than 3 days for: 30 examples using the graphical technique, 28 using the NL technique, 24 using the LOG technique, and 15 using the LSM technique.

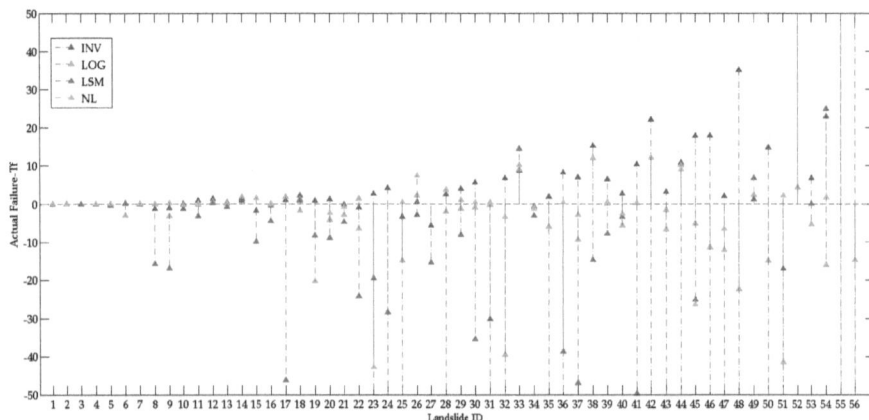

Figure 4. Error diagram obtained for the actual monitoring datasets. The colours identify the different failure forecasting methods (FFMs).

3.2. Sampled Datasets

For the application of forecasting methods, a minimum of three monitoring points within the accelerating phase is required. Under these conditions, we can build a function that can provide information on the increasing displacement velocity and, thus, we are able to apply semi-empirical models. Based on the TC duration and on the revisit time of the different satellites, the landslides with at least three data points within the TC phase have been defined as compatible for the application of the forecasting methods (Table 3).

Table 3. Analysis results for the sampled datasets related to different RTs. The number of compatible landslides and the number of satisfactory predictions (error lower than ±5 days) for each satellite are reported. The compatible landslides include at least three data samples in the TC phase. For each forecasting method (inverse velocity method—INV, log rate vs. log acceleration technique—LOG, linearized least-square technique—LSM, and non-linear least-square technique—NL), the number of landslides with a prediction error lower than ±5 days is reported. The best performances obtained in terms of prediction accuracy are underlined.

Period	Satellite	RT (Days)	Band	λ (mm)	Compatible Landslides (Revisit Time)	E < |5| Days INV	E < |5| Days LOG	E < |5| Days LSM	E < |5| Days NL
						Number of Landslides			
Past	ALOS PALSAR 1	46	L	236	14	1	<u>2</u>	1	1
	J-ERS	44	L	235	14	<u>2</u>	1	0	1
	ERS 1/2-Envisat	35	C	56	14	1	0	1	0
Present	COSMO-SkyMed	4	X	31	43	<u>17</u>	11	11	11
		8	X	31	35	8	5	<u>10</u>	<u>10</u>
		16	X	31	26	4	2	<u>7</u>	<u>7</u>
	Sentinel-1A	12	C	56	30	<u>7</u>	2	<u>7</u>	4
	Sentinel 1A + 1B	6	C	56	38	<u>13</u>	3	5	9
	ALOS PALSAR 2	14	L	236	29	4	<u>6</u>	3	<u>6</u>
	TerraSAR-X	11	X	31	30	<u>6</u>	2	4	3
Future	Radarsat Constellation	4	C	56	43	<u>17</u>	11	11	11

Considering the past satellites (ALOS PALSAR, J-ERS, ERS, and Envisat) characterized by the longer revisit times ranging from 35 to 46 days, 14 compatible landslides (i.e., those that can be monitored using satellite InSAR) have been identified out of the 56 collected in the database. For the present and the future satellite generations, a maximum of 43 compatible landslides have been identified based on the revisit time of the Sentinel 1A-1B, Cosmo Sky-Med (theoretical RT of 4 days), and Radarsat Constellation.

Considering the present satellites' features, the number of compatible landslides ranges from 26 to 43, i.e., from 25% to 86% of the landslides collected in the database. This means that, at present, considering only the revisit time constraint, we could observe a change in the displacement trend via satellite with at least three data points for more than 25% of the collected landslides, reaching 86% of the landslides considering the Sentinel-1 characteristics.

The four FFMs have been applied to every compatible landslide based on each satellite's revisit time. Figure 5 shows the analysis performed considering the shorter revisit time (namely, 4 days), belonging to the future Radarsat Constellation and, theoretically, to the COSMO-SkyMed Constellation. For the forecasting analysis applied to the actual monitoring data, the prediction error increases toward the higher ID values. Specifically, the errors derived from the INV application are small (<±5 days) up to ID 25 and start to increase after ID 25.

Table 3 and Figure 6 show that the number of satisfactory predictions increases toward the shorter revisit times. A prediction is considered satisfactory if the prediction error is within ±5 days. The INV method shows the strongest relationship between the number of successful forecasts and the RT (Figure 6b), because a shorter RT results in a higher number of accurate predictions.

The analysis shows that the INV method results in a better accuracy for the TF estimation. The INV technique has a prediction error of less than 5 days for 17 landslides based on an RT of 4 days, and for 13 landslides based on the RT of the Sentinel-1 satellite.

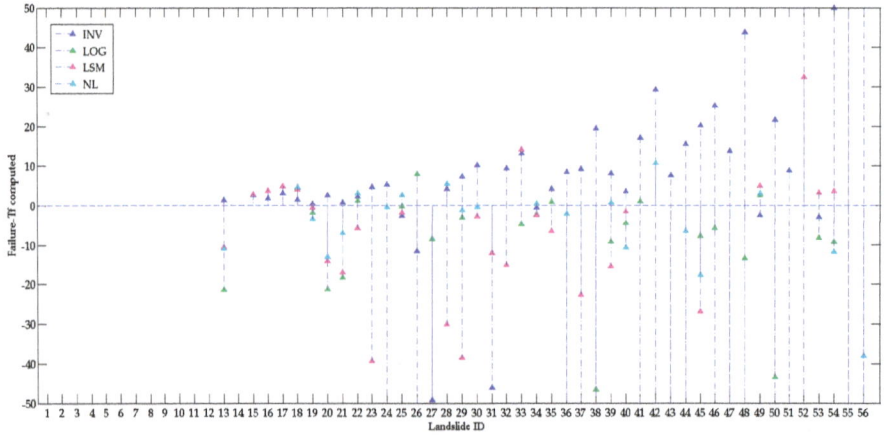

Figure 5. Diagram of the prediction error for the compatible landslides based on a RT of 4 days.

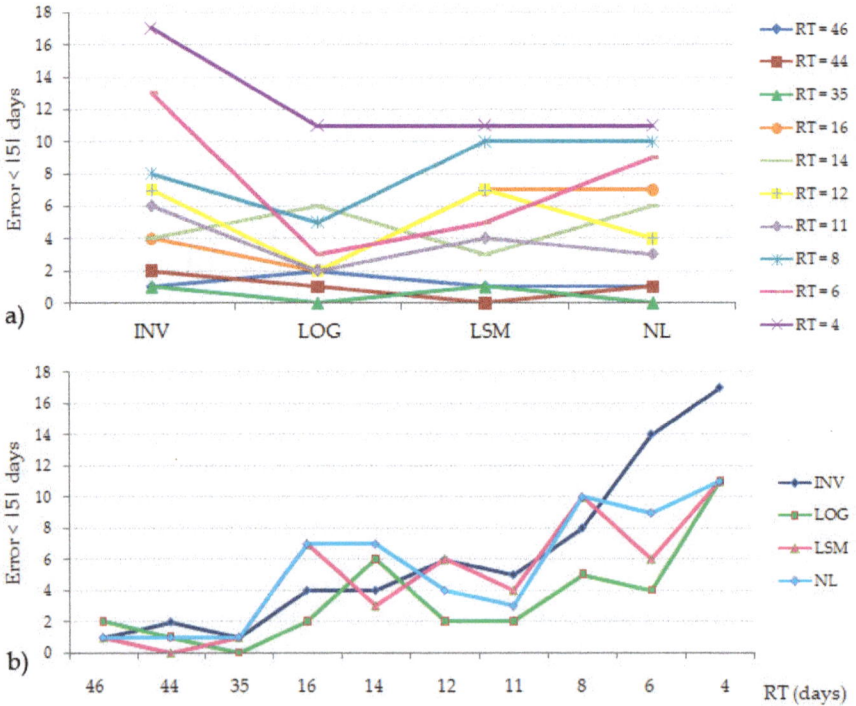

Figure 6. (a) Relationship between the number of successful forecasts and the forecasting method considering the different revisit times; (b) Relationship between the revisit time and the number of satisfactory predictions based on each FFM.

3.3. Simulated Time Series

Considering the phase ambiguity constraints, a maximum of 24 compatible landslides out of the 56 have been identified based on the Radarsat Constellation features and characterized by the best association between the revisit time and the sensor's wavelength. Currently, the Sentinel-1 constellation would have observed 17 landslides with at least three measurement points inside the acceleration phase preceding the slope collapse. By introducing the phase ambiguity, the suitability of the failure forecasting methods decrease. For the simulated satellite datasets, only a maximum of four satisfactory predictions have been obtained using the inverse velocity method (Table 4).

Table 4. Analysis results introducing the phase ambiguity constraint based on the different satellite's features. "Compatible Landslides (Phase ambiguity)" represents the number of landslides with at least three data points in the TC phase accounting for the phase ambiguity limit. For each forecasting method, the number of landslides with a prediction error lower than 5 days is reported.

Period	Satellite	RT (days)	Band	λ (mm)	Compatible Landslides (RT)	Compatible Landslides (Phase Ambiguity)	E < |5| days INV	E < |5| days LOG	E < |5| days LSM	E < |5| days NL
						Number of Landslides				
Past	ALOS PALSAR 1	46	L	236	14	7	0	1	0	0
	J-ERS	44	L	235	14	8	0	0	0	0
	ERS ½-Envisat	35	C	56	14	4	1	0	1	0
Present	COSMO-SkyMed	4	X	31	43	19	3	0	3	1
		8	X	31	35	11	0	0	2	0
		16	X	31	26	9	0	0	0	0
	Sentinel-1A	12	C	56	30	11	1	0	0	0
	Sentinel 1A + 1B	6	C	56	38	17	4	3	2	1
	ALOS PALSAR 2	14	L	236	29	14	1	2	1	1
	TerraSAR-X	11	X	31	30	9	0	0	0	0
Future	Radarsat Constellation	4	C	56	43	24	4	1	4	2

4. Discussion

The forecasting methods have great potential for landslide hazard management [16,18–20]. By applying FFMs to actual monitoring data, we have obtained a prediction error lower than 3 days in 53.7% of the examples using the INV method, 51.7% using the LOG technique, 48.2% using the NL technique, and 27.8% using the LSM technique. The percentage of successful predictions increases when considering landslides with shorter TC phases and datasets characterized by a higher sampling frequency, namely, the landslide database presented here with lower ID values. When considering only the first 21 landslides (TC < 15 days), an error lower than 3 days has been obtained in 100% of the examples using the INV method, 85.7% using the LOG technique, 76.2% using the NL technique, and 57.1% using the LSM method.

For the sampled and simulated datasets, the number of successful predictions decreases (Tables 3 and 4). Only a maximum of four satisfactory predictions have been obtained with the inverse velocity method including the ambiguity phase constraints. The principal reason is because the FFMs have been applied to only the landslides with a TC phase long enough to be observed with the satellite's RT (high ID values). The predictions obtained using the actual and simulated datasets are similar (Figure 7) when the FFMs are accurate using the actual monitoring data (ID 23, 27, 35, 49, 53). A reliable prediction can also be obtained using the simulated datasets; however, if the prediction error is high using the actual time series, the FFMs fail using the simulated satellite data as well.

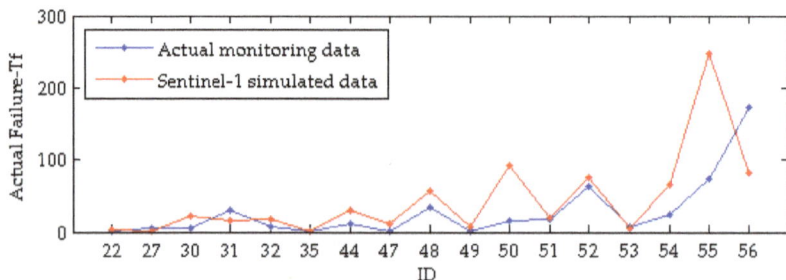

Figure 7. Comparison between the INV method applied to the actual monitoring data and to the simulated datasets. The results of the simulated datasets are referenced to the Sentinel 1A + 1B features.

The error growth towards higher ID values could be related to both the decrease of the data sampling frequency and to the increase of the TC phase duration. Because of the inverse relationship between the TC length and data sampling frequency (Figure 8), we are not able to clearly determine which parameter primarily affects the success or the failure of the forecasting analyses. A higher sampling frequency produces more robust datasets for the application of forecasting methods, especially near the time of slope collapse; however, a very long TC phase can lead to complex deformational behaviours due to the variation of the stress conditions, which has a negative impact on the FFM stability. The difference in the prediction errors obtained with the actual and simulated time series data is related to the loss of the last monitoring data caused by the sampling and ambiguity phase reproduction. It is assumed that the last monitoring data before the collapse significantly affects the forecasting analyses results because they include the highest velocity of the phenomenon. Under equal conditions, the closer the monitoring data are to the actual slope failure time, the more accurate the forecast. This can be proven by comparing the predictions obtained using the entire time series (datasets from the beginning of TC until the last monitoring datum) and only a part of the time series (i.e., removing the last measurement points) (Figure 9). Based on our results, there is a connection between the time intervals from the last available measurement point to the actual failure (GAP) and the accuracy of the forecasting methods. The relationship between the GAP and the prediction accuracy can be observed for the majority of the landslides collected (Figure 9a,b). In Figure 9c, an example is shown where the prediction accuracy increases when approaching the actual failure based on the Preonzo landslide monitoring data.

The GAP reduction capability is related to the sampling frequency; the higher the sampling frequency, the greater the ability to obtain measurements closer to the actual failure. Figure 10 shows the relationship between the GAP and the sampling frequency for the time series of our database.

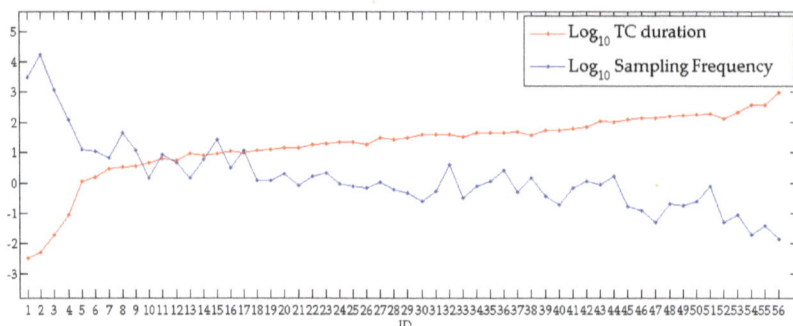

Figure 8. Inverse correlation between the TC's length and data sampling frequency.

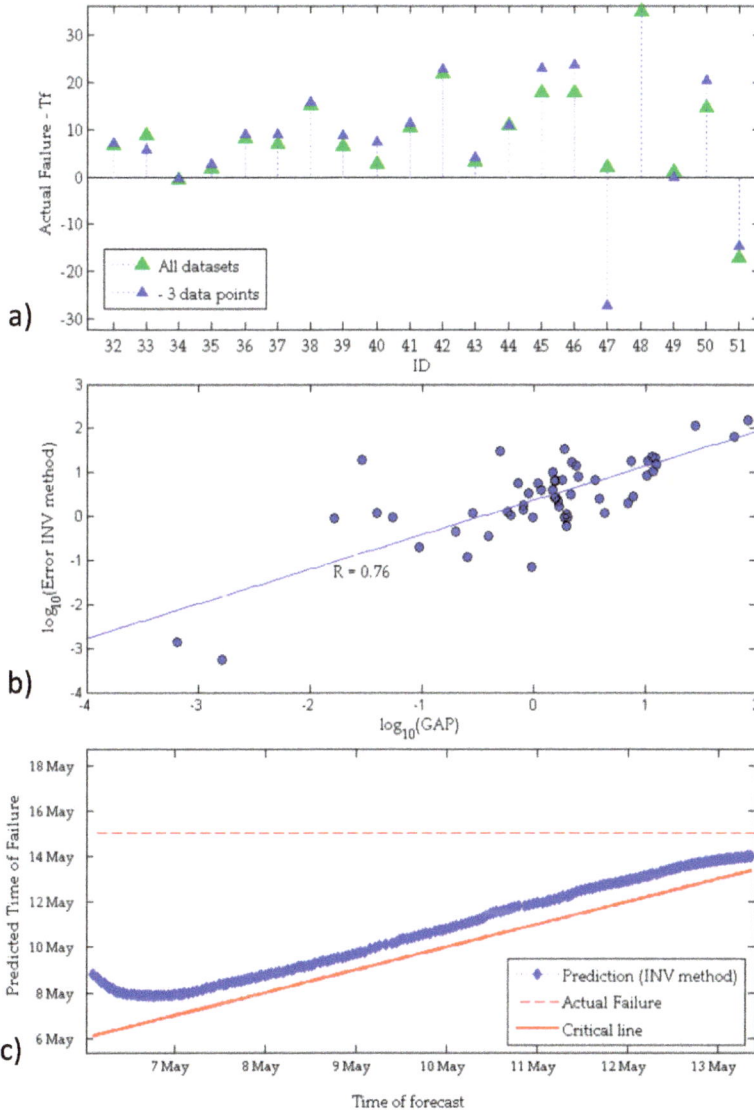

Figure 9. Prediction errors. (**a**) Errors using the INV method to: (i) the entire available time series (green predictions) and (ii) removing the last three measurement points (blue predictions); (**b**) Prediction error vs. GAP (time interval between the last datum and the actual failure); (**c**) INV iterative forecasting analysis for the Preonzo Landslide (ID 12) (modified after [66]). The predictions have been updated by simulating the data acquisition over time. The time of the actual failure (15 May) is displayed as a red dashed line. The critical line (red line) corresponds to the day the forecast was made. Consequently, it represents the lower boundary of a reasonable prediction. The analysis shows that the closer the last available datum is to the actual failure, the more accurate the prediction. When approaching the actual failure, the prediction error decreases, in this case, following a linear trend for the INV methods.

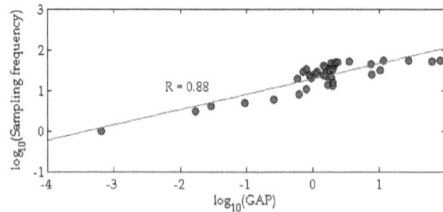

Figure 10. Relationship between the data sampling frequency and the GAP for the collected landslide database.

5. Conclusions

From our analyses, the following conclusions can be drawn:

1. Monitoring sampling frequency is important for the application of forecasting methods. For some events, an accurate prediction using FFMs on simulated satellite data has been obtained. However, satellite interferometry cannot provide, at present, the sampling frequency required for a reliable FFM for early warning purposes.

2. Satellite interferometry has potential for monitoring landslide precursors and for detecting the slopes affected by a change in their deformational behaviour. According to the deformational behaviours of our landslide database, at present, 30% of the events could have been monitored during the acceleration phase with at least three data points, even considering the phase ambiguity constraints. Thus, based on the Sentinel 1A + 1B features, it is possible to monitor 17/56 landslides using satellite SAR interferometry.

3. Considering both the RT and ambiguity phase limits, it would have been possible to forecast a maximum of 4 landslides with an error smaller than 5 days, based on the Sentinel-1 features. This is due to the strong constraints related to the phase ambiguity, which does not allow the monitoring of the high displacement rate occurring during the tertiary creep phase. Notably, it is sometimes possible to monitor consecutive displacements greater than $\lambda/4$. If one direction of motion can be assumed ('upward' or 'downward', as for the majority of landslides), the maximum measurable displacement between a pair of scenarios becomes one-half of the wavelength [28,29]. However, considering $\lambda/2$ as the maximum detectable displacement between two consecutive acquisitions, we achieved a similar result: six successful forecasts were obtained based on the Sentinel-1 features.

4. If a landslide process is monitored using a suitable technique (high spatial and temporal resolution), forecasting methods can be a powerful tool for risk management and for early-warning systems.

Acknowledgments: The present work has been carried out in the framework of the project *"Grandi Frane in roccia e Frane Superficiali a Cinematica Rapida in Aree Montane: Metodi Per La Previsione Temporale e Spaziale (Prediction and Susceptibility)"*, supported by the Dipartimento per gli Affari Regionali e le Autonomie della Presidenza del Consiglio dei Ministri and the Università di Roma «La Sapienza». We would like to thank the two anonymous reviewers and the Editor for their helpful and constructive comments.

Author Contributions: Francesca Bozzano and Paolo Mazzanti conceived the study. Francesca Bozzano, Carlo Esposito and Paolo Mazzanti coordinated the research and contributed to the article organization. Serena Moretto and Alfredo Rocca built the landslides database. Serena Moretto designed and performed the forecasting analyses. Serena Moretto and Francesca Bozzano drafted the manuscript which was revised by all Authors.

Conflicts of Interest: The authors declare no conflict of interest.

References

1. Heim, A. *Bergsturz und Menschenleben*; Fretz. & Wasmuth: Zürich, Switzerland, 1932. (In Germany)
2. Terzaghi, K. Mechanism of landslides. In *Application of Geology to Engineering Practice (Berkeley Volume)*; Geological Society of America: Washington, DC, USA, 1950; pp. 83–123.

3. Saito, M.; Uezawa, H. Failure of Soil due to Creep. In Proceedings of the 5th International Conference on Soil Mechanics and Foundation Engineering (ICSMFE), Paris, France, 17–22 July 1961; Volume 1, pp. 315–318.

4. Saito, M. Forecasting the Time of Occurrence of a Slope Failure. In Proceedings of the 6th International Conference on Soil Mechanics and Foundatin Engineering, Montreal, QC, Canada, 8–15 September 1965; pp. 537–541.

5. Voight, B.; Kennedy, B.A. Slope Failure of 1967–1969, Chuquicamata Mine, Chile. *Dev. Geotech. Eng.* **1979**, *4*, 595–632.

6. Fukuzono, T. A new method for predicting the failure time of a slope. In Proceedings of the 4th International Conference and Field Workshop on Landslides, Tokyo, Japan, 23–31 August 1985; pp. 145–150.

7. Azimi, C.; Biarez, J.; Desvarreux, P.; Keime, F. Forecasting time of failure for a rockslide in gypsum. In Proceedings of the 5th international symposium on landslides, Lausanne, Switherland, 10–15 July 1988; pp. 531–536.

8. Voight, B. A relation to describe rate-dependent material failure. *Science* **1989**, *243*, 200–203. [CrossRef] [PubMed]

9. Voight, B. A method for prediction of volcanic eruptions. *Nature* **1988**, *332*, 125–130. [CrossRef]

10. Fukuzono, T. A simple method for predicting the failure time of slope using reciprocal of velocity. *Technol. Disaster Prev. Sci. Technol. Agency Jpn. Int. Coop. Agency* **1989**, *13*, 111–128.

11. Cornelius, R.R.; Voight, B. Graphical and PC-software analysis of volcano eruption precursors according to the materials failure forecast method (FFMS). *J. Volcanol. Geotherm. Res.* **1995**, *64*, 295–320. [CrossRef]

12. Kilburn, C.R.; Petley, D.N. Forecasting giant, catastrophic slope collapse: Lessons from Vajont, Northern Italy. *Geomorphology* **2003**, *54*, 21–32. [CrossRef]

13. Petley, D.N.; Higuchi, T.; Petley, D.J.; Bulmer, M.H.; Carey, J. Development of progressive landslide failure in cohesive materials. *Geology* **2005**, *33*, 201–204. [CrossRef]

14. Petley, D.N.; Petley, D.J.; Allison, R.J. Temporal prediction in landslides. Understanding the Saito effect. In *Landslides and Engineered Slopes: From the Past to the Future*; Taylor & Francis: Xi'an, China, 2008; pp. 794–800.

15. Rose, N.D.; Hungr, O. Forecasting potential rock slope failure in open pit mines using the inversevelocity method. *Int. J. Rock Mech. Min. Sci.* **2007**, *44*, 308–320. [CrossRef]

16. Gigli, G.; Fanti, R.; Canuti, P.; Casagli, N. Integration of advanced monitoring and numerical modeling techniques for the complete risk scenario analysis of rockslides: The case of Mt. Beni (Florence, Italy). *Eng. Geol.* **2011**, *120*, 48–59. [CrossRef]

17. Bozzano, F.; Cipriani, I.; Mazzanti, P.; Prestininzi, A. A field experiment for calibrating landslide time-of-failure prediction functions. *Int. J. Rock Mech. Min. Sci.* **2014**, *67*, 69–77. [CrossRef]

18. Mazzanti, P.; Bozzano, F.; Cipriani, I.; Prestininzi, A. New insights into the temporal prediction of landslides by a terrestrial SAR interferometry monitoring case study. *Landslides* **2015**, *12*, 55. [CrossRef]

19. Manconi, A.; Giordan, D. Landslide early warning based on failure forecast models: The example of the Mt. de La Saxe rockslide, northern Italy. *Nat. Hazards Earth Syst. Sci.* **2015**, *15*, 1639–1644. [CrossRef]

20. Sättele, M.; Krautblatter, M.; Bründl, M.; Straub, D. Forecasting rock slope failure: How reliable and effective are warning systems? *Landslide* **2016**, *13*, 737. [CrossRef]

21. Loew, S.; Gschwind, S.; Keller-Signer, A.; Valenti, G. Monitoring and Early Warning of the 2012 Preonzo Catastrophic Rockslope Failure. *Landslides* **2017**, *14*, 141. [CrossRef]

22. Ferretti, A.; Prati, C.; Rocca, F. Nonlinear subsidence rate estimation using permanent scatterers in differential SAR interferometry. *IEEE Trans. Geosci. Remote Sens.* **2000**, *38*, 2202–2212. [CrossRef]

23. Ferretti, A.; Prati, C.; Rocca, F. Permanent scatterers in SAR interferometry. *IEEE Trans. Geosc. Remote Sens.* **2001**, *39*, 8–20. [CrossRef]

24. Berardino, P.; Fornaro, G.; Lanari, R.; Sansosti, E. A new algorithm for surface deformation monitoring based on small baseline differeential SAR interferograms. *IEEE Trans. Geosc. Remote Sens.* **2002**, *40*, 2375–2383. [CrossRef]

25. Perissin, D.; Wang, T. Repeat-Pass SAR Interferometry with Partially Coherent Targets. *IEEE Trans. Geosci. Remote Sens.* **2012**, *50*, 271–280. [CrossRef]

26. Early Warning from Space of Homes on the Slide. Available online: http://www.esa.int/Our_Activities/ Observing_the_Earth/Early_warning_from_space_of_homes_on_the_slide (accessed on 9 May 2017).

27.   Mazzanti, P.; Rocca, A.; Bozzano, F.; Cossu, R.; Floris, M. Landslides Forecasting Analysis by Time Series Displacement Derived from Satellite InSAR Data: Preliminary Results. In Proceedings of the Fringe 2011, Frascati, Italy, 19–23 September 2011.

28.   Ferretti, A.; Prati, C.; Rocca, F.; Nicola, C.; Farina, P.; Young, B. Permanent scatterers technology: A powerful state of the art tool for historic and future monitoring of landslides and other terrain instability phenomena. In *International Conference on Landslide Risk Management*; Balkema: Vancouver, BC, Canada, 2005.

29.   Crosetto, M.; Monserrat, O.; Iglesias, R.; Crippa, B. Persistent scatterer interferometry: Potential, limits and initial cand X-band comparison. *Photogramm. Eng. Remote Sens.* **2010**, *76*, 1061–1069. [CrossRef]

30.   Fukuzono, T. Experimental Study of Slope Failure Caused by Heavy Rainfall. *IAHS-AISH Publ.* **1987**, *165*, 133–134.

31.   Moriwaki, H.; Inokuchi, T.; Hattanji, T.; Sassa, K.; Ochiai, H.; Wang, G. Failure processes in a full-scale landslide experiment using a rainfall simulator. *Landslides* **2012**, *1*, 277–288. [CrossRef]

32.   Mufundirwa, A.; Fujii, Y.; Kodama, J. A new practical method for prediction of geomechanical failure-time. *Int. J. Rock Mech. Min. Sci.* **2010**, *47*, 1079–1090. [CrossRef]

33.   Kayesa, G. Prediction of slope failure at Letlhakane mine with the geomos slope monitoring system. In *International Symposium on Stability of Rock Slopes in Open Pit Mining and Civil Engineering*; The South African Institute of Mining and Metallurgy: Johannesburg, South Africa, 2006; pp. 606–622.

34.   Saito, M. Forecasting Time of Slope Failure by Tertiary Creep. In Proceedings of the 7th International Conference on Soil Mechanics and Foundation Engineering; Sociedad Mexicana de Mecanica: Mexico City, Mexico, 1969; pp. 677–683.

35.   Manconi, A.; Giordan, D. Landslide failure forecast in near-real-time. *Geomat. Nat. Hazard. Risk* **2014**, *7*, 639–648. [CrossRef]

36.   Kent, A. Coal mine waste dumps in British Columbia stability issues and recent developments. *Int. Mine Waste Manag.* **1992**, 10–18.

37.   Iovine, G.; Petrucci, O.; Rizzo, V.; Tansi, C. The march 7th 2005 Cavallerizzo (Cerzeto) Landslide in Calabria—Southern Italy. In Proceedings of the 10th IAEG Congress, Nottingham, UK, 6–10 September 2006; pp. 1–12.

38.   Okamoto, T.; Larsen, J.O.; Matsuura, S.; Asano, S.; Takeuchi, Y.; Grande, L. Displacement properties of landslide masses at the initiation of failure in quick clay deposits and the effects of meteorological and hydrological factors. *Eng. Geol.* **2004**, *72*, 233–251. [CrossRef]

39.   Thompson, P.W. Confirmation of a Failure Mechanism Using Open Pit Monitoring Methods. In Proceedings of the Australian Conference on Geotechnical Instrumentation and Monitoring in Open Pit and Underground Mining, Kalgoorlie, Australia, 21–23 June 1993; pp. 483–489.

40.   Glastonbury, B.; Fell, R. *Report on the Analysis of the Deformation Behaviour of Excavated Rock Slopes*; School of Civil & Environmental Engineering, University of New South Wales: Sydney, Australia, 2002.

41.   GEOPRAEVENT. Available online: www.geopraevent.ch (accessed on10 May 2017).

42.   Metodi Innovativi per il Controllo e la Gestione dei Fenomeni Franosi. Available online: http://www.eventi.saie.bolognafiere.it/media/saie/presentazioni%20sba/sala%20A/Matteo_Berti.pdf (accessed on 10 May 2017). (In Italian)

43.   Ulusay, R.; Aksoy, H. Assessment of the failure mechanism of a highwall slope under spoil pile loadings at a coal mine. *Eng. Geol.* **1994**, *38*, 117–134. [CrossRef]

44.   Ryan, T.M.; Call, R.D. Applications of Rock Mass Monitoring for Stability Assessment of Pit Slope Failure. In Proceedings of the 33th U.S. Symposium on Rock Mechanics (USRMS), Santa Fe, NM, USA, 3–5 June 1992; Balkema: Rotterdam, The Netherlands, 1992.

45.   Cahill, J.; Lee, M. Ground control at Leinster nickel operations. In *International Symposium on Stability of Rock Slopes in Open Pit Mining and Civil. Engineering*; The South African Institute of Mining and Metallurgy: Johannesburg, South Africa, 2006; pp. 321–334.

46.   Salt, G. Landslide Mobility and Remedial Measures. In Proceedings of the 5 International Symposium on Landslides, Rotterdam, The Netherlands, 1988; pp. 757–762.

47.   Carey, J.M.; Moore, R.; Petley, D.N.; Siddle, H.J. Pre-failure behaviour of slope materials and their significance in the progressive failure of landslides. In *Landslides and Climate Change: Challenges and Solutions*; Taylor & Francis Group: London, UK, 2007; pp. 207–215.

48. Urcioli, G.; Picarelli, L. Interaction between landslides and man-made works. In *Landslides and Engineered Slopes*; Taylor & Francis Group: London, UK, 2008.

49. Borsetto, M.; Frassoni, A.; La Barbera, G.; Fanelli, M.; Giuseppetti, G.; Mazzà, G. An application of Voight empirical model for the prediction of soil and rock instabilities. In Proceedings of the 7th International Symposium on Landslides; Bell, D.H., Ed.; Balkema: Rotterdam, The Netherlands; Christchurch, New Zealand, 1991; pp. 335–341.

50. Petley, D.N. The evolution of slope failures: Mechanisms of rupture propagation. *Nat. Hazard. Earth Sys. Sci.* **2004**, *4*, 147–152. [CrossRef]

51. Bai, J.G.; Lu, S.D.; Han, J.S. Importance of study of creep sliding mechanism to prevention and treatment of reservoir landslide. In *Landslides and Engineered Slopes: From the Past to the Future*; Taylor & Francis Group: Xi'an, China, 2008; pp. 1071–1076.

52. Bhandari, R.K. Special lecture—Some practical lessons in the investigation and field monitoring of landslides. *Landslides* **1988**, *1–3*, 1435–1457.

53. Zavodni, Z.M.; Broadbent, C.D. Slope failure kinematics. *CIM Bull.* **1980**, 69–74.

54. Brawner, C.O.; Stacey, P.F. Hogarth pit slope failure, Ontario, Canada. In *Rockslides and Avalanches, B*; Elsevier: Amsterdam, The Netherlands, 1979; pp. 691–707.

55. Glastonbury, B.; Fell, R. *Report On the Analysis of Rapid Natural Rock Slope*. R-390; School of Civil & Environmental, University of New South Wales (UNSW): Sydney, Australia, 2001.

56. Brown, I.; Hittinger, M.; Goodman, R. Finite element study of the Nevis Bluff (New Zealand) rock slope failure. *Rock Mech.* **1980**, *12*, 231–245. [CrossRef]

57. Keqiang, H.; Sijing, W. Double-parameter threshold and its formation mechanism of the colluvial landslide: Xintan landslide, China. *Environ. Geol.* **2006**, *49*, 696–707. [CrossRef]

58. Schumm, S.A.; Chorley, R.J. The fall of Threatening Rock. *Am. J. Sci.* **1964**, *262*, 1041–1054. [CrossRef]

59. D'Elia, B.; Picarelli, L.; Leroueil, S.; Vaunat, J. Geotechnical characterisation of slope movements in structurally complex clay soils and stiff jointed clays. *Riv. Ital. Geotec.* **1998**, *3*, 5–47.

60. Zabusky, L.; Swidzinski, W.; Kulczykowski, M.; Mrozek, T.; Laskowicz, I. Monitoring of landslides in the Brada river valley in Koronowo (Polish Lowlands). *Environ. Earth Sci.* **2015**, *73*, 8609–8619. [CrossRef]

61. Boyd, J.M.; Hinds, D.V.; Moy, D.; Rogers, C. Two simple devices for monitoring movements in rock slopes. *Quart. J. Eng. Geol. Hydrogeol.* **1973**, *6*, 295–302. [CrossRef]

62. Siqing, Q.; Sijing, W. A homomorphic model for identifying abrupt abnormalities of landslide forerunners. *Eng. Geol.* **2000**, *57*, 163–168. [CrossRef]

63. Royán, M.J.; Abellán, A.; Vilaplana, J.M. Progressive failure leading to the 3 December 2013 rockfall at Puigcercós scarp (Catalonia, Spain). *Landslides* **2015**, *12*, 585–595. [CrossRef]

64. Muller, L. The rock slide in the Vaiont valley. *Felsmech. Ing.* **1964**, *42*, 148–212.

65. Bozzano, F.; Mazzanti, P.; Esposito, C.; Moretto, S.; Rocca, A. Potential of satellite InSAR monitoring for landslide Failure Forecasting. In *Landslides and Engineered Slopes. Experience, Theory and Practice*; CRC Press: Boca Raton, FL, USA, 2016; Volume 2, pp. 523–530.

66. Moretto, S.; Bozzano, F.; Esposito, C.; Mazzanti, P. Lesson learned from the pre-collapse time series of displacement of the Preonzo landslide (Switzerland). *Rend. Online Soc. Geol. Ital.* **2016**, *41*, 247–250. [CrossRef]

MDPI AG

St. Alban-Anlage 66

4052 Basel, Switzerland

Tel. +41 61 683 77 34

Fax +41 61 302 89 18

http://www.mdpi.com

Geosciences Editorial Office

E-mail: geosciences@mdpi.com

http://www.mdpi.com/journal/geosciences

www.ingramcontent.com/pod-product-compliance
Lightning Source LLC
Chambersburg PA
CBHW051728210326
41597CB00032B/5642